Cartilage

Susanne Grässel • Attila Aszódi
Editors

Cartilage

Volume 2: Pathophysiology

Springer

Editors
Susanne Grässel
Orthopädische Klinik für die Universität Regensburg
Universitätsklinikum Regensburg
Regensburg
Germany

Attila Aszódi
Department of General, Trauma, and Reconstruction Surgery
Ludwig-Maximilians-University
Munich
Germany

ISBN 978-3-319-45801-4   ISBN 978-3-319-45803-8 (eBook)
DOI 10.1007/978-3-319-45803-8

Library of Congress Control Number: 2017931314

Printed on acid-free paper

This Springer imprint is published by Springer Nature
The registered company is Springer International Publishing AG
The registered company address is: Gewerbestrasse 11, 6330 Cham, Switzerland

Preface

Volume two of this book series comprised of three volumes is dedicated to provide an overview about the pathophysiology of cartilage, joint tissue, and intervertebral disks.

The text is designed to be of use to multiple medical and basic science disciplines as orthopedics, rheumatology, and trauma surgery and all basic investigators working in the field of cartilage, joint, and intervertebral disk pathophysiology.

This volume focuses on the major cartilage pathophysiologies which include osteoarthritis and rheumatoid arthritis, degeneration of intervertebral disks, and genetic skeletal diseases as cartilage collagenopathies and other hereditary chondrodysplasias resulting from mutations in structural cartilage proteins.

Chapter 1 provides an overview about *osteoarthritis (OA)* which is the most common joint disorder and known as a leading cause of disability in the adult population. It is now appreciated that all components of the joint, including the cartilage, calcified cartilage, synovial joint lining, and periarticular bone, undergo pathological changes during the initiation and progression of OA. Some of these alterations can be attributed to direct injury and mechanical disruption of the tissues, but in general the mechanisms are dependent on active cell-mediated processes that occur during the long time course of the disease. A deeper understanding of the specific and unique roles of complex signaling networks and their downstream targets involving biochemical crosstalk among the cartilage, synovium, bone, and other joint tissues will provide mechanistic insights into the pathologic processes that affect the cartilage and other joint tissues in OA, but also may identify potential therapeutic targets for treatment of this debilitating disease. Chapter 2 provides insight into mechanical stress as an obligatory etiological factor in the development of OA. Understanding how tissues of the joint respond to mechanical injury is likely to inform our understanding of pathogenesis. Articular cartilage is avascular yet responds rapidly and strongly to a range of mechanical stresses. It does so by activating a number of mechanosensitive pathways mediated by release of molecules trapped within the pericellular matrix as well as by triggering mechanoreceptors at the cell surface. These pathways appear to be relevant to the in vivo response to mechanical disruption and affect the course of experimental OA.

The gradual loss of articular cartilage from the surface of articulating joints is a feature of OA. It is marked by degradation of the cartilage matrix, including the large aggregating proteoglycan aggrecan, the small leucine-rich proteoglycans known as SLRPs, and the fibrillar type II collagen. Chapter 3 discusses the major families of cartilage-degrading enzymes, the matrix metalloproteinases (MMPs), and a disintegrin and metalloproteinase with thrombospondin motifs (ADAMTS) families. Factors that regulate MMP and ADAMTS activity, with a focus on MMP-13, ADAMTS-4, and ADAMTS-5 as the major protagonists of cartilage degradation, are discussed. The important role of degraded matrix fragments in regulating inflammation in osteoarthritis, via Toll-like receptor signaling, is highlighted. Chapter 4 puts emphasis on the functions of proteoglycans as one of the main components of the articular cartilage ECM. Proteoglycans bind water and provide the basis for absorbing high compressive loads. Additionally, they bind cytokines, chemokines, growth factors, and morphogens, thereby protecting these factors against proteolysis and/or acting as a depot of regulatory factors when matrix degradation occurs. They also modulate signaling pathways and create morphogen gradients by immobilization of ligands in the ECM and regulation of the turnover of ligands. Given these important roles of proteoglycans in regulating cell functions, it is well understandable that the loss of ECM and degradation of proteoglycans during OA induce severe changes in cartilage homeostasis.

The presence and production of soluble factors in the osteoarthritic joint have always been a focus of research, as they are assumed to play a role in the initiation and/or progression of the disease. Chapter 5 reviews research data which assign an important role to chemokines, growth factors, and adipokines in OA; however it also emphasizes on a traditionally studied subset of inflammatory, anti-inflammatory, and modulatory cytokines. Differential profiles of these factors compared to healthy joints were found in the knee and other OA affected joints, whereby joint damage itself induces a specific change in the secretory pattern of diverse soluble factors.

Genetic skeletal diseases are a diverse and complex group of over 450 rare diseases that affect the development and homeostasis of the skeleton. Although individually rare, as a group of related genetic skeletal diseases, they have an overall prevalence of at least 1 per 4,000 children, which extrapolates to a minimum of 225,000 people in the European Union, and this extensive burden in pain and disability leads to poor quality of life and high healthcare costs. Dominant-negative (qualitative) defects in numerous cartilage structural proteins result in a broad range of genetic skeletal diseases. Chapter 6 will focus on mutations in fibrillar and fibril-associated collagen genes which cause a wide range of chondrodysplasias, ranging from premature arthritis to severe early lethal disorders. Mutations of cartilage-specific collagens can cause cartilage tissue dysfunction by reducing synthesis of structurally normal protein or through protein misfolding which leads to intracellular retention and degradation and consequent secretion of reduced amounts of structurally abnormal protein. In addition, collagen misfolding mutations can induce a cellular unfolded protein response which ultimately promote apoptosis and thus contribute to the pathology. Chapter 7 will focus on a disease spectrum resulting from mutations in the glycoproteins, cartilage oligomeric matrix protein (COMP), type IX collagen, and matrilin-3, which together cause a continuum of phenotypes that are among the most common of the autosomal dominant genetic skeletal diseases. Pseudoachondroplasia (PSACH) and autosomal dominant multiple epiphyseal dysplasia (MED) define a disease spectrum typified by varying degrees of short-limbed dwarfism, joint pain with stiffness, and early-onset OA. New insight into disease-related musculoskeletal complications such as myopathy, ligamentous laxity, and tendinopathy has been gained through the analysis of mouse models of the PSACH and MED disease spectrum.

Chapter 8 will summarize and discuss the role of integrins in the physiology and pathophysiology of the growth plate and articular cartilage. Integrins are membrane receptors responsible for bidirectional communication between the cells and the surrounding by transmitting physicochemical signals through adhesion complexes. In addition, integrins are involved in sensing mechanical stress signals generated by the extracellular matrix and transduce them into the cell interior converting physical stimuli to biochemical signaling. Chondrocyte integrins have thus indispensable roles in cartilage development, skeletal growth, and articular cartilage function.

Chapter 9 will focus on the peripheral nervous system which is critically involved in the metabolism of joint tissue and intervertebral disks (IVD). Nerve fibers of sympathetic and sensory origin innervate synovial tissue and subchondral bone of diarthrodial joints. During endochondral ossification in embryonic limb development, sensory and sympathetic neurotransmitters modulate osteo-chondrogenic differentiation of mesenchymal progenitor cells, vascularization, and matrix differentiation indicating a distinct role in skeletal growth and possible limb regeneration processes. In adults, sensory and sympathetic neurotransmitters are involved in the pathology of inflammatory diseases as rheumatoid arthritis which manifests mainly in joints. In addition, they might play a role in the pathogenesis of a priori degenerative joint disorders, as OA and intervertebral disk degeneration.

Tissues of intervertebral disks share similarities to those of diarthrodial joints, such as a thin layer of cartilage that lines the interface between the joint and the bony elements and a central space rich in extracellular matrix molecules that promotes lubrication and maintains osmotic pressure. Like the pathophysiology of other cartilaginous joints, intervertebral disks undergo biomechanical and structural changes as a result of aging and mechanical insults. Due to higher mechanical loading, lumbar disks are more susceptible to degeneration, which can lead to symptomatic outcomes such as low back pain, sciatica, and other physical disabilities. These affect the quality of life as we age and present a significant burden to the healthcare system globally. Chapter 10 will provide an overview of the intervertebral disk in health and disease.

Bringing together international experts from diverse fields of musculoskeletal research was a demanding task requiring patience and persistence not only for volume one of this book series but also for this volume. For that we are very grateful to our authors of this volume who managed to complete their chapters and who dedicated their spare free time to writing their reviews.

Regensburg, Germany Susanne Grässel
Munich, Germany Attila Aszódi

Contents

Contributors

Attila Aszódi, PhD Laboratory of Experimental Surgery and Regenerative Medicine, Clinic for General, Trauma and Reconstruction Surgery, Ludwig-Maximilians-University, Munich, Germany

Yvonne Bastiaansen-Jenniskens, PhD University Medical Center Utrecht, Department of Orthopaedics, Utrecht, The Netherlands

John F. Bateman, PhD Murdoch Childrens Research Institute, Royal Children's Hospital, Parkville, Australia

Biochemistry and Molecular Biology, University of Melbourne, Parkville, Australia

Peter Bell, PhD Institute of Genetic Medicine, Newcastle University, International Centre for Life, Newcastle upon Tyne, UK

Jessica Bertrand, PhD Otto-von-Guericke University Magdeburg, Department of Orthopaedic Surgery, Magdeburg, Germany

Michael D. Briggs, PhD Institute of Genetic Medicine, Newcastle University, International Centre for Life, Newcastle upon Tyne, UK

Trevor L. Cameron, PhD Murdoch Childrens Research Institute, Royal Children's Hospital, Parkville, Australia

Danny Chan, PhD School of Biomedical Sciences, The University of Hong Kong, Pokfulam, Hong Kong

Wilson C.W. Chan, PhD School of Biomedical Sciences, The University of Hong Kong, Pokfulam, Hong Kong

Laura B. Creemers, PhD University Medical Center Utrecht, Department of Orthopaedics, Utrecht, The Netherlands

Kirsty L. Culley, PhD Hospital for Special Surgery Research Institute and Weill Cornell Medical College, New York, NY, USA

Amanda J. Fosang, PhD University of Melbourne Department of Paediatrics and Murdoch Childrens Research Institute, Royal Children's Hospital, Parkville, VIC, Australia

Stephanie J. Gauci, PhD University of Melbourne Department of Paediatrics and Murdoch Childrens Research Institute, Royal Children's Hospital, Parkville, VIC, Australia

Mary B. Goldring, PhD Hospital for Special Surgery Research Institute and Weill Cornell Medical College, New York, NY, USA

Susanne Grässel, PhD Experimental Orthopaedics, Department of Orthopaedic Surgery, ZMB/BioPark 1, University of Regensburg, Regensburg, Germany

Annelena Held, PhD Otto-von-Guericke University Magdeburg, Department of Orthopaedic Surgery, Magdeburg, Germany

Heba M. Ismail, PhD Arthritis Research UK Centre for OA Pathogenesis, Kennedy Institute of Rheumatology, NDORMS, University of Oxford, Oxford, UK

Zsuzsa Jenei-Lanzl, PhD Experimental Orthopaedics, Department of Orthopaedic Surgery and Laboratory of Experimental Rheumatology and Neuroendocrine Immunology, Department of Internal Medicine I, University of Regensburg, Regensburg, Germany

Sarah Yoon Ji Kwon, PhD School of Biomedical Sciences, The University of Hong Kong, Pokfulam, Hong Kong

Shireen R. Lamandé, PhD Murdoch Childrens Research Institute, Royal Children's Hospital, Parkville, Australia

Departments of Paediatrics, University of Melbourne, Parkville, Australia

Christopher B. Little, PhD Raymond Purves Bone and Joint Research Laboratories, Kolling Institute, Institute of Bone and Joint Research, Sydney Medical School Northern, University of Sydney, St. Leonards, NSW, Australia

Miguel Otero, PhD Hospital for Special Surgery Research Institute and Weill Cornell Medical College, New York, NY, USA

Katarzyna A. Piróg, PhD Institute of Genetic Medicine, Newcastle University, International Centre for Life, Newcastle upon Tyne, UK

Daniel Saris, PhD University Medical Center Utrecht, Department of Orthopaedics, Utrecht, The Netherlands

Ravi Savarirayan, PhD Murdoch Childrens Research Institute, Royal Children's Hospital, Parkville, Australia

Departments of Paediatrics, University of Melbourne, Parkville, Australia

Victorian Clinical Genetics Service, Parkville, Australia

Akansha M. Shah, PhD School of Biomedical Sciences, The University of Hong Kong, Pokfulam, Hong Kong

Heather Stanton, PhD University of Melbourne Department of Paediatrics and Murdoch Childrens Research Institute, Royal Children's Hospital, Parkville, VIC, Australia

Rainer H. Straub, MD Laboratory of Experimental Rheumatology and Neuroendocrine Immunology, Department of Internal Medicine I, University of Regensburg, Regensburg, Germany

Tonia L. Vincent, FRCP, PhD Arthritis Research UK Centre for OA Pathogenesis, Kennedy Institute of Rheumatology, NDORMS, University of Oxford, Oxford, UK

Pathogenesis of Osteoarthritis in General

1

Mary B. Goldring, Kirsty L. Culley, and Miguel Otero

Abstract

Osteoarthritis (OA) is the most common joint disorder and is a leading cause of disability in the adult population. It is now appreciated that all components of the joint, including the cartilage, calcified cartilage, synovial joint lining, and periarticular bone, undergo pathological changes during the initiation and progression of OA. Some of these alterations can be attributed to direct injury and mechanical disruption of the tissues, but in general the mechanisms are dependent on active cell-mediated processes that occur during the long time course of the disease. Based on clinical observations and experimental studies, it is now recognized that it is possible for individual patients to exhibit common sets of symptoms and structural abnormalities due to distinct pathophysiological pathways that act independently or in combination. Recent research focusing on the underlying pathological mechanisms has identified complex signaling networks involving biochemical cross talk among the cartilage, synovium, bone, and other joint tissues. These complex networks involve interplay among anabolic, catabolic, and inflammatory signals within a background of poorly characterized genetic factors. A deeper understanding of the specific and unique roles of these mediators and their downstream targets will provide mechanistic insights into the pathologic processes that affect the cartilage and other joint tissues in OA but also may identify potential therapeutic targets for treatment of this debilitating disease.

M.B. Goldring (✉) • K.L. Culley • M. Otero
Hospital for Special Surgery Research Institute and Weill Cornell Medical College, New York, NY 10021, USA
e-mail: goldringm@hss.edu; culleyk@hss.edu; oterom@hss.edu

S. Grässel, A. Aszódi (eds.), *Cartilage*, DOI 10.1007/978-3-319-45803-8_1

1.1 Introduction

Osteoarthritis (OA) is the most common joint disorder and the major cause of disability in the adult population. The pathophysiology of the disease is characterized by progressive loss of articular cartilage, cartilage calcification, osteophyte formation, subchondral bone remodeling, and mild to moderate inflammation of the synovial lining. Although cartilage destruction is the hallmark of OA disease, the involvement of changes in the periarticular tissues, including the subchondral bone, ligaments, tendons, menisci, and synovial membrane, is now well recognized (Loeser et al. 2012a; Goldring and Goldring 2007). For example, ligaments and menisci are important for maintaining biomechanical stability in the joint, and their injury can lead eventually to cartilage loss. In addition, multiple factors are involved in the pathogenesis of OA, including genetic susceptibility, biomechanics of the affected joint, and the presence and extent of inflammation. It has been difficult, therefore, to identify specific targets for therapy.

Investigations in various in vitro models and preclinical in vivo models during the past decades have focused primarily on cartilage degradation or repair as a therapeutic target and more recently on how biomechanical and cellular responses in chondrocytes are modified by interactions with other joint tissues, in particular, the synovium and bone (Goldring et al. 2011; Goldring and Otero 2011; Goldring and Berenbaum 2015). This chapter focuses on the role of the chondrocyte in maintaining cartilage homeostasis and responding to adverse events in the whole joint that modify cartilage integrity and result in the initiation and progress of the osteoarthritis disease program.

1.2 Disease Etiologies and Therapeutic Prospects

Studies in patients that undergo operative procedures indicate that there are different etiologies and time courses that result in the initiation and development of OA. These subsets represent a continuum of early, progressive, and end-stage OA and include (1) anterior cruciate ligament (ACL) injury (<35 years of age), (2) acute meniscal injury (26–40 years of age), (3) degenerative meniscus (40–65 years of age), and (4) total joint replacement (>50 years of age). Epidemiologic studies have established that there is a strong relationship between ACL disruption and the risk for subsequent development of OA (Segawa et al. 2001; Buckwalter and Brown 2004; Roos 2005; Lohmander et al. 2007; Meunier et al. 2007). Studies of populations with meniscal injury have also been useful for identifying OA risk factors (Englund and Lohmander 2004). Meniscal injuries are commonly seen in association with ACL injury (Jones et al. 2003; Louboutin et al. 2009).

Current understanding indicates that the aberrant distribution of forces in cartilage leads to altered mechanotransduction in the chondrocytes and subsequent activation of catabolic and inflammatory genes, deregulated matrix synthesis, and decreased repair capacity (Lotz and Kraus 2010; de Lange-Brokaar et al. 2012) (see also Chap. 2). The development of posttraumatic cartilage pathology may in turn adversely impact the structural and functional properties of periarticular bone. The damaged meniscus is an additional source of inflammatory cytokines, chemokines, and reactive oxygen species that could promote expression and activation of proteolytic enzymes and adversely affect cell survival and synthetic activity of chondrocytes and cells of other joint tissues (Englund et al. 2009; Rai et al. 2013).

In addition to trauma or injury, there are other factors that influence the disease process, including aging, genetic predisposition, abnormal biomechanics, obesity, and comorbidities such as cardiovascular disease, metabolic syndrome, and diabetes. However, OA clinical trial designs have not accounted for these multifactorial aspects in disease subsets, but rather have selected subjects based on joint location or diagnosis as primary OA or secondary to other types of arthritis (Punzi et al. 2010; Kloppenburg 2014; Stiebel et al. 2014; Detert et al. 2014). There is a need, therefore, for optimization of cohort selection based on classifying OA patients according to disease phenotypes related to distinct pathophysiological pathways (Bijlsma et al. 2011; Blanco and Ruiz-Romero 2012; Conaghan 2013; Henrotin 2014).

Given the complexity of OA, it is not surprising that there is no structure-modifying agent for OA and the available pain therapies have limited efficacy and associated toxicities (Matthews and Hunter 2011). However, it is unlikely that a single therapy will be effective against both symptoms and structural changes in the entire spectrum of OA patients (Mobasheri 2013a, b; Pulsatelli et al. 2013; Thakur et al. 2014). Most trials have addressed function and pain at later stages of disease when there is already radiographic evidence of disease progression such as joint space narrowing and osteophytes. The development of validated diagnostic and prognostic molecular biomarkers (Lafeber and van Spil 2013; Hsueh et al. 2014; Lotz et al. 2013; Tonge et al. 2014), as well as novel imaging biomarkers (Hunter et al. 2013), that could be used for evaluating pre-symptomatic early-stage disease, may permit therapeutic interventions to halt or slow OA progression prior to irreversible joint damage. There is therefore a need for deeper understanding of the structure and function of articular cartilage and other joint tissues and how they interact and respond to adverse environmental insults in ways that disrupt normal joint homeostasis.

1.3 Articular Cartilage Physiology

Adult articular cartilage is composed of a specialized matrix of type II collagen and the large aggregating proteoglycan, aggrecan, along with several "minor" collagens and small proteoglycans. Its unique structural organization provides tensile strength via the collagen network and compressive resistance via the proteoglycans, which contribute to the capacity of the matrix to accommodate more than 70 % water (Heinegard and Saxne 2011; Onnerfjord et al. 2012; Hunziker et al. 2014) (for further details, see Chaps. 3 and 4). Cartilage is relatively hypocellular compared with other tissues, with the chondrocytes constituting only 1–2 % of the total cartilage volume, and it lacks a vascular supply and innervation. In normal adult articular cartilage, the chondrocyte has limited proliferative capacity and its ability to perform low-turnover repair declines with age. Since the half-life of type II collagen within the cartilage collagen network is more than 100 years, whereas the proteoglycan components have half-lives of weeks to years, the chondrocyte is involved mostly in replacing the glycosaminoglycans on the aggrecan and other small proteoglycan core proteins. The importance of these matrix proteins in determining the structural and functional properties of the articular cartilage can be observed in chondrodysplasias and other heritable disorders where mutations or deficiencies in cartilage matrix genes result in altered skeletal development often associated with the premature development of OA (Sandell 2012) (see also Chap. 6) (Fig. 1.1).

Chondrocytes in articular cartilage exist in lacunae as single cells encased in a pericellular matrix (PCM) consisting of collagen VI microfibrils, hyaluronan, perlecan, biglycan, aggrecan as monomers or small aggregates, and type IX collagen but virtually no type II collagen (Wilusz et al. 2014). The PCM helps to maintain the

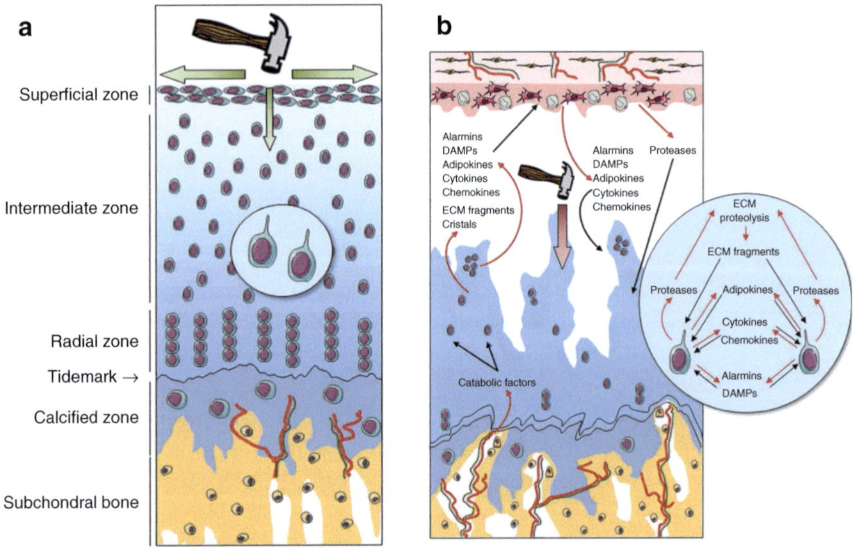

Fig. 1.1 Schematic representation of cartilage organization in the healthy joint and in osteoarthritis. (**a**) Each of the four different zones of healthy articular cartilage, the superficial, intermediate, radial, and calcified zones, is characterized by distinct chondrocyte morphology and extracellular matrix organization and composition. The calcified zone differs from the three other zones by the mineralization of its extracellular matrix, by the presence of vessels (*red*), and by nerve fibers (*green*) that originate from the subchondral bone. The calcified zone interfaces with the non-mineralized cartilage, from which it is separated by the tidemark, and the subchondral bone. (**b**) In OA, there is progressive loss of cartilage matrix from the superficial zone associated with chondrocyte phenotypic modifications, including the formation of clusters, catabolic activation, and hypertrophic differentiation. Cytokines, chemokines, alarmins, DAMPs, adipokines, and other mediators released into the synovial fluid from the cartilage, synovium, and other joint tissues amplify a vicious circle of cartilage damage. In addition to cartilage damage, remodeling of the subchondral bone occurs with the development of vessels (*red*) located in vascular channels, which also contain osteoblasts, osteoclasts, and sensory nerves (*green*) (From Houard et al. (2013) with permission)

chondrocyte in a low-turnover, survival state by protecting it from interacting with molecules in the interterritorial cartilage matrix via cell-surface receptors such as integrins, cell determinant 44 (CD44), annexins, syndecans, and discoidin domain receptor 2 (DDR2) (Loeser 2014; Pap and Bertrand 2013; Xu et al. 2011, 2014). Chondrocytes exist in a low-oxygen tension environment and intracellular survival factors such as HIF-1α are required for maintenance of homeostasis and adaptation to the mechanical environment (Maes et al. 2012). Primary cilia located on the chondrocyte surface are required for chondrocytes to respond to mechanical forces and to maintain hedgehog signaling (Wann et al. 2012; Ruhlen and Marberry 2014; Ho et al. 2013; Thompson et al. 2016). These organelles contain mechanosensitive receptors, including the transient receptor potential vanilloid 4 (TRPV4), piezo channels, and connexin 43, as well as integrins, which permit the chondrocytes to sense and adapt their metabolic activity in response to physical forces (Loeser 2014; O'Conor et al. 2014; Knight et al. 2009; Mayan et al. 2015).

Cartilage also provides a unique articulating surface with a very low coefficient of friction, facilitated by a boundary layer of lubricants, including lubricin, encoded by the PRG4 gene, and hyaluronic acid that are produced by chondrocytes and synovial cells (Waller et al. 2013; Jay and Waller 2014). Joint motion and mechanical loading induce fluid movement between the cartilage and the synovial fluid, facilitating the diffusion of molecules across cartilage and thus providing nutrition. Soluble products transferred from the underlying subchondral bone could also be an important source of nutrients, especially for the deeper layers of the articular cartilage.

1.4 Articular Cartilage Pathology in OA

Of clinical importance is the ability of the chondrocyte to react to mechanical stimuli and to structural changes in the surrounding cartilage matrix through responding to and producing anabolic and catabolic cytokines, which may act to influence cartilage homeostasis in a positive or negative manner (Goldring and Otero 2011; Goldring et al. 2011) (for further details, see Chaps. 2 and 5). Cartilage destruction with loss of proteoglycans and the degradation of the collagen network are the pivotal events that determine the irreversible progression of OA disease. Cartilage destruction in OA is believed to be chondrocyte mediated in response to biomechanical insults and may occur directly or indirectly through the production of cytokines and cartilage matrix-degrading proteinases in cartilage and other joint tissues (Anderson et al. 2011). Joint injury alters the biomechanical dynamics of the joint and joint kinematics, thereby stimulating chondrocyte mechanotransduction, activation of catabolic and inflammatory genes, deregulated matrix synthesis, and decreased repair capacity (Lohmander et al. 2007; Lotz and Kraus 2010; Louboutin et al. 2009) (Fig. 1.1).

Direct analysis of cartilage or chondrocytes from OA patients undergoing joint replacement has yielded evidence that chondrocytes respond not only to proinflammatory cytokines but also inhibitory and anabolic cytokines that modulate responses. The impact of cytokines on chondrocyte function, particularly with respect to their various roles in cartilage destruction, has been reviewed extensively (Goldring and Otero 2011; Goldring et al. 2011; Loeser et al. 2012a) (see also Chap. 5). Studies of cartilage degradation in genetically modified mouse strains that are resistant to or accelerate the development of OA spontaneously with aging or when subjected to trauma and biomechanical injury have provided information about the molecular effectors of the disease and potential targets for therapy (Little and Hunter 2013; Vincent et al. 2012; Malfait et al. 2013; Fang and Beier 2014) (for details, see Chap. 2).

The sequence of pathological changes in the cartilage matrix associated with OA development has been elucidated by histological and biochemical analyses of cartilage from human subjects with OA and various animal models of OA (Table 1.1) (Goldring and Goldring 2007; Houard et al. 2013; Loeser et al. 2012a). The earliest changes include the loss of negatively charged glycosaminoglycans resulting in increased water content associated with swelling of the cartilage matrix. Macroscopic changes in the cartilage matrix composition are accompanied by the appearance of surface fibrillations, characterized by microscopic cracks in the superficial zone. As the disease progresses, exfoliation of fragments of cartilage and deep fissures extending into the deeper cartilage layers lead to exposure of the underlying zones of calcified cartilage and subchondral bone.

Accompanying or possibly preceding these matrix changes, chondrocytes undergo marked phenotypic changes, which have been attributed to the adverse effects of excessive mechanical loading that directly injures the cells or modulates chondrocyte function via cell-surface mechanosensors. A requisite early event is the disruption of the PCM associated with the biomechanically induced release of inductive factors such as transforming growth factor (TGF)-β from the matrix and action of proteinases such as HTRA1. This results in the exposure of receptors such as DDR2 and syndecan-4 to extracellular matrix ligands, which are not normally in contact with the chondrocyte. Activation of these receptors and downstream signaling then induces gene expression of proteinases that can directly degrade the major interterritorial matrix components, type II collagen, and aggrecan and produce fragments that can further amplify the degradative process through binding to integrins (Pap and Bertrand 2013; Xu et al. 2014; Loeser 2014) (for further details, see Chap. 8). There is also evidence of increased chondrocyte synthetic activity in regions of enhanced pericellular staining for proteoglycans, reflecting attempts at repair. With progressive matrix depletion, the chondrocytes appear to form clonal clusters suggesting earlier proliferative events. Many of the differentially expressed genes identified in genomic analyses encode proteins that can be detected in these clonal clusters in human OA cartilage (Sandell 2012). There are also vast regions of decreased chondrocyte density containing cell membrane "ghosts" or fragmented chondrocyte nuclei consistent with apoptosis. The accumulation of advanced glycation end products, a feature of aging cartilage that alters the composition and material properties of the matrix, can also be observed. Chondrocytes with features of a senescence-associated secretory phenotype have also been noted in aging cartilage predisposed to OA development (Loeser 2013).

Table 1.1 Changes in the cartilage, synovium, calcified cartilage, and bone during the evolution of osteoarthritis

Alterations in cartilage matrix and chondrocyte function in OA	Increased water content with swelling of the cartilage matrix
	Increased chondrocyte anabolic and catabolic activities (MMPs, ADAMTSs, other proteases)
	Decreased proteoglycan content: increased degradation and decreased synthesis
	Development of surface fibrillations
	Disruption of the collagen network (increased breakdown)
	Fissuring and fragmentation of cartilage matrix
	Chondrocyte death and development of chondrocyte clusters
	Chondrocyte hypertrophy
Contributions of synovium and synovial fluid	Low-grade synovitis associated with the infiltration of mononuclear cells, e.g., activated B and T lymphocytes
	Production of reactive oxygen species, nitric oxide, DAMPs, alarmins, and proinflammatory cytokines
	Production of IL-2, IL-7, IL-15, and IL-21 involved in recruitment, survival, and activity of lymphocytes
	Novel chemokine signature: IL-8, CCL5, CCL19, and its receptor CCR7, as well as MCP-1 and MIP-1β and γ
	Induction of Toll-like receptors by DAMPs and alarmins and activation of NF-kB signaling
	Induction of complement cascade
Alterations in the calcified cartilage and osteochondral junction in OA	Penetration of the calcified cartilage by vascular channels with neural elements
	Expansion of the zone of calcified cartilage with advancement into the articular cartilage
	Duplication of the tidemark
	New bone formation at the osteochondral junction
Bone changes in OA	Increased cortical plate and subchondral cancellous bone remodeling
	Increased cortical plate thickness (initial increase in porosity and reduced stiffness with progression to increased cortical plate stiffness)
	Decreased subchondral cancellous bone mass and alterations in architecture
	Formation of osteophytes
	Development of bone marrow lesions
	Formation of bone cysts
	Bone attrition (altered bone contour)

Proteoglycan degradation is mediated by aggrecanases of the a disintegrin and metalloproteinase with thrombospondin motifs (ADAMTS) family that cleave the aggrecan core protein (Fosang and Beier 2011) (see Chap. 3). The collagen network also becomes susceptible to disruption either by physical forces or by increased activity of matrix metalloproteinases (MMPs) (Wang et al. 2013). The increased activities of degradative enzymes produced by chondrocytes not only disrupt the integrity of the cartilage matrix but also lead to the generation of cartilage matrix breakdown products and secreted damage-associated molecular patterns (DAMPs) and alarmins that contribute to further deregulation of chondrocyte function through Toll-like receptors, integrins, and other cell-surface receptors (Houard et al. 2013; Loeser et al. 2012a; Liu-Bryan and Terkeltaub 2015) (see also Chap. 8). They also act on the adjacent synovial tissue to induce inflammation and the release of proinflammatory products, including cytokines and reactive oxygen species that feed back on the chondrocytes to enhance the catabolic state.

Many of the signaling events activated in chondrocytes by excessive mechanical loading are similar to those induced by inflammatory and oxidative stress (Perera et al. 2010; Nam et al. 2009). During oxidative stress, HIF-2α, induced by NF-kB signaling, directly targets hypoxia-responsive DNA elements in the promoter regions of MMP13, COL10A1, VEGFA, and other genes that are involved in chondrocyte dysfunction in OA (Saito et al. 2010; Yang et al. 2010; Hashimoto et al. 2013). In contrast, the upregulation of HIF-3α by lubricin is associated with downregulation of HIF-1α and HIF-2α target genes as part of a mechanism that protects against aging-related or posttraumatic OA development in mouse models (Ruan et al. 2013). Hypoxia and oxidative stress via induction of the metal regulatory transcription factor-1 (MTF1) and the Zn^{2+} transporter, ZIP8, also regulate the so-called Zinc-ZIP8-MTF1 axis to promote cartilage destructive proteinases in vitro and in vivo (Kim et al. 2014).

Taken together, all of the changes observed in the cartilage matrix and resident chondrocytes over the course of OA initiation and progression have been interpreted as reflecting the influences of local biomechanical conditions to produce both dysfunctional catabolic and anabolic cellular responses. However, these changes in the cartilage cannot be considered in isolation, as the adjacent synovium and subchondral bone play significant roles in OA pathology (Loeser et al. 2012a; Scanzello and Goldring 2012; Goldring and Goldring 2010).

1.5 Influence of Synovitis and Inflammation on OA Pathology

The cartilage is particularly responsive to factors released by the synovium, and the alterations in the cartilage composition and properties induced by the synovial products may in turn adversely affect the adjacent subchondral and periarticular bone. Inflammation, characterized by synovitis, is one of the risk factors for the development of OA (Scanzello and Goldring 2012), and several studies have shown that the presence of synovitis is correlated with OA symptoms and progression of OA joint pathology (Ayral et al. 2005; Hill et al. 2007; Guermazi et al. 2011). The inflamed synovium releases cytokines, chemokines, and other soluble mediators that can be detected in OA synovial fluids and influence chondrocyte functions via cell-surface receptors to induce catabolic and inflammatory mediators (Fig. 1.1). The synovium is also a source of degradative enzymes, including MMPs and aggrecanases that can directly degrade the cartilage matrix (Loeser et al. 2012a), although this occurs more characteristically in RA where there is direct interaction between tissues at the cartilage-synovial pannus junction (Otero and Goldring 2007; Goldring and Marcu 2009).

Synovitis is also observed frequently following joint injury and may be a prognostic indicator of the rate of development of posttraumatic OA (PTOA) (Scanzello et al. 2009, 2011, 2013). Magnetic resonance imaging (MRI) findings correlate with microscopic and macroscopic evaluations of synovitis and suggest that synovitis is often present soon after a traumatic event (Loeuille et al. 2005). The low-grade synovitis following injury is associated with the infiltration of mononuclear cells, including activated B and T lymphocytes, and with the production of inflammatory cytokines (de Lange-Brokaar et al. 2012; Benito et al. 2005; Pearle et al. 2007). Of interest, there is evidence that cartilage damage may not progress rapidly without inflammation in the synovial compartment (Scanzello et al. 2011, 2013; Scanzello and Goldring 2012).

Many proinflammatory cytokines, adipokines, and chemokines have been implicated in the initiation and development of OA joint pathology, including the perpetuation of synovitis and cartilage damage (Wojdasiewicz et al. 2014). In contrast to the findings in rheumatoid arthritis, the involvement of IL-1 and TNF-α in the pathogenesis of OA synovitis and pathology remains controversial. There is evidence, however, implicating other cytokines, including the common γ-chain family of cytokines, IL-2, IL-7, IL-15, and IL-21, which are involved in recruitment, survival, and activity of lymphocytes (Scanzello et al. 2009).

Multiple chemokines have also been identified in synovial tissues and synovial fluid from patients with OA, including IL-8, CCL5, CCL19, and its receptor CCR7, as well as MCP-1 and MIP-1β and γ (Scanzello et al. 2009, 2011; Haringman et al. 2006; Gobezie et al. 2007; Endres et al. 2010; Ritter et al. 2013) (see also Chap. 5). Of interest is the finding that several of the chemokines isolated from synovial fluid are chemotactic for human subchondral bone progenitor cells (Endres et al. 2010).

There is increasing evidence implicating a role for innate immunity in the pathogenesis of the synovial inflammation in OA (de Lange-Brokaar et al. 2012; Scanzello and Goldring 2012; Berenbaum 2013; Haseeb and Haqqi 2013). In the context of OA, activation of the innate response is initiated by stimulation of TLRs, which are pattern-recognition receptors that were identified initially as receptors for microbial products generated during infection. These receptors are now known to play key roles during cellular stress and are responsive to products of extracellular matrix damage, including cartilage degradation products, during sterile tissue injury (Piccinini and Midwood 2010; Liu-Bryan and Terkeltaub 2015). TLRs have been detected in both chondrocytes and synovial cells in tissues from OA patients where they can be activated by the endogenous DAMPS released from injured cartilage and other joint tissues, including the bone (Schelbergen et al. 2012; Sohn et al. 2012; Nair et al. 2012; Gomez et al. 2014). Ligand-mediated TLR activation in chondrocytes and synovial cells is an important stimulus for canonical NF-kB signaling, the major pathway mediating the gene expression of a broad spectrum of proinflammatory mediators and products, including chemokines and cytokines (e.g., IL-1, IL-6, and TNF-α) (Marcu et al. 2010).

Gene profiling studies of cartilage or whole joints from animal models of OA have helped to identify pathways that are involved in deregulation of chondrocyte function in OA pathogenesis (Appleton et al. 2007; Olex et al. 2014). Both age-related and injury-induced OA pathologies (Loeser et al. 2012b, 2013), as well as spontaneous OA models such as the Str/ort mouse (Poulet et al. 2012), are associated with upregulation of genes of the senescence-associated secretory phenotype, including inflammatory cytokines, chemokines, MMPs, and immune and defense response genes (Loeser et al. 2012b). The prominence of the NF-kB pathway as a common thread among the different gene signatures has been emphasized in these models, suggesting a key regulatory role for stress and inflammatory signaling via canonical NF-kB signaling in human OA (Goldring et al. 2011; Marcu et al. 2010). The findings that mechanical stimuli modulate NF-kB signaling (Nam et al. 2009) provide an explanation for why NF-kB-related gene signatures may be upregulated in mouse models of PTOA in the absence of overt signs of inflammation such as synovitis and immune cell infiltration. Profiling studies have revealed that inflammatory signatures are present before the appearance of overt OA and, in some models, are associated with increased numbers of activated T and B lymphocytes in the spleens of the mice destined to develop OA (Poulet et al. 2012).

The complement cascade is one of the major effector mechanisms of the immune system, and activation of the complement cascade plays an important role in the pathogenesis of a broad spectrum of inflammatory and autoimmune disorders (Holers 2014). There is evidence that both chondrocytes and synovial macrophages can produce complement components and inhibitors, and previous studies have demonstrated increased synovial complement component deposition in the setting of acute flare-ups of OA and in synovial fluids from patients with OA (Konttinen et al. 1996; Wang et al. 2011). A recent study by Wang et al. (2011) showed that mice with impaired ability to activate the complement system were partially protected from the development of OA, providing direct evidence for a role of the complement system in OA pathogenesis, at least in murine models. A role for the complement system in human subjects requires further study.

1.6 Osteophytes

Osteophytes do not originate from the cartilage, but are bony outgrowths that grow by a process of endochondral ossification involving TGF-β and bone morphogenetic protein-2 (BMP-2), through proliferation of periosteal cells at the joint margin, differentiation into cells with morphological features of hypertrophic chondrocytes, mineralization, and replacement of the endochondral tissue with the bone (van der

Kraan and van den Berg 2007). Osteophytes are localized to the joint margins and are a radiographic hallmark of OA. However, rather than contributing to joint dysfunction and OA progression, they may serve to stabilize the joint (Guyton and Brand 2002; Messent et al. 2007; van der Kraan and van den Berg 2007), based on the finding that removal of osteophytes from the knee joint in animal models induces joint instability (Pottenger et al. 1990). In addition, there is no relationship between osteophyte size and the risk for development of structural progression of knee OA in human subjects (Felson et al. 2005).

1.7 Cross Talk Between Articular Cartilage and Subchondral Bone

The articular cartilage, calcified cartilage, and subchondral cortical and trabecular bone in diarthrodial joints form a biocomposite that is uniquely adapted to allow the transfer of load. The concept that an impermeable barrier consisting of the calcified cartilage and tidemark, the histologically defined demarcation between the deep articular and calcified cartilage, prevents the transfer of solutes and soluble products between the articular cartilage and subchondral bone has been challenged by studies showing direct contact between bone and cartilage cells and tissues in localized regions in the osteochondral junction (Brower et al. 1962; Imhof et al. 2000; Lyons et al. 2006; Pan et al. 2009). Microcracks produced by mechanical stress and exacerbation of naturally occurring pores in the subchondral bone plate provide conduits for vascular invasion into the calcified cartilage and enable cross talk through diffusion of small molecules (Lane et al. 1977; Burr and Schaffler 1997; Bullough 2004; Imhof et al. 2000). TGF-β-mediated angiogenesis, potentially among the earliest events driving OA (Zhen et al. 2013), is followed by vascular invasion into the osteochondral junction (Pan et al. 2012; Saito et al. 2012). Examination of OA mouse models by microCT shows early and temporal subchondral plate porosity and increased perforation with enhanced biochemical and mechanical interactions among the subchondral trabeculae, bone marrow cells, and articular cartilage (Botter et al. 2011; Gu et al. 2012) (Fig. 1.1).

The expansion of the calcified cartilage, associated with tidemark duplication, into the deep zones of the articular cartilage leads to local cartilage thinning, thereby compromising the mechanical function of the osteochondral unit (Burr and Gallant 2012). This process may be mediated by chondrocytes that appear in the deep zone OA cartilage contiguous with the calcified cartilage and have phenotypic features that recapitulate a developmental molecular program, including expression of genes encoding type X collagen, MMP13, and transcription factors such as Runx2 (van der Kraan and van den Berg 2012). These hypertrophic-like chondrocytes also are a source of proangiogenic factors, including VEGF, that play a role in promoting vascular penetration of the calcified cartilage at the osteochondral junction through channels originating at sites of microcracks and fissures, accompanied by sensory and sympathetic nerves that may play a role in joint pain (Ashraf et al. 2011a, b; Stoppiello et al. 2014) (for details, see Chap. 9). Cuffs of new bone form around these channels involving increased osteoclast activity, infiltration of inflammatory cells into the marrow spaces, increased endothelial cell proliferation and vascular density, and localized bone marrow replacement by fibrovascular tissue expressing VEGF (Suri and Walsh 2012; Walsh et al. 2010).

Additional in vitro findings provide direct evidence that interactions between bone cells and chondrocytes can modulate phenotypic expression in each cell type and also demonstrate the differential effects of osteoblasts from normal versus OA bone. Coculture of bovine articular cartilage explants with fragments of subchondral bone can increase chondrocyte survival compared to explants cultured alone (Amin et al. 2009). In cocultures of chondrocytes and osteoblasts separated by semipermeable membranes to allow transfer of soluble mediators, osteoblasts derived from the "sclerotic" bone from patients with OA have the capacity to reduce the expression of genes encoding aggrecan (ACAN), collagen type II (COL2A1), and the transcription factor Sox9 and to increase expression of MMP3 and MMP13 compared to osteoblasts from normal, "nonsclerotic" bone. Pretreatment of the osteoblasts with IL-1, IL-6, or oncostatin M can further enhance MMP3 and MMP13 production (Sanchez et al. 2008). Coculture also upregulates MMP13, COL2A1, osteopontin, and alkaline phosphatase in osteoblasts. Gene expression profiling of osteoblasts cultured from OA bone specimens compared to osteoblasts obtained from comparable regions of normal bone shows differential gene expression that may account for the findings obtained in coculture studies (Hilal et al. 1998; Zhang et al. 2012; Chou et al. 2013).

Among the important molecular networks that play roles in cartilage-bone cross talk in OA are the TGF-β/BMP, fibroblast growth factor (FGF), and Wnt/β-catenin pathways. All three pathways are involved in regulating cartilage and bone development, are dysregulated during OA pathogenesis, and have been used as additive factors in tissue engineering strategies (Baron and Kneissel 2013; Lories et al. 2013).

TGF-β signaling has roles in cartilage and bone homeostasis, as well as in OA pathogenesis (Bush and Beier 2013). This may be explained partly by the availability of receptors and downstream intracellular signaling molecules. In normal articular chondrocytes, the canonical receptor, ALK5, activates Smad2/3 to inhibit chondrocyte hypertrophy, whereas in OA chondrocytes the decreased ratio of ALK5 to ALK1, which signals through BMP-related Smad1/5/8, results in induction of MMP13 and increased cartilage catabolism (Blaney Davidson et al. 2009; van der Kraan et al. 2012; van den Bosch et al. 2014; van der Kraan 2014), associated with increased nerve growth factor (NGF) (Blaney Davidson et al. 2014). These aberrant chondrocyte responses may account for the off-target effects of TGF-β injections in joints that result in osteophyte formation and synovial fibrosis. Studies in animal models of OA also show the differential effects of TGF-ß inhibitors depending upon the dose and mode of administration. For example, the implantation of a TGF-ß-specific antibody in the subchondral bone attenuates both cartilage loss and OA bone changes. Furthermore, systemic administration of low doses of a TGF-ß inhibitor can prevent the migration of mesenchymal stem cells in the subchondral bone of operated limbs and attenuate neovascularization and cartilage loss (Zhen et al. 2013).

FGF signaling is important in skeletal development and cartilage and bone homeostasis (Vincent 2012), where the differential use of receptors determines a temporal and spatial program of cellular differentiation. FGF-2, for example, preserves chondrocyte phenotype through FGFR3 but promotes cartilage destruction via FGFR1 (Yan et al. 2011). Of the four receptors, FGFR1 and 3 are the most abundant, and OA cartilage has a reduced ratio of FGFR3 to FGFR1. In vitro and in vivo studies have shown that FGF-18 via FGFR3 protects against cartilage damage (Barr et al. 2014; Mori et al. 2014). The potential of FGF-18 and other anabolic factors to promote cartilage regeneration tissue engineering approaches is a subject of intensive study (Ellman et al. 2013). Proof-of-concept clinical trials in humans to address structural end points by MRI and radiographic joint space narrowing and pain are needed (Lohmander et al. 2014).

Wnt signaling in the cartilage and bone is also complex and is determined by the availability of Wnt ligands in different tissues and their use of distinct, canonical versus noncanonical pathways during skeletal development and growth versus OA pathogenesis (Baron and Kneissel 2013; Lories et al. 2013). A number of Wnt antagonists have been identified in the bone and cartilage and explored as OA therapeutics or targets. These include dickkopf-related protein-1 (DKK-1), secreted frizzled-related protein 1 (sFRP1), and sclerostin (SOST). DKK-1 and sFRP1, in addition to affecting bone homeostasis, also can have profound affects in cartilage by modulating hypertrophic differentiation (Leijten et al. 2012; van den Bosch et al. 2014) and expression of proteinases (Oh et al. 2012; Bougault et al. 2014). DKK-1 has differential effects in the cartilage and bone depending on dosage and mode of delivery. The inhibitory effect of DKK-1 on cartilage degradation is related to its capacity to inhibit VEGF production by osteocytes and osteoblasts in the bone (Funck-Brentano et al. 2014).

SOST, identified originally as specific to osteocytes, can be detected in cartilage and in osteochondral tissues, plasma, and synovial fluid in OA patients and animal models (Chan et al. 2011; Mabey et al. 2014). SOST inhibitors increase bone formation (Lewiecki 2014), but their effects in cartilage and OA are controversial (Roudier et al. 2013; Lories et al. 2013). Conflicting findings may be due partially to cross talk between Wnt signaling and TGF-β/BMP pathways that disrupt homeostasis by inducing endochondral ossification during osteophyte formation (Gelse et al. 2012) or promoting the switch toward BMP-related ALK1 and Smad1/5/8 (van den Bosch et al. 2014). In addition, TGF-β is required for the suppression of SOST by mechanical load (Nguyen et al. 2013).

The differential capacity of the bone and cartilage to adapt to changes in local biomechanics and injury is an important factor in the development of OA joint pathology and has implications for designing therapeutic interventions. In animal models of traumatic joint injury, the onset of bone pathology can be linked to a specific initiating event (Anderson et al. 2011; Christiansen et al. 2015). Importantly all of the studies demonstrate a strong association between local changes in subchondral bone and cartilage pathology, thereby supporting the concept that the cellular and morphological adaptations to local mechanical factors occur in both tissues.

There is, however, considerable controversy about whether OA pathology begins in the subchondral bone or the overlying articular cartilage. Although the articular cartilage is designed to deform under compressive loads without failure, it has a lower capacity to withstand tension or shear stresses that occur during joint motion (Burr and Gallant 2012). Under normal conditions, the subchondral bone can protect the overlying articular cartilage from damage by distributing loads. During OA development, alterations in the contour of the subchondral bone or heterogeneity in bone density or stiffness create local shear stresses that increase deformation of the cartilage and predispose it to splitting and fibrillation, since it is not able to adapt as rapidly as the bone by modifying its matrix composition and mechanical properties (Burr and Gallant 2012). The initial fibrillation and splitting of the articular cartilage reflect the acute effects of the loading and are followed by alterations in cartilage proteoglycan turnover and progressive activation of a catabolic program in chondrocytes. The increased thickness of the zone of calcified cartilage also contributes to the thinning of the articular cartilage (Radin et al. 1984; Radin and Rose 1986).

Insights into the sequential changes in both articular cartilage and periarticular bone have been obtained in noninvasive tibial loading models in mice in which the magnitude and duration of cyclic compressive loading can be defined (Poulet et al. 2011; Ko et al. 2013). The bone changes involving increased subchondral cortical bone plate thickness occurs in the predicted region of increased load transfer and corresponds to the region with cartilage pathology. In contrast to the effects on the cortical bone, the trabecular epiphyseal bone mass decreases, consistent with stress shielding of this region by the thickened subchondral plate (Ko et al. 2013). These findings are consistent with the observations in patients with established OA, in whom there is loss of subchondral cancellous bone associated with trabecular thinning (Karvonen et al. 1998; Burr and Gallant 2012).

In contrast, a single loading session, which does not acutely alter chondrocyte viability or articular cartilage or bone integrity, induces proteoglycan loss and cartilage thinning consistent with activation of a progressive catabolic chondrocyte response (Ko et al. 2016); however, bone remodeling increases only transiently and begins to resolve at later time points. These findings provide evidence that a single event of traumatic loading produces sustained adverse effects on chondrocyte function that are not reversible and lead to progressive cartilage loss, whereas the bone responds rapidly to the initial loading by changing cellular remodeling activities and eventually reestablishes a physiological state.

Further evidence of an association between the cartilage and bone in the response to trauma is provided by the association of bone marrow lesions with regions of OA cartilage pathology and their persistence as predictive of progressive cartilage loss and joint pain (Kothari et al. 2010; Roemer et al. 2009, 2011, 2015; Hunter et al. 2006). Originally termed "bone marrow edema" by Wilson and colleagues in 1988 (Wilson et al. 1988), they can be detected using fluid-sensitive MRI sequences in localized regions of increased signal intensity in the subchondral bone in human subjects with OA. However, histological analyses of regions corresponding to the anatomical sites of the bone marrow lesions have revealed the presence of fat necrosis, localized marrow fibrosis, and vascular changes associated with microfractures of the trabecular bone at various stages of healing (Zanetti et al. 2000; Taljanovic et al. 2008; Leydet-Quilici et al. 2010). These findings suggest that the MRI signal is not generated by actual "edema" but rather by an active cellular process associated with local regions of bone damage. Bone marrow lesions are especially common in regions of denuded cartilage (Bowes et al. 2015), indicating that the overlying cartilage has a role in distributing mechanical forces and protecting the subchondral bone from the adverse effects of excessive load. The anatomic association of the cartilage and bone pathology provides further evidence that both tissues are responsive to local environmental conditions and that interactions between them affect their composition and structural properties in a reciprocal manner.

References

Amin AK, Huntley JS, Simpson AH, Hall AC (2009) Chondrocyte survival in articular cartilage: the influence of subchondral bone in a bovine model. J Bone Joint Surg Br 91(5):691–699. doi:10.1302/0301-620X.91B5.21544

Anderson DD, Chubinskaya S, Guilak F, Martin JA, Oegema TR, Olson SA, Buckwalter JA (2011) Post-traumatic osteoarthritis: improved understanding and opportunities for early intervention. J Orthop Res 29(6):802–809. doi:10.1002/jor.21359

Appleton CT, Pitelka V, Henry J, Beier F (2007) Global analyses of gene expression in early experimental osteoarthritis. Arthritis Rheum 56(6):1854–1868. doi:10.1002/art.22711

Ashraf S, Mapp PI, Walsh DA (2011a) Contributions of angiogenesis to inflammation, joint damage, and pain in a rat model of osteoarthritis. Arthritis Rheum 63(9):2700–2710. doi:10.1002/art.30422

Ashraf S, Wibberley H, Mapp PI, Hill R, Wilson D, Walsh DA (2011b) Increased vascular penetration and nerve growth in the meniscus: a potential source of pain in osteoarthritis. Ann Rheum Dis 70(3):523–529. doi:10.1136/ard.2010.137844

Ayral X, Pickering EH, Woodworth TG, Mackillop N, Dougados M (2005) Synovitis: a potential predictive factor of structural progression of medial tibiofemoral knee osteoarthritis – results of a 1 year longitudinal arthroscopic study in 422 patients. Osteoarthritis Cartilage 13(5):361–367

Baron R, Kneissel M (2013) WNT signaling in bone homeostasis and disease: from human mutations to treatments. Nat Med 19(2):179–192. doi:10.1038/nm.3074

Barr L, Getgood A, Guehring H, Rushton N, Henson FM (2014) The effect of recombinant human fibroblast growth factor-18 on articular cartilage following single impact load. J Orthop Res 32(7):923–927. doi:10.1002/jor.22622

Benito MJ, Veale DJ, FitzGerald O, van den Berg WB, Bresnihan B (2005) Synovial tissue inflammation in early and late osteoarthritis. Ann Rheum Dis 64(9):1263–1267. doi:10.1136/ard.2004.025270

Berenbaum F (2013) Osteoarthritis as an inflammatory disease (osteoarthritis is not osteoarthrosis!). Osteoarthritis Cartilage 21(1):16–21. doi:10.1016/j.joca.2012.11.012

Bijlsma JW, Berenbaum F, Lafeber FP (2011) Osteoarthritis: an update with relevance for clinical practice. Lancet 377(9783):2115–2126. doi:10.1016/S0140-6736(11)60243-2, S0140-6736(11)60243-2 [pii]

Blanco FJ, Ruiz-Romero C (2012) Osteoarthritis: metabolomic characterization of metabolic phenotypes in OA. Nat Rev Rheumatol 8(3):130–132. doi:10.1038/nrrheum.2012.11

Blaney Davidson EN, Remst DF, Vitters EL, van Beuningen HM, Blom AB, Goumans MJ, van den Berg WB, van der Kraan PM (2009) Increase in ALK1/ALK5 ratio as a cause for elevated

MMP-13 expression in osteoarthritis in humans and mice. J Immunol 182(12):7937–7945. doi:10.4049/jimmunol.0803991
Blaney Davidson EN, van Caam AP, Vitters EL, Bennink MB, Thijssen E, van den Berg WB, Koenders MI, van Lent PL, van de Loo FA, van der Kraan PM (2014) TGF-beta is a potent inducer of Nerve Growth Factor in articular cartilage via the ALK5-Smad2/3 pathway. Potential role in OA related pain? Osteoarthritis Cartilage. doi:10.1016/j.joca.2014.12.005
Botter SM, van Osch GJ, Clockaerts S, Waarsing JH, Weinans H, van Leeuwen JP (2011) Osteoarthritis induction leads to early and temporal subchondral plate porosity in the tibial plateau of mice: an in vivo microfocal computed tomography study. Arthritis Rheum 63(9):2690–2699. doi:10.1002/art.30307
Bougault C, Priam S, Houard X, Pigenet A, Sudre L, Lories RJ, Jacques C, Berenbaum F (2014) Protective role of frizzled-related protein B on matrix metalloproteinase induction in mouse chondrocytes. Arthritis Res Ther 16(4):R137. doi:10.1186/ar4599
Bowes MA, McLure SW, Wolstenholme CB, Vincent GR, Williams S, Grainger A, Conaghan PG (2015) Osteoarthritic bone marrow lesions almost exclusively colocate with denuded cartilage: a 3D study using data from the Osteoarthritis Initiative. Ann Rheum Dis. doi:10.1136/annrheumdis-2015-208407
Brower TD, Akahoshi Y, Orlic P (1962) The diffusion of dyes thorugh articular cartilage in vivo. J Bone Joint Surg 44-A:456–463
Buckwalter JA, Brown TD (2004) Joint injury, repair, and remodeling: roles in post-traumatic osteoarthritis. Clin Orthop Relat Res 423:7–16
Bullough PG (2004) The role of joint architecture in the etiology of arthritis. Osteoarthritis Cartilage 12 Suppl A:S2–S9
Burr DB, Gallant MA (2012) Bone remodelling in osteoarthritis. Nat Rev Rheumatol 8(11):665–673. doi:10.1038/nrrheum.2012.130
Burr DB, Schaffler MB (1997) The involvement of subchondral mineralized tissues in osteoarthrosis: quantitative microscopic evidence. Microsc Res Tech 37(4):343–357
Bush JR, Beier F (2013) TGF-beta and osteoarthritis – the good and the bad. Nat Med 19(6):667–669. doi:10.1038/nm.3228
Chan BY, Fuller ES, Russell AK, Smith SM, Smith MM, Jackson MT, Cake MA, Read RA, Bateman JF, Sambrook PN, Little CB (2011) Increased chondrocyte sclerostin may protect against cartilage degradation in osteoarthritis. Osteoarthritis Cartilage 19(7):874–885. doi:10.1016/j.joca.2011.04.014
Chou CH, Wu CC, Song IW, Chuang HP, Lu LS, Chang JH, Kuo SY, Lee CH, Wu JY, Chen YT, Kraus VB, Lee MT (2013) Genome-wide expression profiles of subchondral bone in osteoarthritis. Arthritis Res Ther 15(6):R190. doi:10.1186/ar4380
Christiansen BA, Guilak F, Lockwood KA, Olson SA, Pitsillides AA, Sandell LJ, Silva MJ, van der Meulen MC, Haudenschild DR (2015) Non-invasive mouse models of post-traumatic osteoarthritis. Osteoarthritis Cartilage 23(10):1627–1638. doi:10.1016/j.joca.2015.05.009
Conaghan PG (2013) Osteoarthritis in 2012: parallel evolution of OA phenotypes and therapies. Nat Rev Rheumatol 9(2):68–70. doi:10.1038/nrrheum.2012.225
de Lange-Brokaar BJ, Ioan-Facsinay A, van Osch GJ, Zuurmond AM, Schoones J, Toes RE, Huizinga TW, Kloppenburg M (2012) Synovial inflammation, immune cells and their cytokines in osteoarthritis: a review. Osteoarthritis Cartilage 20(12):1484–1499. doi:10.1016/j.joca.2012.08.027
Detert J, Klaus P, Listing J, Hohne-Zimmer V, Braun T, Wassenberg S, Rau R, Buttgereit F, Burmester GR (2014) Hydroxychloroquine in patients with inflammatory and erosive osteoarthritis of the hands (OA TREAT): study protocol for a randomized controlled trial. Trials 15:412. doi:10.1186/1745-6215-15-412
Ellman MB, Yan D, Ahmadinia K, Chen D, An HS, Im HJ (2013) Fibroblast growth factor control of cartilage homeostasis. J Cell Biochem 114(4):735–742. doi:10.1002/jcb.24418
Endres M, Andreas K, Kalwitz G, Freymann U, Neumann K, Ringe J, Sittinger M, Haupl T, Kaps C (2010) Chemokine profile of synovial fluid from normal, osteoarthritis and rheumatoid

arthritis patients: CCL25, CXCL10 and XCL1 recruit human subchondral mesenchymal progenitor cells. Osteoarthritis Cartilage 18(11):1458–1466. doi:10.1016/j.joca.2010.08.003
Englund M, Lohmander LS (2004) Risk factors for symptomatic knee osteoarthritis fifteen to twenty-two years after meniscectomy. Arthritis Rheum 50(9):2811–2819
Englund M, Guermazi A, Roemer FW, Aliabadi P, Yang M, Lewis CE, Torner J, Nevitt MC, Sack B, Felson DT (2009) Meniscal tear in knees without surgery and the development of radiographic osteoarthritis among middle-aged and elderly persons: the Multicenter Osteoarthritis Study. Arthritis Rheum 60(3):831–839. doi:10.1002/art.24383
Fang H, Beier F (2014) Mouse models of osteoarthritis: modelling risk factors and assessing outcomes. Nat Rev Rheumatol 10(7):413–421. doi:10.1038/nrrheum.2014.46
Felson DT, Gale DR, Elon Gale M, Niu J, Hunter DJ, Goggins J, Lavalley MP (2005) Osteophytes and progression of knee osteoarthritis. Rheumatology (Oxford) 44(1):100–104. doi:10.1093/rheumatology/keh411
Fosang AJ, Beier F (2011) Emerging Frontiers in cartilage and chondrocyte biology. Best Pract Res Clin Rheumatol 25(6):751–766. doi:10.1016/j.berh.2011.11.010
Funck-Brentano T, Bouaziz W, Marty C, Geoffroy V, Hay E, Cohen-Solal M (2014) Dkk1-mediated inhibition of Wnt signaling in bone ameliorates osteoarthritis. Arthritis Rheum. doi:10.1002/art.38799
Gelse K, Ekici AB, Cipa F, Swoboda B, Carl HD, Olk A, Hennig FF, Klinger P (2012) Molecular differentiation between osteophytic and articular cartilage – clues for a transient and permanent chondrocyte phenotype. Osteoarthritis Cartilage 20(2):162–171. doi:10.1016/j.joca.2011.12.004
Gobezie R, Kho A, Krastins B, Sarracino DA, Thornhill TS, Chase M, Millett PJ, Lee DM (2007) High abundance synovial fluid proteome: distinct profiles in health and osteoarthritis. Arthritis Res Ther 9(2):R36. doi:10.1186/ar2172
Goldring MB, Berenbaum F (2015) Emerging targets in osteoarthritis therapy. Curr Opin Pharmacol 22:51–63. doi:10.1016/j.coph.2015.03.004
Goldring MB, Goldring SR (2007) Osteoarthritis. J Cell Physiol 213(3):626–634. doi:10.1002/jcp.21258
Goldring MB, Goldring SR (2010) Articular cartilage and subchondral bone in the pathogenesis of osteoarthritis. Ann N Y Acad Sci 1192:230–237. doi:10.1111/j.1749-6632.2009.05240.x
Goldring MB, Marcu KB (2009) Cartilage homeostasis in health and rheumatic diseases. Arthritis Res Ther 11(3):224. doi:10.1186/ar2592
Goldring MB, Otero M (2011) Inflammation in osteoarthritis. Curr Opin Rheumatol 23(5):471–478. doi:10.1097/BOR.0b013e328349c2b1
Goldring MB, Otero M, Plumb DA, Dragomir C, Favero M, El Hachem K, Hashimoto K, Roach HI, Olivotto E, Borzi RM, Marcu KB (2011) Roles of inflammatory and anabolic cytokines in cartilage metabolism: signals and multiple effectors converge upon MMP-13 regulation in osteoarthritis. Eur Cell Mater 21:202–220
Gomez R, Villalvilla A, Largo R, Gualillo O, Herrero-Beaumont G (2014) TLR4 signalling in osteoarthritis-finding targets for candidate DMOADs. Nat Rev Rheumatol. doi:10.1038/nrrheum.2014.209
Gu XI, Palacio-Mancheno PE, Leong DJ, Borisov YA, Williams E, Maldonado N, Laudier D, Majeska RJ, Schaffler MB, Sun HB, Cardoso L (2012) High resolution micro arthrography of hard and soft tissues in a murine model. Osteoarthritis Cartilage 20(9):1011–1019. doi:10.1016/j.joca.2012.05.004
Guermazi A, Roemer FW, Hayashi D, Crema MD, Niu J, Zhang Y, Marra MD, Katur A, Lynch JA, El-Khoury GY, Baker K, Hughes LB, Nevitt MC, Felson DT (2011) Assessment of synovitis with contrast-enhanced MRI using a whole-joint semiquantitative scoring system in people with, or at high risk of, knee osteoarthritis: the MOST study. Ann Rheum Dis 70(5):805–811. doi:10.1136/ard.2010.139618
Guyton GP, Brand RA (2002) Apparent spontaneous joint restoration in hip osteoarthritis. Clin Orthop Relat Res 404:302–307

Haringman JJ, Smeets TJ, Reinders-Blankert P, Tak PP (2006) Chemokine and chemokine receptor expression in paired peripheral blood mononuclear cells and synovial tissue of patients with rheumatoid arthritis, osteoarthritis, and reactive arthritis. Ann Rheum Dis 65(3):294–300. doi:10.1136/ard.2005.037176

Haseeb A, Haqqi TM (2013) Immunopathogenesis of osteoarthritis. Clin Immunol 146(3):185–196. doi:10.1016/j.clim.2012.12.011

Hashimoto K, Otero M, Imagawa K, de Andres MC, Coico JM, Roach HI, Oreffo RO, Marcu KB, Goldring MB (2013) Regulated transcription of human matrix metalloproteinase 13 (MMP13) and interleukin-1beta (IL1B) genes in chondrocytes depends on methylation of specific proximal promoter CpG sites. J Biol Chem 288(14):10061–10072. doi:10.1074/jbc.M112.421156

Heinegard D, Saxne T (2011) The role of the cartilage matrix in osteoarthritis. Nat Rev Rheumatol 7(1):50–56. doi:10.1038/nrrheum.2010.198

Henrotin Y (2014) Does signaling pathway inhibition hold therapeutic promise for osteoarthritis? Joint Bone Spine 81(4):281–283. doi:10.1016/j.jbspin.2014.03.002

Hilal G, Martel-Pelletier J, Pelletier JP, Ranger P, Lajeunesse D (1998) Osteoblast-like cells from human subchondral osteoarthritic bone demonstrate an altered phenotype in vitro: possible role in subchondral bone sclerosis. Arthritis Rheum 41(5):891–899. doi:10.1002/1529-0131(199805)41:5<891::AID-ART17>3.0.CO;2-X

Hill CL, Hunter DJ, Niu J, Clancy M, Guermazi A, Genant H, Gale D, Grainger A, Conaghan P, Felson DT (2007) Synovitis detected on magnetic resonance imaging and its relation to pain and cartilage loss in knee osteoarthritis. Ann Rheum Dis 66(12):1599–1603. doi:10.1136/ard.2006.067470

Ho L, Ali SA, Al-Jazrawe M, Kandel R, Wunder JS, Alman BA (2013) Primary cilia attenuate hedgehog signalling in neoplastic chondrocytes. Oncogene 32(47):5388–5396. doi:10.1038/onc.2012.588

Holers VM (2014) Complement and its receptors: new insights into human disease. Annu Rev Immunol 32:433–459. doi:10.1146/annurev-immunol-032713-120154

Houard X, Goldring MB, Berenbaum F (2013) Homeostatic mechanisms in articular cartilage and role of inflammation in osteoarthritis. Curr Rheumatol Rep 15(11):375. doi:10.1007/s11926-013-0375-6

Hsueh MF, Onnerfjord P, Kraus VB (2014) Biomarkers and proteomic analysis of osteoarthritis. Matrix Biol 39:56–66. doi:10.1016/j.matbio.2014.08.012

Hunter DJ, Zhang Y, Niu J, Goggins J, Amin S, LaValley MP, Guermazi A, Genant H, Gale D, Felson DT (2006) Increase in bone marrow lesions associated with cartilage loss: a longitudinal magnetic resonance imaging study of knee osteoarthritis. Arthritis Rheum 54(5):1529–1535. doi:10.1002/art.21789

Hunter DJ, Eckstein F, Kraus VB, Losina E, Sandell L, Guermazi A (2013) Imaging biomarker validation and qualification report: sixth OARSI Workshop on Imaging in Osteoarthritis combined with third OA Biomarkers Workshop. Osteoarthritis Cartilage 21(7):939–942. doi:10.1016/j.joca.2013.04.014

Hunziker EB, Lippuner K, Shintani N (2014) How best to preserve and reveal the structural intricacies of cartilaginous tissue. Matrix Biol 39:33–43. doi:10.1016/j.matbio.2014.08.010

Imhof H, Sulzbacher I, Grampp S, Czerny C, Youssefzadeh S, Kainberger F (2000) Subchondral bone and cartilage disease: a rediscovered functional unit. Invest Radiol 35(10):581–588

Jay GD, Waller KA (2014) The biology of lubricin: near frictionless joint motion. Matrix Biol 39:17–24. doi:10.1016/j.matbio.2014.08.008

Jones HP, Appleyard RC, Mahajan S, Murrell GA (2003) Meniscal and chondral loss in the anterior cruciate ligament injured knee. Sports Med 33(14):1075–1089

Karvonen RL, Miller PR, Nelson DA, Granda JL, Fernandez-Madrid F (1998) Periarticular osteoporosis in osteoarthritis of the knee. J Rheumatol 25(11):2187–2194

Kim JH, Jeon J, Shin M, Won Y, Lee M, Kwak JS, Lee G, Rhee J, Ryu JH, Chun CH, Chun JS (2014) Regulation of the catabolic cascade in osteoarthritis by the zinc-ZIP8-MTF1 axis. Cell 156(4):730–743. doi:10.1016/j.cell.2014.01.007

Kloppenburg M (2014) Hand osteoarthritis-nonpharmacological and pharmacological treatments. Nat Rev Rheumatol 10(4):242–251. doi:10.1038/nrrheum.2013.214

Knight MM, McGlashan SR, Garcia M, Jensen CG, Poole CA (2009) Articular chondrocytes express connexin 43 hemichannels and P2 receptors – a putative mechanoreceptor complex involving the primary cilium? J Anat 214(2):275–283. doi:10.1111/j.1469-7580.2008.01021.x

Ko FC, Dragomir C, Plumb DA, Goldring SR, Wright TM, Goldring MB, van der Meulen MC (2013) In vivo cyclic compression causes cartilage degeneration and subchondral bone changes in mouse tibiae. Arthritis Rheum 65(6):1569–1578. doi:10.1002/art.37906

Ko FC, Dragomir CL, Plumb DA, Hsia AW, Adebayo OO, Goldring SR, Wright TM, Goldring MB, van der Meulen MC (2016) Progressive cell-mediated changes in articular cartilage and bone in mice are initiated by a single session of controlled cyclic compressive loading. J Orthop Res. doi:10.1002/jor.23204

Konttinen YT, Ceponis A, Meri S, Vuorikoski A, Kortekangas P, Sorsa T, Sukura A, Santavirta S (1996) Complement in acute and chronic arthritides: assessment of C3c, C9, and protectin (CD59) in synovial membrane. Ann Rheum Dis 55(12):888–894

Kothari A, Guermazi A, Chmiel JS, Dunlop D, Song J, Almagor O, Marshall M, Cahue S, Prasad P, Sharma L (2010) Within-subregion relationship between bone marrow lesions and subsequent cartilage loss in knee osteoarthritis. Arthritis Care Res (Hoboken) 62(2):198–203. doi:10.1002/acr.20068

Lafeber FP, van Spil WE (2013) Osteoarthritis year 2013 in review: biomarkers; reflecting before moving forward, one step at a time. Osteoarthritis Cartilage 21(10):1452–1464. doi:10.1016/j.joca.2013.08.012

Lane LB, Villacin A, Bullough PG (1977) The vascularity and remodelling of subchondrial bone and calcified cartilage in adult human femoral and humeral heads. An age- and stress-related phenomenon. J Bone Joint Surg Br 59(3):272–278

Leijten JC, Emons J, Sticht C, van Gool S, Decker E, Uitterlinden A, Rappold G, Hofman A, Rivadeneira F, Scherjon S, Wit JM, van Meurs J, van Blitterswijk CA, Karperien M (2012) Gremlin 1, frizzled-related protein, and Dkk-1 are key regulators of human articular cartilage homeostasis. Arthritis Rheum 64(10):3302–3312. doi:10.1002/art.34535

Lewiecki EM (2014) Role of sclerostin in bone and cartilage and its potential as a therapeutic target in bone diseases. Ther Adv Musculoskelet Dis 6(2):48–57. doi:10.1177/1759720X13510479

Leydet-Quilici H, Le Corroller T, Bouvier C, Giorgi R, Argenson JN, Champsaur P, Pham T, de Paula AM, Lafforgue P (2010) Advanced hip osteoarthritis: magnetic resonance imaging aspects and histopathology correlations. Osteoarthritis Cartilage 18(11):1429–1435. doi:10.1016/j.joca.2010.08.008

Little CB, Hunter DJ (2013) Post-traumatic osteoarthritis: from mouse models to clinical trials. Nat Rev Rheumatol 9(8):485–497. doi:10.1038/nrrheum.2013.72

Liu-Bryan R, Terkeltaub R (2015) Emerging regulators of the inflammatory process in osteoarthritis. Nat Rev Rheumatol 11(1):35–44. doi:10.1038/nrrheum.2014.162

Loeser RF (2013) Aging processes and the development of osteoarthritis. Curr Opin Rheumatol 25(1):108–113. doi:10.1097/BOR.0b013e32835a9428

Loeser RF (2014) Integrins and chondrocyte-matrix interactions in articular cartilage. Matrix Biol 39:11–16. doi:10.1016/j.matbio.2014.08.007

Loeser RF, Goldring SR, Scanzello CR, Goldring MB (2012a) Osteoarthritis: a disease of the joint as an organ. Arthritis Rheum 64(6):1697–1707. doi:10.1002/art.34453

Loeser RF, Olex AL, McNulty MA, Carlson CS, Callahan MF, Ferguson CM, Chou J, Leng X, Fetrow JS (2012b) Microarray analysis reveals age-related differences in gene expression during the development of osteoarthritis in mice. Arthritis Rheum 64(3):705–717. doi:10.1002/art.33388

Loeser RF, Olex AL, McNulty MA, Carlson CS, Callahan M, Ferguson C, Fetrow JS (2013) Disease progression and phasic changes in gene expression in a mouse model of osteoarthritis. PLoS One 8(1):e54633. doi:10.1371/journal.pone.0054633

Loeuille D, Chary-Valckenaere I, Champigneulle J, Rat AC, Toussaint F, Pinzano-Watrin A, Goebel JC, Mainard D, Blum A, Pourel J, Netter P, Gillet P (2005) Macroscopic and microscopic features of synovial membrane inflammation in the osteoarthritic knee: correlating magnetic resonance imaging findings with disease severity. Arthritis Rheum 52(11):3492–3501. doi:10.1002/art.21373

Lohmander LS, Englund PM, Dahl LL, Roos EM (2007) The long-term consequence of anterior cruciate ligament and meniscus injuries: osteoarthritis. Am J Sports Med 35(10):1756–1769

Lohmander LS, Hellot S, Dreher D, Krantz EF, Kruger DS, Guermazi A, Eckstein F (2014) Intraarticular sprifermin (recombinant human fibroblast growth factor 18) in knee osteoarthritis: a randomized, double-blind, placebo-controlled trial. Arthritis Rheum 66(7):1820–1831. doi:10.1002/art.38614

Lories RJ, Corr M, Lane NE (2013) To Wnt or not to Wnt: the bone and joint health dilemma. Nat Rev Rheumatol 9(6):328–339. doi:10.1038/nrrheum.2013.25

Lotz MK, Kraus VB (2010) New developments in osteoarthritis. Posttraumatic osteoarthritis: pathogenesis and pharmacological treatment options. Arthritis Res Ther 12(3):211. doi:10.1186/ar3046, ar3046 [pii]

Lotz M, Martel-Pelletier J, Christiansen C, Brandi ML, Bruyere O, Chapurlat R, Collette J, Cooper C, Giacovelli G, Kanis JA, Karsdal MA, Kraus V, Lems WF, Meulenbelt I, Pelletier JP, Raynauld JP, Reiter-Niesert S, Rizzoli R, Sandell LJ, Van Spil WE, Reginster JY (2013) Value of biomarkers in osteoarthritis: current status and perspectives. Ann Rheum Dis 72(11):1756–1763. doi:10.1136/annrheumdis-2013-203726

Louboutin H, Debarge R, Richou J, Selmi TA, Donell ST, Neyret P, Dubrana F (2009) Osteoarthritis in patients with anterior cruciate ligament rupture: a review of risk factors. Knee 16(4):239–244. doi:10.1016/j.knee.2008.11.004

Lyons TJ, McClure SF, Stoddart RW, McClure J (2006) The normal human chondro-osseous junctional region: evidence for contact of uncalcified cartilage with subchondral bone and marrow spaces. BMC Musculoskelet Disord 7:52. doi:10.1186/1471-2474-7-52

Mabey T, Honsawek S, Tanavalee A, Wilairatana V, Yuktanandana P, Saetan N, Zhan D (2014) Plasma and synovial fluid sclerostin are inversely associated with radiographic severity of knee osteoarthritis. Clin Biochem 47(7–8):547–551. doi:10.1016/j.clinbiochem.2014.03.011

Maes C, Carmeliet G, Schipani E (2012) Hypoxia-driven pathways in bone development, regeneration and disease. Nat Rev Rheumatol 8(6):358–366. doi:10.1038/nrrheum.2012.36

Malfait AM, Little CB, McDougall JJ (2013) A commentary on modelling osteoarthritis pain in small animals. Osteoarthritis Cartilage 21(9):1316–1326. doi:10.1016/j.joca.2013.06.003

Marcu KB, Otero M, Olivotto E, Borzi RM, Goldring MB (2010) NF-kappaB signaling: multiple angles to target OA. Curr Drug Targets 11(5):599–613

Matthews GL, Hunter DJ (2011) Emerging drugs for osteoarthritis. Expert Opin Emerg Drugs 16(3):479–491. doi:10.1517/14728214.2011.576670

Mayan MD, Gago-Fuentes R, Carpintero-Fernandez P, Fernandez-Puente P, Filgueira-Fernandez P, Goyanes N, Valiunas V, Brink PR, Goldberg GS, Blanco FJ (2015) Articular chondrocyte network mediated by gap junctions: role in metabolic cartilage homeostasis. Ann Rheum Dis 74(1):275–284. doi:10.1136/annrheumdis-2013-204244

Messent EA, Ward RJ, Tonkin CJ, Buckland-Wright C (2007) Osteophytes, juxta-articular radiolucencies and cancellous bone changes in the proximal tibia of patients with knee osteoarthritis. Osteoarthritis Cartilage 15(2):179–186

Meunier A, Odensten M, Good L (2007) Long-term results after primary repair or non-surgical treatment of anterior cruciate ligament rupture: a randomized study with a 15-year follow-up. Scand J Med Sci Sports 17(3):230–237. doi:10.1111/j.1600-0838.2006.00547.x

Mobasheri A (2013a) The future of osteoarthritis therapeutics: emerging biological therapy. Curr Rheumatol Rep 15(12):385. doi:10.1007/s11926-013-0385-4

Mobasheri A (2013b) The future of osteoarthritis therapeutics: targeted pharmacological therapy. Curr Rheumatol Rep 15(10):364. doi:10.1007/s11926-013-0364-9

Mori Y, Saito T, Chang SH, Kobayashi H, Ladel CH, Guehring H, Chung UI, Kawaguchi H (2014) Identification of fibroblast growth factor-18 as a molecule to protect adult articular cartilage by gene expression profiling. J Biol Chem 289(14):10192–10200. doi:10.1074/jbc.M113.524090

Nair A, Kanda V, Bush-Joseph C, Verma N, Chubinskaya S, Mikecz K, Glant TT, Malfait AM, Crow MK, Spear GT, Finnegan A, Scanzello CR (2012) Synovial fluid from patients with early osteoarthritis modulates fibroblast-like synoviocyte responses to toll-like receptor 4 and toll-like receptor 2 ligands via soluble CD14. Arthritis Rheum 64(7):2268–2277. doi:10.1002/art.34495

Nam J, Aguda BD, Rath B, Agarwal S (2009) Biomechanical thresholds regulate inflammation through the NF-kappaB pathway: experiments and modeling. PLoS One 4(4):e5262. doi:10.1371/journal.pone.0005262

Nguyen J, Tang SY, Nguyen D, Alliston T (2013) Load regulates bone formation and Sclerostin expression through a TGFbeta-dependent mechanism. PLoS One 8(1):e53813. doi:10.1371/journal.pone.0053813

O'Conor CJ, Leddy HA, Benefield HC, Liedtke WB, Guilak F (2014) TRPV4-mediated mechanotransduction regulates the metabolic response of chondrocytes to dynamic loading. Proc Natl Acad Sci U S A 111(4):1316–1321. doi:10.1073/pnas.1319569111

Oh H, Chun CH, Chun JS (2012) Dkk-1 expression in chondrocytes inhibits experimental osteoarthritic cartilage destruction in mice. Arthritis Rheum 64(8):2568–2578. doi:10.1002/art.34481

Olex AL, Turkett WH, Fetrow JS, Loeser RF (2014) Integration of gene expression data with network-based analysis to identify signaling and metabolic pathways regulated during the development of osteoarthritis. Gene 542(1):38–45. doi:10.1016/j.gene.2014.03.022

Onnerfjord P, Khabut A, Reinholt FP, Svensson O, Heinegard D (2012) Quantitative proteomic analysis of eight cartilaginous tissues reveals characteristic differences as well as similarities between subgroups. J Biol Chem 287(23):18913–18924. doi:10.1074/jbc.M111.298968

Otero M, Goldring MB (2007) Cells of the synovium in rheumatoid arthritis. Chondrocytes. Arthritis Res Ther 9(5):220. doi:10.1186/ar2292

Pan J, Zhou X, Li W, Novotny JE, Doty SB, Wang L (2009) In situ measurement of transport between subchondral bone and articular cartilage. J Orthop Res 27(10):1347–1352. doi:10.1002/jor.20883

Pan J, Wang B, Li W, Zhou X, Scherr T, Yang Y, Price C, Wang L (2012) Elevated cross-talk between subchondral bone and cartilage in osteoarthritic joints. Bone 51(2):212–217. doi:10.1016/j.bone.2011.11.030

Pap T, Bertrand J (2013) Syndecans in cartilage breakdown and synovial inflammation. Nat Rev Rheumatol 9(1):43–55. doi:10.1038/nrrheum.2012.178

Pearle AD, Scanzello CR, George S, Mandl LA, DiCarlo EF, Peterson M, Sculco TP, Crow MK (2007) Elevated high-sensitivity C-reactive protein levels are associated with local inflammatory findings in patients with osteoarthritis. Osteoarthritis Cartilage 15(5):516–523. doi:10.1016/j.joca.2006.10.010

Perera PM, Wypasek E, Madhavan S, Rath-Deschner B, Liu J, Nam J, Rath B, Huang Y, Deschner J, Piesco N, Wu C, Agarwal S (2010) Mechanical signals control SOX-9, VEGF, and c-Myc expression and cell proliferation during inflammation via integrin-linked kinase, B-Raf, and ERK1/2-dependent signaling in articular chondrocytes. Arthritis Res Ther 12(3):R106. doi:10.1186/ar3039

Piccinini AM, Midwood KS (2010) DAMPening inflammation by modulating TLR signalling. Mediators Inflamm. doi:10.1155/2010/672395

Pottenger LA, Phillips FM, Draganich LF (1990) The effect of marginal osteophytes on reduction of varus-valgus instability in osteoarthritic knees. Arthritis Rheum 33(6):853–858

Poulet B, Hamilton RW, Shefelbine S, Pitsillides AA (2011) Characterizing a novel and adjustable noninvasive murine joint loading model. Arthritis Rheum 63(1):137–147. doi:10.1002/art.27765

Poulet B, Ulici V, Stone TC, Pead M, Gburcik V, Constantinou E, Palmer DB, Beier F, Timmons JA, Pitsillides AA (2012) Time-series transcriptional profiling yields new perspectives on susceptibility to murine osteoarthritis. Arthritis Rheum 64(10):3256–3266. doi:10.1002/art.34572
Pulsatelli L, Addimanda O, Brusi V, Pavloska B, Meliconi R (2013) New findings in osteoarthritis pathogenesis: therapeutic implications. Ther Adv Chronic Dis 4(1):23–43. doi:10.1177/2040622312462734
Punzi L, Frigato M, Frallonardo P, Ramonda R (2010) Inflammatory osteoarthritis of the hand. Best Pract Res Clin Rheumatol 24(3):301–312. doi:10.1016/j.berh.2009.12.007
Radin EL, Rose RM (1986) Role of subchondral bone in the initiation and progression of cartilage damage. Clin Orthop Relat Res 213:34–40
Radin EL, Martin RB, Burr DB, Caterson B, Boyd RD, Goodwin C (1984) Effects of mechanical loading on the tissues of the rabbit knee. J Orthop Res 2(3):221–234. doi:10.1002/jor.1100020303
Rai MF, Patra D, Sandell LJ, Brophy RH (2013) Transcriptome analysis of injured human meniscus reveals a distinct phenotype of meniscus degeneration with aging. Arthritis Rheum 65(8):2090–2101. doi:10.1002/art.37984
Ritter SY, Subbaiah R, Bebek G, Crish J, Scanzello CR, Krastins B, Sarracino D, Lopez MF, Crow MK, Aigner T, Goldring MB, Goldring SR, Lee DM, Gobezie R, Aliprantis AO (2013) Proteomic analysis of synovial fluid from the osteoarthritic knee: comparison with transcriptome analyses of joint tissues. Arthritis Rheum 65(4):981–992. doi:10.1002/art.37823
Roemer FW, Hunter DJ, Guermazi A (2009) MRI-based semiquantitative assessment of subchondral bone marrow lesions in osteoarthritis research. Osteoarthritis Cartilage 17(3):414–415. doi:10.1016/j.joca.2008.07.019; author reply 416–417
Roemer FW, Guermazi A, Felson DT, Niu J, Nevitt MC, Crema MD, Lynch JA, Lewis CE, Torner J, Zhang Y (2011) Presence of MRI-detected joint effusion and synovitis increases the risk of cartilage loss in knees without osteoarthritis at 30-month follow-up: the MOST study. Ann Rheum Dis 70(10):1804–1809. doi:10.1136/ard.2011.150243
Roemer FW, Kwoh CK, Hannon MJ, Hunter DJ, Eckstein F, Fujii T, Boudreau RM, Guermazi A (2015) What comes first? Multitissue involvement leading to radiographic osteoarthritis: magnetic resonance imaging-based trajectory analysis over four years in the osteoarthritis initiative. Arthritis Rheum 67(8):2085–2096. doi:10.1002/art.39176
Roos EM (2005) Joint injury causes knee osteoarthritis in young adults. Curr Opin Rheumatol 17(2):195–200
Roudier M, Li X, Niu QT, Pacheco E, Pretorius JK, Graham K, Yoon BR, Gong J, Warmington K, Ke HZ, Black RA, Hulme J, Babij P (2013) Sclerostin is expressed in articular cartilage but loss or inhibition does not affect cartilage remodeling during aging or following mechanical injury. Arthritis Rheum 65(3):721–731. doi:10.1002/art.37802
Ruan MZ, Erez A, Guse K, Dawson B, Bertin T, Chen Y, Jiang MM, Yustein J, Gannon F, Lee BH (2013) Proteoglycan 4 expression protects against the development of osteoarthritis. Sci Transl Med 5(176):176ra134. doi:10.1126/scitranslmed.3005409
Ruhlen R, Marberry K (2014) The chondrocyte primary cilium. Osteoarthritis Cartilage 22(8):1071–1076. doi:10.1016/j.joca.2014.05.011
Saito T, Fukai A, Mabuchi A, Ikeda T, Yano F, Ohba S, Nishida N, Akune T, Yoshimura N, Nakagawa T, Nakamura K, Tokunaga K, Chung UI, Kawaguchi H (2010) Transcriptional regulation of endochondral ossification by HIF-2alpha during skeletal growth and osteoarthritis development. Nat Med 16(6):678–686. doi:10.1038/nm.2146
Saito M, Sasho T, Yamaguchi S, Ikegawa N, Akagi R, Muramatsu Y, Mukoyama S, Ochiai N, Nakamura J, Nakagawa K, Nakajima A, Takahashi K (2012) Angiogenic activity of subchondral bone during the progression of osteoarthritis in a rabbit anterior cruciate ligament transection model. Osteoarthritis Cartilage 20(12):1574–1582. doi:10.1016/j.joca.2012.08.023
Sanchez C, Deberg MA, Bellahcene A, Castronovo V, Msika P, Delcour JP, Crielaard JM, Henrotin YE (2008) Phenotypic characterization of osteoblasts from the sclerotic zones of osteoarthritic subchondral bone. Arthritis Rheum 58(2):442–455. doi:10.1002/art.23159

Sandell LJ (2012) Etiology of osteoarthritis: genetics and synovial joint development. Nat Rev Rheumatol 8(2):77–89. doi:10.1038/nrrheum.2011.199

Scanzello CR, Goldring SR (2012) The role of synovitis in osteoarthritis pathogenesis. Bone. doi:10.1016/j.bone.2012.02.012, S8756-3282(12)00068-3 [pii]

Scanzello CR, Umoh E, Pessler F, Diaz-Torne C, Miles T, Dicarlo E, Potter HG, Mandl L, Marx R, Rodeo S, Goldring SR, Crow MK (2009) Local cytokine profiles in knee osteoarthritis: elevated synovial fluid interleukin-15 differentiates early from end-stage disease. Osteoarthritis Cartilage 17(8):1040–1048. doi:10.1016/j.joca.2009.02.011, S1063-4584(09)00057-0 [pii]

Scanzello CR, McKeon B, Swaim BH, DiCarlo E, Asomugha EU, Kanda V, Nair A, Lee DM, Richmond JC, Katz JN, Crow MK, Goldring SR (2011) Synovial inflammation in patients undergoing arthroscopic meniscectomy: molecular characterization and relationship to symptoms. Arthritis Rheum 63(2):391–400. doi:10.1002/art.30137

Scanzello CR, Albert AS, DiCarlo E, Rajan KB, Kanda V, Asomugha EU, Swaim BH, Katz JN, Goldring SR, Richmond JC, McKeon B (2013) The influence of synovial inflammation and hyperplasia on symptomatic outcomes up to 2 years post-operatively in patients undergoing partial meniscectomy. Osteoarthritis Cartilage 21(9):1392–1399. doi:10.1016/j.joca.2013.05.011

Schelbergen RF, Blom AB, van den Bosch MH, Sloetjes A, Abdollahi-Roodsaz S, Schreurs BW, Mort JS, Vogl T, Roth J, van den Berg WB, van Lent PL (2012) Alarmins S100A8 and S100A9 elicit a catabolic effect in human osteoarthritic chondrocytes that is dependent on Toll-like receptor 4. Arthritis Rheum 64(5):1477–1487. doi:10.1002/art.33495

Segawa H, Omori G, Koga Y (2001) Long-term results of non-operative treatment of anterior cruciate ligament injury. Knee 8(1):5–11

Sohn DH, Sokolove J, Sharpe O, Erhart JC, Chandra PE, Lahey LJ, Lindstrom TM, Hwang I, Boyer KA, Andriacchi TP, Robinson WH (2012) Plasma proteins present in osteoarthritic synovial fluid can stimulate cytokine production via Toll-like receptor 4. Arthritis Res Ther 14(1):R7. doi:10.1186/ar3555

Stiebel M, Miller LE, Block JE (2014) Post-traumatic knee osteoarthritis in the young patient: therapeutic dilemmas and emerging technologies. Open Access J Sports Med 5:73–79. doi:10.2147/OAJSM.S61865

Stoppiello LA, Mapp PI, Wilson D, Hill R, Scammell BE, Walsh DA (2014) Structural associations of symptomatic knee osteoarthritis. Arthritis Rheum 66(11):3018–3027. doi:10.1002/art.38778

Suri S, Walsh DA (2012) Osteochondral alterations in osteoarthritis. Bone 51(2):204–211. doi:10.1016/j.bone.2011.10.010

Taljanovic MS, Graham AR, Benjamin JB, Gmitro AF, Krupinski EA, Schwartz SA, Hunter TB, Resnick DL (2008) Bone marrow edema pattern in advanced hip osteoarthritis: quantitative assessment with magnetic resonance imaging and correlation with clinical examination, radiographic findings, and histopathology. Skeletal Radiol 37(5):423–431. doi:10.1007/s00256-008-0446-3

Thakur M, Dickenson AH, Baron R (2014) Osteoarthritis pain: nociceptive or neuropathic? Nat Rev Rheumatol 10(6):374–380. doi:10.1038/nrrheum.2014.47

Thompson CL, Wiles A, Poole CA, Knight MM (2016) Lithium chloride modulates chondrocyte primary cilia and inhibits Hedgehog signaling. FASEB J 30(2):716–726. doi:10.1096/fj.15-274944

Tonge DP, Pearson MJ, Jones SW (2014) The hallmarks of osteoarthritis and the potential to develop personalised disease-modifying pharmacological therapeutics. Osteoarthritis Cartilage 22(5):609–621. doi:10.1016/j.joca.2014.03.004

van den Bosch MH, Blom AB, van Lent PL, van Beuningen HM, Blaney Davidson EN, van der Kraan PM, van den Berg WB (2014) Canonical Wnt signaling skews TGF-beta signaling in chondrocytes towards signaling via ALK1 and Smad 1/5/8. Cell Signal 26(5):951–958. doi:10.1016/j.cellsig.2014.01.021

van der Kraan PM (2014) Age-related alterations in TGF beta signaling as a causal factor of cartilage degeneration in osteoarthritis. Biomed Mater Eng 24(1 Suppl):75–80. doi:10.3233/BME-140976

van der Kraan PM, van den Berg WB (2007) Osteophytes: relevance and biology. Osteoarthritis Cartilage 15(3):237–244

van der Kraan PM, van den Berg WB (2012) Chondrocyte hypertrophy and osteoarthritis: role in initiation and progression of cartilage degeneration? Osteoarthritis Cartilage 20(3):223–232. doi:10.1016/j.joca.2011.12.003

van der Kraan PM, Goumans MJ, Blaney Davidson E, ten Dijke P (2012) Age-dependent alteration of TGF-beta signalling in osteoarthritis. Cell Tissue Res 347(1):257–265. doi:10.1007/s00441-011-1194-6

Vincent TL (2012) Explaining the fibroblast growth factor paradox in osteoarthritis: lessons from conditional knockout mice. Arthritis Rheum 64(12):3835–3838. doi:10.1002/art.34648

Vincent TL, Williams RO, Maciewicz R, Silman A, Garside P (2012) Mapping pathogenesis of arthritis through small animal models. Rheumatology 51(11):1931–1941. doi:10.1093/rheumatology/kes035

Waller KA, Zhang LX, Elsaid KA, Fleming BC, Warman ML, Jay GD (2013) Role of lubricin and boundary lubrication in the prevention of chondrocyte apoptosis. Proc Natl Acad Sci U S A 110(15):5852–5857. doi:10.1073/pnas.1219289110

Walsh DA, McWilliams DF, Turley MJ, Dixon MR, Franses RE, Mapp PI, Wilson D (2010) Angiogenesis and nerve growth factor at the osteochondral junction in rheumatoid arthritis and osteoarthritis. Rheumatology (Oxford) 49(10):1852–1861. doi:10.1093/rheumatology/keq188

Wang Q, Rozelle AL, Lepus CM, Scanzello CR, Song JJ, Larsen DM, Crish JF, Bebek G, Ritter SY, Lindstrom TM, Hwang I, Wong HH, Punzi L, Encarnacion A, Shamloo M, Goodman SB, Wyss-Coray T, Goldring SR, Banda NK, Thurman JM, Gobezie R, Crow MK, Holers VM, Lee DM, Robinson WH (2011) Identification of a central role for complement in osteoarthritis. Nat Med 17(12):1674–1679. doi:10.1038/nm.2543

Wang M, Sampson ER, Jin H, Li J, Ke QH, Im HJ, Chen D (2013) MMP13 is a critical target gene during the progression of osteoarthritis. Arthritis Res Ther 15(1):R5. doi:10.1186/ar4133

Wann AK, Zuo N, Haycraft CJ, Jensen CG, Poole CA, McGlashan SR, Knight MM (2012) Primary cilia mediate mechanotransduction through control of ATP-induced Ca2+ signaling in compressed chondrocytes. FASEB J 26(4):1663–1671. doi:10.1096/fj.11-193649

Wilson AJ, Murphy WA, Hardy DC, Totty WG (1988) Transient osteoporosis: transient bone marrow edema? Radiology 167(3):757–760. doi:10.1148/radiology.167.3.3363136

Wilusz RE, Sanchez-Adams J, Guilak F (2014) The structure and function of the pericellular matrix of articular cartilage. Matrix Biol 39:25–32. doi:10.1016/j.matbio.2014.08.009

Wojdasiewicz P, Poniatowski LA, Szukiewicz D (2014) The role of inflammatory and anti-inflammatory cytokines in the pathogenesis of osteoarthritis. Mediators Inflamm 2014:561459. doi:10.1155/2014/561459

Xu L, Polur I, Servais JM, Hsieh S, Lee PL, Goldring MB, Li Y (2011) Intact pericellular matrix of articular cartilage is required for unactivated discoidin domain receptor 2 in the mouse model. Am J Pathol 179(3):1338–1346. doi:10.1016/j.ajpath.2011.05.023

Xu L, Golshirazian I, Asbury BJ, Li Y (2014) Induction of high temperature requirement A1, a serine protease, by TGF-beta1 in articular chondrocytes of mouse models of OA. Histol Histopathol 29(5):609–618

Yan D, Chen D, Cool SM, van Wijnen AJ, Mikecz K, Murphy G, Im HJ (2011) Fibroblast growth factor receptor 1 is principally responsible for fibroblast growth factor 2-induced catabolic activities in human articular chondrocytes. Arthritis Res Ther 13(4):R130. doi:10.1186/ar3441

Yang S, Kim J, Ryu JH, Oh H, Chun CH, Kim BJ, Min BH, Chun JS (2010) Hypoxia-inducible factor-2alpha is a catabolic regulator of osteoarthritic cartilage destruction. Nat Med 16(6):687–693. doi:10.1038/nm.2153

Zanetti M, Bruder E, Romero J, Hodler J (2000) Bone marrow edema pattern in osteoarthritic knees: correlation between MR imaging and histologic findings. Radiology 215(3):835–840. doi:10.1148/radiology.215.3.r00jn05835

Zhang R, Fang H, Chen Y, Shen J, Lu H, Zeng C, Ren J, Zeng H, Li Z, Chen SL, Cai D, Zhao Q (2012) Gene expression analysis of subchondral bone in early experimental osteoarthritis by microarray. PLoS One 7:e32356. dol:32310.31371

Zhen G, Wen C, Jia X, Li Y, Crane JL, Mears SC, Askin FB, Frassica FJ, Chang W, Yao J, Carrino JA, Cosgarea A, Artemov D, Chen Q, Zhao Z, Zhou X, Riley L, Sponseller P, Wan M, Lu WW, Cao X (2013) Inhibition of TGF-[beta] signaling in mesenchymal stem cells of subchondral bone attenuates osteoarthritis. Nat Med 19 (6):704–712. doi:10.1038/nm.3143. http://www.nature.com/nm/journal/v19/n6/abs/nm.3143.html – supplementary-information

Cartilage Injury and Osteoarthritis

2

Heba M. Ismail and Tonia L. Vincent

Abstract

Mechanical stress is an obligatory aetiological factor in the development of OA so understanding how tissues of the joint respond to mechanical injury is likely to inform our understanding of pathogenesis. Much is known about how vascular tissues respond to damage, a process that involves activation of platelets on the exposed endothelium and recruitment of leukocytes to the site of injury. The articular cartilage is avascular yet responds rapidly and strongly to a range of mechanical stresses including cutting, avulsion, impact loading and shearing. It does so by activating a number of mechanosensitive pathways mediated by release of molecules trapped within the pericellular matrix as well as by triggering mechanoreceptors at the cell surface. In this way injury drives a number of intracellular signalling pathways, leading to a broad range of cellular responses. These pathways appear to be relevant to the in vivo response to mechanical disruption and affect the course of experimental OA.

2.1 Introduction

The role of mechanical overload/injury in osteoarthritis development cannot be overstated. The epidemiology of this is compelling; paralysed joints do not develop OA and excessive mechanical strain accelerates OA development. Increased mechanical strain is not only due to increased load traversing the joint (obesity, high impacts, direct cartilage trauma) but also importantly includes loss of joint mechanoprotective mechanisms (through muscle weakness, loss of gait reflexes with age,

H.M. Ismail • T.L. Vincent (✉)
Arthritis Research UK Centre for OA Pathogenesis, Kennedy Institute of Rheumatology, NDORMS, University of Oxford, Roosevelt Drive, Oxford OX3 7FY, UK
e-mail: heba.ismail@keneedy.ox.ac.uk; tonia.vincent@kennedy.ox.ac.uk

S. Grässel, A. Aszódi (eds.), *Cartilage*, DOI 10.1007/978-3-319-45803-8_2

Table 2.1 How mechanical factors contribute to OA risk

Increased load through the joint	Intra-articular fracture
	High-impact injuries
	Obesity
	Joint malalignment
	Developmental shape abnormalities, e.g. chondrodysplasias
Loss of joint mechanoprotective mechanisms	Joint destabilising injuries, e.g. meniscal and cruciate ligament injury
	Loss of gait reflexes upon ageing
	Muscle atrophy (age, immobilisation, genetic)

tendon/meniscal injuries leading to joint destabilisation, etc.) (reviewed in Brandt et al. 2009) (Table 2.1). A somewhat outdated explanation for this association is that simple 'wear and tear' leads to passive attrition of the articular surface with excessive use. This theory has largely been discounted by showing that disease is highly dependent upon the induction and regulation of specific matrix-degrading enzymes (see Chap. 3) and that mice in which these enzymes have been deleted genetically are substantially protected from developing OA despite the fact that their joints have been surgically destabilised (Glasson et al. 2005; Little et al. 2009). Our subsequent observation that induction of inflammatory response genes (including metalloproteinases) in surgically destabilised joints is highly mechanosensitive allowed us to speculate that excessive mechanical stress in vivo is sufficient to activate pathways that drive matrix degradation and initiate disease (Burleigh et al. 2012).

2.2 Articular Cartilage: Structure and Mechano-Sensing Function

The articular cartilage is a paucicellular type II collagen- and proteoglycan-rich tissue. It is a highly lubricated connective tissue with exceptionally low surface friction that allows smooth articulation of the opposing joint surfaces. Lubrication is achieved by a synergistic relationship between the proteoglycan, the lubricin and the glycosaminoglycan hyaluronan, together with phosphatidylcholine lipids (Seror et al. 2015). Intact lubrication protects the surface from shear stress. Loss of lubricin leads to increased shear stress and superficial chondrocyte apoptosis (Waller et al. 2013). The type II collagen fibres are highly cross-linked and provide the matrix with tensile strength. They are also oriented to aid articulation and reduce shear-associated damage, lying parallel to the articular surface in the most superficial regions of the matrix and perpendicular to the surface in deeper regions. The collagenous component of the matrix is laid down around the time of skeletal maturity and is not thought to turn over in later adult life. Although osteoarthritic chondrocytes do secrete type II collagen, it is not clear whether this can become successfully incorporated into the damaged matrix. Proteoglycan turnover, on the other hand, is relatively rapid, and it is this component of the matrix that can change rapidly as the cartilage volume shrinks with loss of weightbearing activity (so-called atrophy) (Palmoski et al. 1979; Vanwanseele et al. 2002).

The chondrocyte is regarded as the single cell of the cartilage although there are different morphological and molecular features of chondrocytes through the different layers of the tissue. These cells are responsible for developing and maintaining the matrix. Although they rarely divide in the healthy tissue and appear relatively 'quiescent', chondrocytes are exquisitely mechanosensitive. The chondrocyte is a spiculated cell that extends multiple short processes into the immediate pericellular matrix (PCM). The PCM, just 3–5 μm wide, is rich in type VI collagen, perlecan and other minor proteoglycans such as decorin and biglycan (Kvist et al. 2008; Soder et al. 2002; Tesche and Miosge 2004). As it is devoid of type II collagen, it exhibits a much reduced Young's modulus (stiffness) compared with the further remote matrix, and this is thought to allow selective compression of the PCM upon cartilage loading (Guilak et al. 2006). Both cell surface 'mechanoreceptors' and mechanisms controlled directly by the PCM drive mechano-responses in chondrocytes. These will be considered next.

2.3 Chondrocyte Injury Assays

A number of different methods for applying injurious mechanical stress to a chondrocyte in vitro have been described (see Table 2.2). These include injury to the chondrocyte embedded within its native matrix (cartilage explant), e.g. by cutting with a scalpel, explantation of the cartilage from the intact joint or compression (Dell'accio et al. 2008; Gruber et al. 2004; Quinn et al. 1998; Redman et al. 2004). Cartilage injury models using murine tissues allow to interrogate the cellular response to injury in a pathway-dependent manner and have been done by 'avulsing' the immature femoral condyle of 4–6-week-old mice (Chong et al. 2013). Alternatively, mechanical stress can be applied to isolated cells either grown in monolayer or

Table 2.2 In vitro cartilage injury models

In vitro cartilage injury models	Tissue	Injury stimulus
Cutting injury	Articular cartilage explants	Explantation (cutting the tissue from the intact joint surface)
	Cultured articular cartilage explants	Recutting explants that have been cultured for >24 h post explantation
Injurious loading	Articular cartilage explants	Injurious load (excessive magnitude of load or high velocity (impact))
	Isolated cells grown in artificial matrix (alginate, agarose)	Injurious shear or compressive stress
Avulsion injury (murine)	Whole femoral heads	Avulsion – shearing through the growth plate
Stretch?	Cells in monolayer	Stretch activation (? injurious)

embedded within three-dimensional artificial matrices, e.g. by stretching or mechanical compression (Salter et al. 2002; Wright et al. 1996; Vincent et al. 2007). The frequency of load and speed of load application can also be altered. High-velocity loads create impact-associated cartilage injuries in vitro (Aspden et al. 2002).

2.4 Cellular Response to Cartilage Injury: Activation of Intracellular Signalling

Most common cell signalling systems have been described in the context of cartilage/chondrocyte injury. For the purposes of this chapter, we will restrict our précis to those studies that have specifically addressed the response to injurious mechanical stress rather than covering mechano-responses of chondrocytes to modest cyclic and static load that are likely to be regarded as non-injurious (Fig. 2.1).

2.4.1 Focal Adhesions: Src, FAK and Integrin Signalling

Integrins have been defined as 'mechanoreceptors' in other cell types, triggered by perturbations in the extracellular matrix and transducing extracellular signals to the cell (Huveneers and Danen 2009; Wang et al. 1993). Upon activation, integrins

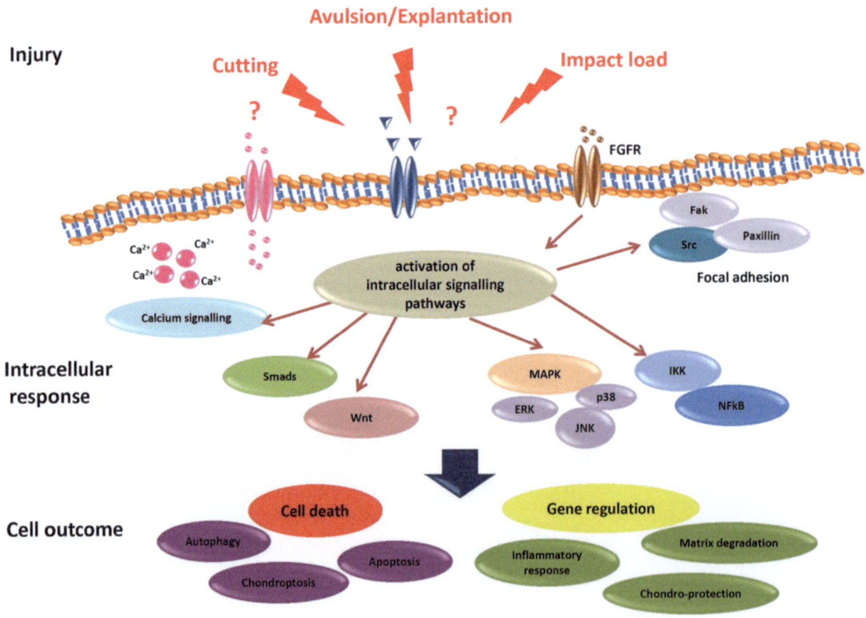

Fig. 2.1 Intracellular responses to cartilage injury. Multiple pathways are activated by a variety of injurious stimuli. Some involve known receptor systems on the cell membrane, e.g. FGF receptor. Others are unknown. Cellular outcomes that have been investigated largely include mechanisms of cell death upon injury or regulation of genes that determine tissue health

cluster in focal adhesions to create transmembrane protein complexes that form tyrosine kinase signalling hubs. Articular chondrocytes express multiple integrins for binding to type II collagen, fibronectin and other ECM molecules (Vinall et al. 2002) (for further details, see Chap. 8). α5β1-integrin is responsible for cell attachment to fibronectin and plays a role in proliferation and adhesion in chondrocytes (Enomoto-Iwamoto et al. 1997). The interaction between α5β1-integrin, stretch-activated ion channels and interleukin-4 in response to cyclic mechanical strain has been reported in chondrocytes in a number of studies (reviewed in Chen et al. 2013). Integrin-cytoskeleton interactions are implicated in death-signalling pathways in chondrocytes following impact injury. Cytoskeletal dissolution by treatment with cytochalasin B or nocodazole considerably decreased impact-induced chondrocyte death, which suggests that the cytoskeleton is required for impact-induced chondrocyte death (Sauter et al. 2012). Additionally, Jang et al. (2014) found that excessive compression of the articular cartilage led to cytoskeleton-dependent chondrocyte death. This process involved integrin activation and signalling via focal adhesion kinase (FAK) and Src family kinases. Treatment with FAK and Src kinase inhibitors significantly improved chondrocyte viability after impact loads.

Sterile cutting injury to the surface of the intact articular cartilage caused rapid (within seconds) activation of protein tyrosine kinases of the Src kinase family including focal adhesion kinase, paxillin and cortactin (Watt et al. 2013). The Src inhibitor, PP2, blocked injury-induced tyrosine phosphorylation and modulated inflammatory gene expression patterns induced in response to explantation injury. Interestingly Src activation was not seen in response to recutting cartilage explants that had been cultured post explantation (recut) suggesting that injury to the intact joint surface may be different to injuring cartilage in vitro (Fig. 2.1).

2.4.2 MAPK and NFkB Signalling

The mitogen-activated protein (MAP) kinases, p38, extracellularly regulated kinase (ERK) and c-Jun N terminal kinase (JNK) are implicated in multiple cellular processes. In chondrocytes p38 and ERK MAPKs are induced by shear and compressive forces (Fitzgerald et al. 2008, Fanning et al. 2003) as well as by explantation and recutting injury (Gruber et al. 2004; Vincent et al. 2002). Using an osteochondral explant blunt injury model, Ding et al. (2010) demonstrated enhanced phosphorylation of the three MAPKs 20 min post impaction. The activation of MAP kinases has been observed in various osteoarthritic-cartilage injury models also (Chowdhury et al. 2008; Ding et al. 2008; Guo et al. 2009). Cartilage explantation and murine femoral condyle avulsion are also able to activate NFkB signalling although, interestingly, these pathways do not appear to be activated upon recutting injury (Chong et al. 2013; Gruber et al. 2004).

Treatment with specific p38 and ERK MAP kinase inhibitors partly reversed loss of cell viability and proteoglycan content in impacted explants. Similar findings were reported by Rosenzweig et al. (2012). In this study, mechanical injury to bovine osteochondral explants led to MAPK activation and decreased chondrocyte viability especially in the superficial zone. Reduction in cell viability post-injury correlated with increased caspase-3 activity and could be reversed by ERK and p38 inhibition. Inhibition of p38 in vitro also caused a reduction in metalloprotease activation, although inactivation of p38 in vivo was not chondroprotective. Indeed, genetic loss of function of p38 (by overexpressing a dominant negative form of p38) was associated with higher grades of osteoarthritis in the knee (Namdari et al. 2008).

Sustained ERK activation caused by excising or cutting the cartilage is at least partly attributed to the release of FGF2 from the pericellular matrix (PCM) of chondrocytes (Vincent et al. 2002). FGF2 is sequestered on the heparan sulphate chains of perlecan and released within seconds of injury or compression (Vincent et al. 2004, 2007). Interestingly, neither NFkB nor JNK activation upon injury is due to a soluble factor released from the injured cartilage (our unpublished observations) (Fig. 2.1).

2.4.3 Smad and Wnt Signalling

Other factors such as TGFβ are also released from the matrix upon mechanical compression and drive phosphorylation of Smad2/Smad3 (Madej et al. 2016). This pathway becomes less responsive with age perhaps explaining why cartilage health declines over time. Activation of BMP and Wnt signalling has also been reported in response to cartilage injury, and this drives gene regulation in injured explants (see below) (Dell'Accio et al. 2006, 2008) (Fig. 2.1).

2.4.4 Ion Channel Signalling

Mechanically induced Ca^{2+} signalling has been studied extensively in chondrocytes. Han et al. (2012) developed a system to measure Ca^{2+} signalling of chondrocytes in situ and found that Ca^{2+} signalling is regulated not only by mechanical force but also by the cellular environment, such as ECM topography and temperature.

A number of mechanosensitive Ca^{2+} channels have been identified in chondrocytes including the intracellular PLC-inositol 1,4,5-trisphosphate pathway, stretch-activated ion channels, the transient receptor potential vanilloid 4 (TRPV4) cation channel (reviewed in Chen et al. 2013) and the newly identified family of cation-permeable, directly mechanically activated (MA) ion channels, named 'Piezo' (Lee et al. 2014). TRPV4 is responsible for mediating the anabolic response of chondrocytes to osmotic or mechanical stress (Phan et al. 2009; O'Conor et al. 2014) and may therefore respond to physiological rather than injurious mechanical stress. Blocking TRPV4 in vitro makes chondrocytes less responsive to hyposmotic stress and impairs intracellular calcium flux (Phan et al. 2009). TRPV4 knockout mice show a loss of the Ca^{2+} response to hypo-osmotic challenge and have significantly increased OA indicating that this pathway is chondroprotective in vivo (Clark et al. 2010). Chondrocytes express the two channels Piezo 1 and Piezo 2, which respond synergistically to injurious mechanical strain. Inhibition of Piezos by GsMTx4, a biologically derived peptide that specifically inhibits mechanically activated cation channels, caused inhibition of the chondrocyte's response to injury. GsMTx4 significantly protected chondrocytes from cell death following injury (Lee et al. 2014) (Fig. 2.1).

2.5 Cellular Response to Cartilage Injury: Cell Death

Mechanical injury to the articular cartilage can cause chondrocyte cell death (reviewed in Kuhn et al. 2004). Death occurs through two principal mechanisms, necrosis or apoptosis; the latter being more likely to occur if the injury is sharp rather than blunt (Redman et al. 2007; Tew et al. 2000). Since chondrocytes represent only 1–10 % of the cartilage tissue volume and have low regenerative capacity (Stockwell 1978), cell death is considered a potentially important pathological event in OA. The use of inhibitors of caspases, key mediators of apoptosis, has been used in OA and osteochondral injury models in rabbits to significantly reduce chondrocyte cell death (D'Lima et al. 2006; Dang et al. 2006; Costouros and Kim 2007). Goodwin et al. (2010) demonstrated that the enhanced release of mitochondrial reactive oxygen species (ROS) following mechanical injury to the articular cartilage would induce cell death. Suppressing the ROS production either by rotenone, an electron transport chain inhibitor, or by N-acetylcysteine (Martin et al. 2009), an oxidant scavenger, dramatically reduced the rates of cell death. This suggests that oxidative stress is responsible for chondrocyte death following cartilage mechanical injury. Similarly, Killian et al. (2014) have shown that cartilage trauma caused by high impaction triggered acute cell death in rabbit articular chondrocytes in addition to an increase in nitric oxide release by joint tissues.

Bartell et al. (2015) developed a technique to measure cartilage mechanics at microscales of strain fields and rapid time frames. In response to injurious load, the effect on chondrocytes can be measured as early as 3 ms following localised impact. The authors found that the high rates of strain were highly correlated with chondrocyte cell death. Interestingly, chondrocyte cell death was only detected after 2 h following impact and the removal of the superficial zone increased the penetration of cell death to deeper zones suggesting a protective role for the surface layer.

Another form of cell death has also been reported in OA chondrocytes called chondroptosis. Chondroptosis, first described by Roach et al. (2004), includes some elements of the apoptotic and autophagic processes. Chondroptosis involves significant morphological changes within the endoplasmic reticulum and Golgi apparatus, which functions to segment the cytoplasm into several smaller autophagic-like vacuolised compartments (Roach et al. 2004). Studies of growth plate chondrocytes provide evidence that this process may help prevent a chondrocyte from entering a committed stage of apoptosis (Bohensky et al. 2007a, b) and may explain the long life of a chondrocyte.

Almonte-Becerril et al. (2010) have used a Wistar rats OA model followed by a high-impact exercise. They showed that following injury, apoptotic cell death combined with autophagy was induced in chondrocytes. At early-stage OA, apoptosis was induced in the superficial and midzones of the cartilage, whilst an increase in both apoptosis and autophagy (as measured by active caspase-3 and LC3II expression) was observed at later stages of OA.

On the other hand, Carames et al. (2012) reported that mechanical impact injury to the bovine and human cartilage causes inhibition of autophagy (decreased expression of the autophagic markers ULK1, Beclin1 and LC3) in the cartilage superficial zone, and this was associated with increased release of the sulphated glycosaminoglycan (GAG) indicating cartilage damage. Reactivation of the autophagic pathway by rapamycin resulted in inhibition of cell death and a reduction in GAG loss in response to mechanical injury (Fig. 2.1).

2.6 Cellular Response to Cartilage Injury: Gene Regulation

Several groups have studied gene regulation in response to mechanical cartilage injury in vitro. Dell'Accio et al. performed a microarray study of the recut human articular cartilage and identified a key role for the Wnt pathway in the injury response. Injury was associated with the up-regulation of a number of Wnt activators and their inhibitors as well as leading to activation of the canonical pathway mediated by cytoplasmic accumulation of β-catenin (Dell'accio et al. 2008). Several studies have suggested a proteolytic cartilage phenotype following injury. Lee et al. studied the regulation of selective proteases following explantation injury, either alone or with the addition of a traumatic compressive load. Both explantation and compression regulated protease gene expression over the first 24 h of injury (Lee et al. 2005). Injurious compression also caused a rapid early release of glycosaminoglycan that was deemed to be due to matrix damage. This was followed by an active release of GAG that could be suppressed by metalloproteinase inhibition (DiMicco et al. 2004). Explantation also regulates IL1α and IL1β mRNA levels and induces pro-protein production by the chondrocyte, albeit not causing its secretion (Gruber et al. 2004).

By injecting joints with specific inhibitors, it has been possible to investigate the role of individual pathways/molecules in the explantation injury response. Such studies have revealed that several injury genes such as TIMP1, activin A and TSG-6 are FGF2 dependent (Chong et al. 2013). Importantly, FGF2-dependent genes identified in vitro were also regulated upon joint destabilisation in vivo (Chong et al. 2013). Regulation of inflammatory response genes upon explantation such as inflammatory cytokines and metalloproteinases is largely Src kinase dependent, although some genes are also FGF2 dependent (Chong et al. 2013; Watt et al. 2013). Some of these molecules likely contribute to promoting tissue catabolism upon cartilage injury, whilst others appear to be anti-catabolic or repair-promoting. Both may contribute to pathology in OA (Fig. 2.1).

2.7 In Vivo Models of Cartilage Injury

Increasingly, we have seen an attempt to validate injury pathways in vivo (Table 2.2). In vivo models of injury have been around for many decades, and the response to injury has been described in detail histologically and metabolically (reviewed in Campbell 1969). For instance, experiments of rabbit articular cartilage scarification by Meachim showed that chondrocytes around lesions become metabolically more active, increasing their proteoglycan production and clonal proliferation. Longer-term outcome of in vivo injury is inconsistent, with some studies showing repair of focal lesions and others showing limited repair responses. Successful outcome is in part thought to be due to whether the osteochondral junction has been breached (Meachim 1963). Eltawil et al., more recently, suggest that genetic strain and age are other modifiable factors that dictate successful repair in the mouse (Eltawil et al. 2009). In these studies they created full-thickness cartilage defects in the intercondylar groove of the femur and measured filling of the defect after 8 weeks. They also assessed the quality of repair tissue, as well as neo-epitopes of matrix turnover, and showed that non-repairing strains demonstrated evidence of OA changes in the matrix around the lesion. Similarly, Rai et al. demonstrated a strain-dependent ability of focal cartilage lesions to repair, a response that could be partly recapitulated in the injured auricular cartilage (Rai et al. 2012).

Surgically induced models of cartilage injury through joint destabilisation offer a model to study chronic cartilage injury in vivo. Microarray studies of genes regulated in the articular cartilage within the first few weeks following surgery have identified many genes and pathways previously identified in in vitro cartilage injury studies and in OA (Appleton et al. 2007; Gardiner et al. 2015). These models of post-traumatic OA have been used in multiple studies to investigate candidate pathways in the disease (Vincent et al. 2012).

Another experimental system that has been utilised to investigate the immediate and longer-term response to acute traumatic injury in vivo is that induced by cyclic forced compression of the hyper-flexed knee joint in mice (Poulet et al. 2011). In these experiments, mice undergo cyclic injurious joint compression under general anaesthetic. This leads to damage to the posterior condylar cartilage and later, osteoarthritic changes more widely within the joint. Other non-surgical models of cartilage injury have also been developed and are reviewed in (Christiansen et al. 2015) (Table 2.3).

Table 2.3 In vivo cartilage injury models

In vivo cartilage injury models	Invasive (surgical)?	Primary cartilage outcome assessed
Focal cartilage defect	Yes	Repair
Articular cartilage scarification	Yes	Repair and degeneration
Destabilisation of the joint (several different models)	Yes	Degeneration
Cyclic tibial compression load	No	Degeneration
Intra-articular tibial fracture	No	Degeneration
ACL rupture by tibial compression overload	No	Degeneration

ACL anterior cruciate ligament

Conclusion

Cartilage injury is arguably the most important aetiological factor in the development of OA, and studying the cellular response of cartilage to injury offers us insights into how mechanical injury might drive pathological processes that characterise disease. It is worth bearing in mind that the injury responses described in these studies are likely to be pertinent to a more generic response of connective tissues to injury, and therefore conclusions drawn from them are likely also to be relevant to other tissues in the damaged joint.

References

Almonte-Becerril M, Navarro-Garcia F, Gonzalez-Robles A, Vega-Lopez MA, Lavalle C, Kouri JB (2010) Cell death of chondrocytes is a combination between apoptosis and autophagy during the pathogenesis of osteoarthritis within an experimental model. Apoptosis: Int J Program Cell Death 15(5):631–638. doi:10.1007/s10495-010-0458-z

Appleton CT, Pitelka V, Henry J, Beier F (2007) Global analyses of gene expression in early experimental osteoarthritis. Arthritis Rheum 56(6):1854–1868. doi:10.1002/art.22711

Aspden RM, Jeffrey JE, Burgin LV (2002) Impact loading of articular cartilage. Osteoarthr Cartil/OARS Osteoarthr Res Soc 10(7):588–589; author reply 590

Bartell LR, Fortier LA, Bonassar LJ, Cohen I (2015) Measuring microscale strain fields in articular cartilage during rapid impact reveals thresholds for chondrocyte death and a protective role for the superficial layer. J Biomech. doi:10.1016/j.jbiomech.2015.05.035

Bohensky J, Shapiro IM, Leshinsky S, Terkhorn SP, Adams CS, Srinivas V (2007a) HIF-1 regulation of chondrocyte apoptosis: induction of the autophagic pathway. Autophagy 3(3):207–214

Bohensky J, Shapiro IM, Leshinsky S, Watanabe H, Srinivas V (2007b) PIM-2 is an independent regulator of chondrocyte survival and autophagy in the epiphyseal growth plate. J Cell Physiol 213(1):246–251. doi:10.1002/jcp.21117

Brandt KD, Dieppe P, Radin EL (2009) Commentary: is it useful to subset "primary" osteoarthritis? A critique based on evidence regarding the etiopathogenesis of osteoarthritis. Semin Arthritis Rheum 39(2):81–95. doi:10.1016/j.semarthrit.2009.06.001

Burleigh A, Chanalaris A, Gardiner MD, Driscoll C, Boruc O, Saklatvala J, Vincent TL (2012) Joint immobilization prevents murine osteoarthritis and reveals the highly mechanosensitive nature of protease expression in vivo. Arthritis Rheum 64(7):2278–2288. doi:10.1002/art.34420

Campbell CJ (1969) The healing of cartilage defects. Clin Orthop Relat Res 64:45–63

Carames B, Taniguchi N, Seino D, Blanco FJ, D'Lima D, Lotz M (2012) Mechanical injury suppresses autophagy regulators and pharmacologic activation of autophagy results in chondroprotection. Arthritis Rheum 64(4):1182–1192. doi:10.1002/art.33444

Chen C, Tambe DT, Deng L, Yang L (2013) Biomechanical properties and mechanobiology of the articular chondrocyte. Am J Physiol Cell Physiol 305(12):C1202–C1208. doi:10.1152/ajpcell.00242.2013

Chong KW, Chanalaris A, Burleigh A, Jin H, Watt FE, Saklatvala J, Vincent TL (2013) Fibroblast growth factor 2 drives changes in gene expression following injury to murine cartilage in vitro and in vivo. Arthritis Rheum 65(9):2346–2355. doi:10.1002/art.38039

Chowdhury TT, Salter DM, Bader DL, Lee DA (2008) Signal transduction pathways involving p38 MAPK, JNK, NFkappaB and AP-1 influences the response of chondrocytes cultured in agarose constructs to IL-1beta and dynamic compression. Inflamm Res: Off J Eur Histamine Res Soc [et al] 57(7):306–313. doi:10.1007/s00011-007-7126-y

Christiansen BA, Guilak F, Lockwood KA, Olson SA, Pitsillides AA, Sandell LJ, Silva MJ, van der Meulen MC, Haudenschild DR (2015) Non-invasive mouse models of post-traumatic osteoarthritis. Osteoarthr Cartil/OARS Osteoarthr Res Soc 23(10):1627–1638. doi:10.1016/j.joca.2015.05.009

Clark AL, Votta BJ, Kumar S, Liedtke W, Guilak F (2010) Chondroprotective role of the osmotically sensitive ion channel transient receptor potential vanilloid 4: age- and sex-dependent progression of osteoarthritis in Trpv4-deficient mice. Arthritis Rheum 62(10):2973–2983. doi:10.1002/art.27624

Costouros JG, Kim HT (2007) Preventing chondrocyte programmed cell death caused by iatrogenic injury. Knee 14(2):107–111. doi:10.1016/j.knee.2006.10.013

D'Lima D, Hermida J, Hashimoto S, Colwell C, Lotz M (2006) Caspase inhibitors reduce severity of cartilage lesions in experimental osteoarthritis. Arthritis Rheum 54(6):1814–1821. doi:10.1002/art.21874

Dang AC, Warren AP, Kim HT (2006) Beneficial effects of intra-articular caspase inhibition therapy following osteochondral injury. Osteoarthr Cartil/OARS Osteoarthr Res Soc 14(6):526–532. doi:10.1016/j.joca.2005.12.010

Dell'Accio F, De Bari C, El Tawil NM, Barone F, Mitsiadis TA, O'Dowd J, Pitzalis C (2006) Activation of WNT and BMP signaling in adult human articular cartilage following mechanical injury. Arthritis Res Ther 8(5):R139. doi:10.1186/ar2029

Dell'accio F, De Bari C, Eltawil NM, Vanhummelen P, Pitzalis C (2008) Identification of the molecular response of articular cartilage to injury, by microarray screening: Wnt-16 expression and signaling after injury and in osteoarthritis. Arthritis Rheum 58(5):1410–1421. doi:10.1002/art.23444

DiMicco MA, Patwari P, Siparsky PN, Kumar S, Pratta MA, Lark MW, Kim YJ, Grodzinsky AJ (2004) Mechanisms and kinetics of glycosaminoglycan release following in vitro cartilage injury. Arthritis Rheum 50(3):840–848. doi:10.1002/art.20101

Ding L, Guo D, Homandberg GA (2008) The cartilage chondrolytic mechanism of fibronectin fragments involves MAP kinases: comparison of three fragments and native fibronectin. Osteoarthr Cartil/OARS Osteoarthr Res Soc 16(10):1253–1262. doi:10.1016/j.joca.2008.02.015

Ding L, Heying E, Nicholson N, Stroud NJ, Homandberg GA, Buckwalter JA, Guo D, Martin JA (2010) Mechanical impact induces cartilage degradation via mitogen activated protein kinases. Osteoarthr Cartil/OARS Osteoarthr Res Soc 18(11):1509–1517. doi:10.1016/j.joca.2010.08.014

Eltawil NM, De Bari C, Achan P, Pitzalis C, Dell'accio F (2009) A novel in vivo murine model of cartilage regeneration. Age and strain-dependent outcome after joint surface injury. Osteoarthr Cartil/OARS Osteoarthr Res Soc 17(6):695–704. doi:10.1016/j.joca.2008.11.003

Enomoto-Iwamoto M, Iwamoto M, Nakashima K, Mukudai Y, Boettiger D, Pacifici M, Kurisu K, Suzuki F (1997) Involvement of alpha5beta1 integrin in matrix interactions and proliferation of chondrocytes. J Bone Min Res: Off J Am Soc Bone Min Res 12(7):1124–1132. doi:10.1359/jbmr.1997.12.7.1124

Fanning PJ, Emkey G, Smith RJ, Grodzinsky AJ, Szasz N, et al. (2003) Mechanical regulation of mitogen-activated protein kinase signaling in articular cartilage. J Biol Chem 278:50940–50948. doi: 10.1074/jbc.m305107200

Fitzgerald JB, Jin M, Chai DH, Siparsky P, Fanning P, Grodzinsky AJ (2008) Shear- and compression-induced chondrocyte transcription requires MAPK activation in cartilage explants. J Biol Chem 283(11):6735–6743. doi:10.1074/jbc.M708670200

Gardiner MD, Vincent TL, Driscoll C, Burleigh A, Bou-Gharios G, Saklatvala J, Nagase H, Chanalaris A (2015) Transcriptional analysis of micro-dissected articular cartilage in post-traumatic murine osteoarthritis. Osteoarthr Cartil/OARS Osteoarthr Res Soc 23(4):616–628. doi:10.1016/j.joca.2014.12.014

Glasson SS, Askew R, Sheppard B, Carito B, Blanchet T, Ma HL, Flannery CR, Peluso D, Kanki K, Yang Z, Majumdar MK, Morris EA (2005) Deletion of active ADAMTS5 prevents cartilage degradation in a murine model of osteoarthritis. Nature 434(7033):644–648. doi:10.1038/nature03369

Goodwin W, McCabe D, Sauter E, Reese E, Walter M, Buckwalter JA, Martin JA (2010) Rotenone prevents impact-induced chondrocyte death. J Orthop Res: Off Publ Orthop Res Soc 28(8):1057–1063. doi:10.1002/jor.21091

Gruber J, Vincent TL, Hermansson M, Bolton M, Wait R, Saklatvala J (2004) Induction of interleukin-1 in articular cartilage by explantation and cutting. Arthritis Rheum 50(8):2539–2546. doi:10.1002/art.20369

Guilak F, Alexopoulos LG, Upton ML, Youn I, Choi JB, Cao L, Setton LA, Haider MA (2006) The pericellular matrix as a transducer of biomechanical and biochemical signals in articular cartilage. Ann N Y Acad Sci 1068:498–512. doi:10.1196/annals.1346.011

Guo D, Ding L, Homandberg GA (2009) Telopeptides of type II collagen upregulate proteinases and damage cartilage but are less effective than highly active fibronectin fragments. Inflamm Res: Off J Eur Histamine Res Soc [et al] 58(3):161–169. doi:10.1007/s00011-009-8090-5

Han SK, Wouters W, Clark A, Herzog W (2012) Mechanically induced calcium signaling in chondrocytes in situ. J Orthop Res: Off Publ Orthop Res Soc 30(3):475–481. doi:10.1002/jor.21536

Huveneers S, Danen EH (2009) Adhesion signaling – crosstalk between integrins, Src and Rho. J Cell Sci 122(Pt 8):1059–1069. doi:10.1242/jcs.039446

Jang KW, Buckwalter JA, Martin JA (2014) Inhibition of cell-matrix adhesions prevents cartilage chondrocyte death following impact injury. J Orthop Res: Off Publ Orthop Res Soc 32(3):448–454. doi:10.1002/jor.22523

Killian ML, Haut RC, Haut Donahue TL (2014) Acute cell viability and nitric oxide release in lateral menisci following closed-joint knee injury in a lapine model of post-traumatic osteoarthritis. BMC Musculoskelet Disord 15:297. doi:10.1186/1471-2474-15-297

Kuhn K, D'Lima DD, Hashimoto S, Lotz M (2004) Cell death in cartilage. Osteoarthr Cartil/OARS Osteoarthr Res Soc 12(1):1–16

Kvist AJ, Nystrom A, Hultenby K, Sasaki T, Talts JF, Aspberg A (2008) The major basement membrane components localize to the chondrocyte pericellular matrix – a cartilage basement membrane equivalent? Matrix Biol: J Int Soc Matrix Biol 27(1):22–33. doi:10.1016/j.matbio.2007.07.007

Lee JH, Fitzgerald JB, Dimicco MA, Grodzinsky AJ (2005) Mechanical injury of cartilage explants causes specific time-dependent changes in chondrocyte gene expression. Arthritis Rheum 52(8):2386–2395. doi:10.1002/art.21215

Lee W, Leddy HA, Chen Y, Lee SH, Zelenski NA, McNulty AL, Wu J, Beicker KN, Coles J, Zauscher S, Grandl J, Sachs F, Guilak F, Liedtke WB (2014) Synergy between Piezo1 and Piezo2 channels confers high-strain mechanosensitivity to articular cartilage. Proc Natl Acad Sci U S A 111(47):E5114–E5122. doi:10.1073/pnas.1414298111

Little CB, Barai A, Burkhardt D, Smith SM, Fosang AJ, Werb Z, Shah M, Thompson EW (2009) Matrix metalloproteinase 13-deficient mice are resistant to osteoarthritic cartilage erosion but not chondrocyte hypertrophy or osteophyte development. Arthritis Rheum 60(12):3723–3733. doi:10.1002/art.25002

Madej W, van Caam A, Blaney Davidson EN, Hannink G, Buma P, van der Kraan PM (2016) Ageing is associated with reduction of mechanically-induced activation of Smad2/3P signaling in articular cartilage. Osteoarthr Cartil/OARS Osteoarthr Res Soc 24(1):146–157. doi:10.1016/j.joca.2015.07.018

Martin JA, McCabe D, Walter M, Buckwalter JA, McKinley TO (2009) N-acetylcysteine inhibits post-impact chondrocyte death in osteochondral explants. J Bone Joint Surg Am 91(8):1890–1897. doi:10.2106/JBJS.H.00545

Meachim G (1963) The effect of scarification on articular cartilage in the rabbit. J Bone Joint Surg Br 45B:150–161

Namdari S, Wei L, Moore D, Chen Q (2008) Reduced limb length and worsened osteoarthritis in adult mice after genetic inhibition of p38 MAP kinase activity in cartilage. Arthritis Rheum 58(11):3520–3529. doi:10.1002/art.23999

O'Conor CJ, Leddy HA, Benefield HC, Liedtke WB, Guilak F (2014) TRPV4-mediated mechanotransduction regulates the metabolic response of chondrocytes to dynamic loading. Proc Natl Acad Sci U S A 111(4):1316–1321. doi:10.1073/pnas.1319569111

Palmoski M, Perricone E, Brandt KD (1979) Development and reversal of a proteoglycan aggregation defect in normal canine knee cartilage after immobilization. Arthritis Rheum 22(5):508–517. doi:10.1002/Art.1780220511

Phan MN, Leddy HA, Votta BJ, Kumar S, Levy DS, Lipshutz DB, Lee SH, Liedtke W, Guilak F (2009) Functional characterization of TRPV4 as an osmotically sensitive ion channel in porcine articular chondrocytes. Arthritis Rheum 60(10):3028–3037. doi:10.1002/art.24799

Poulet B, Hamilton RW, Shefelbine S, Pitsillides AA (2011) Characterizing a novel and adjustable noninvasive murine joint loading model. Arthritis Rheum 63(1):137–147. doi:10.1002/art.27765

Quinn TM, Grodzinsky AJ, Buschmann MD, Kim YJ, Hunziker EB (1998) Mechanical compression alters proteoglycan deposition and matrix deformation around individual cells in cartilage explants. J Cell Sci 111:573–583

Rai MF, Hashimoto S, Johnson EE, Janiszak KL, Fitzgerald J, Heber-Katz E, Cheverud JM, Sandell LJ (2012) Heritability of articular cartilage regeneration and its association with ear wound healing in mice. Arthritis Rheum 64(7):2300–2310. doi:10.1002/art.34396

Redman SN, Dowthwaite GP, Thomson BM, Archer CW (2004) The cellular responses of articular cartilage to sharp and blunt trauma. Osteoarthr Cartil 12(2):106–116. doi:10.1016/J.Joca.2002.12.001

Redman SN, Khan IM, Tew SR, Archer CW (2007) In situ detection of cell death in articular cartilage. Methods Mol Med 135:183–199

Roach HI, Aigner T, Kouri JB (2004) Chondroptosis: a variant of apoptotic cell death in chondrocytes? Apoptosis: Int J Program Cell Death 9(3):265–277

Rosenzweig DH, Djap MJ, Ou SJ, Quinn TM (2012) Mechanical injury of bovine cartilage explants induces depth-dependent, transient changes in MAP kinase activity associated with apoptosis. Osteoarthr Cartil/OARS Osteoarthr Res Soc 20(12):1591–1602. doi:10.1016/j.joca.2012.08.012

Salter DM, Millward-Sadler SJ, Nuki G, Wright MO (2002) Differential responses of chondrocytes from normal and osteoarthritic human articular cartilage to mechanical stimulation. Biorheology 39(1–2):97–108

Sauter E, Buckwalter JA, McKinley TO, Martin JA (2012) Cytoskeletal dissolution blocks oxidant release and cell death in injured cartilage. J Orthop Res: Off Publ Orthop Res Soc 30(4):593–598. doi:10.1002/jor.21552

Seror J, Zhu L, Goldberg R, Day AJ, Klein J (2015) Supramolecular synergy in the boundary lubrication of synovial joints. Nat Commun 6:6497. doi:10.1038/ncomms7497

Soder S, Hambach L, Lissner R, Kirchner T, Aigner T (2002) Ultrastructural localization of type VI collagen in normal adult and osteoarthritic human articular cartilage. Osteoarthr Cartil/OARS Osteoarthr Res Soc 10(6):464–470. doi:10.1053/joca.2002.0512

Stockwell RA (1978) Chondrocytes. J Clin Pathol Suppl 12:7–13

Tesche F, Miosge N (2004) Perlecan in late stages of osteoarthritis of the human knee joint. Osteoarthr Cartil/OARS Osteoarthr Res Soc 12(11):852–862. doi:10.1016/j.joca.2004.07.004

Tew SR, Kwan AP, Hann A, Thomson BM, Archer CW (2000) The reactions of articular cartilage to experimental wounding: role of apoptosis. Arthritis Rheum 43(1):215–225. doi:10.1002/1529-0131(200001)43:1<215::AID-ANR26>3.0.CO;2-X

Vanwanseele B, Eckstein F, Knecht H, Stussi E, Spaepen A (2002) Knee cartilage of spinal cord-injured patients displays progressive thinning in the absence of normal joint loading and movement. Arthritis Rheum 46(8):2073–2078. doi:10.1002/art.10462

Vinall RL, Lo SH, Reddi AH (2002) Regulation of articular chondrocyte phenotype by bone morphogenetic protein 7, interleukin 1, and cellular context is dependent on the cytoskeleton. Exp Cell Res 272(1):32–44. doi:10.1006/excr.2001.5395

Vincent T, Hermansson M, Bolton M, Wait R, Saklatvala J (2002) Basic FGF mediates an immediate response of articular cartilage to mechanical injury. Proc Natl Acad Sci U S A 99(12):8259–8264. doi:10.1073/pnas.122033199

Vincent TL, Hermansson MA, Hansen UN, Amis AA, Saklatvala J (2004) Basic fibroblast growth factor mediates transduction of mechanical signals when articular cartilage is loaded. Arthritis Rheum 50(2):526–533. doi:10.1002/Art.20047

Vincent TL, McLean CJ, Full LE, Peston D, Saklatvala J (2007) FGF-2 is bound to perlecan in the pericellular matrix of articular cartilage, where it acts as a chondrocyte mechanotransducer. Osteoarthr Cartil 15(7):752–763. doi:10.1016/j.joca.2007.01.021

Vincent TL, Williams RO, Maciewicz R, Silman A, Garside P, Working ARUAM (2012) Mapping pathogenesis of arthritis through small animal models. Rheumatology 51(11):1931–1941. doi:10.1093/rheumatology/kes035

Waller KA, Zhang LX, Elsaid KA, Fleming BC, Warman ML, Jay GD (2013) Role of lubricin and boundary lubrication in the prevention of chondrocyte apoptosis. Proc Natl Acad Sci U S A 110(15):5852–5857. doi:10.1073/pnas.1219289110

Wang N, Butler JP, Ingber DE (1993) Mechanotransduction across the cell-surface and through the cytoskeleton. Science 260(5111):1124–1127. doi:10.1126/Science.7684161

Watt FE, Ismail HM, Didangelos A, Peirce M, Vincent TL, Wait R, Saklatvala J (2013) Src and fibroblast growth factor 2 independently regulate signaling and gene expression induced by experimental injury to intact articular cartilage. Arthritis Rheum 65(2):397–407. doi:10.1002/art.37765

Wright M, Jobanputra P, Bavington C, Salter DM, Nuki G (1996) Effects of intermittent pressure-induced strain on the electrophysiology of cultured human chondrocytes: evidence for the presence of stretch-activated membrane ion channels. Clin Sci 90(1):61–71

Proteoglycan and Collagen Degradation in Osteoarthritis

3

Stephanie J. Gauci, Heather Stanton, Christopher B. Little, and Amanda J. Fosang

Abstract

The gradual loss of articular cartilage from the surface of articulating joints is a feature of osteoarthritis. It is marked by degradation of the cartilage matrix, including the large aggregating proteoglycan aggrecan, the small leucine-rich proteoglycans known as SLRPs and the fibrillar type II collagen. Aggrecan provides the water-holding capacity of cartilage, while the collagen II scaffold provides elastic restraint, aided by a protective coat of small leucine-rich proteoglycans. Damaged aggrecan is readily replaced by synthesis of new aggrecan; however, type II collagen can resist only a limited amount of proteolysis before cartilage function is compromised. In this review the major enzyme families of cartilage-degrading enzymes, the matrix metalloproteinases (MMPs) and a disintegrin and metalloproteinase with thrombospondin motifs (ADAMTS) families, are discussed. We examine factors that regulate MMP and ADAMTS activity, with a focus on MMP-13, ADAMTS-4 and ADAMTS-5 as the major protagonists of cartilage degradation. We also compare the effects of blocking aggrecanolysis and collagenolysis separately, or together, on cartilage erosion in a mouse model of osteoarthritis. The role of degraded matrix fragments in regulating inflammation in osteoarthritis, via Toll-like receptor signalling, is also discussed.

S.J. Gauci, PhD • H. Stanton, PhD • A.J. Fosang, PhD (✉)
University of Melbourne Department of Paediatrics and Murdoch Childrens Research Institute, Royal Children's Hospital, Parkville 3052, VIC, Australia
e-mail: stephanie.gauci@mcri.edu.au; heather.stanton@mcri.edu.au; amanda.fosang@mcri.edu.au

C.B. Little, PhD
Raymond Purves Bone and Joint Research Laboratories, Kolling Institute, Institute of Bone and Joint Research, Sydney Medical School Northern, University of Sydney, Level 10 Kolling Building – B6, Royal North Shore Hospital, St. Leonards 2065, NSW, Australia
e-mail: christopher.little@sydney.edu.au

S. Grässel, A. Aszódi (eds.), *Cartilage*, DOI 10.1007/978-3-319-45803-8_3

3.1 Cartilage Matrix: Composition and Function

Articular cartilage is predominantly an extracellular matrix with no nerves or blood vessels and only a limited population of sparsely distributed chondrocytes. This small proportion of cells combined with the lack of vasculature means that once it is damaged, cartilage has a poor capacity for self-repair. The cartilage matrix has two key functions in the diarthrodial joint. One function is to provide a smooth, frictionless surface to facilitate movement; this function is conferred mostly by the lubricating glycoprotein, lubricin, secreted at the cartilage surface. The second function of cartilage is to distribute mechanical load across joint surfaces; this function is critically dependent upon water, which is attracted by the high concentration of negatively charged aggrecan trapped within a network of fibrous type II collagen. The type II collagen fibres are also decorated with minor collagens and small leucine-rich proteoglycans (SLRPs). Proteinase-mediated degradation of aggrecan, collagens and SLRPs severely compromises the mechanical properties of articular cartilage in osteoarthritis (OA) and other cartilage pathologies. These proteinases are produced by chondrocytes and also by synovial fibroblasts and infiltrating macrophages.

3.1.1 Aggrecan Structure and Function

Aggrecan is a large proteoglycan containing chondroitin sulphate (CS) and keratan sulphate (KS) glycosaminoglycan side chains. The aggrecan core protein comprises three globular domains, including G1 and G2 at the N-terminus and a C-terminal G3 globular domain separated from G2 by an extended CS-attachment region (Fig. 3.1). Multimolecular aggregates of aggrecan containing as many as 100 aggrecan monomers per aggregate ($MW_{aggregate}$~200×10^6 Da) are immobilised in cartilage by binding to hyaluronan via their G1 domains (Fig. 3.1). The hyaluronan-aggrecan complex is further stabilised by binding of link protein to form a trimeric complex. The vast size of the aggrecan aggregates, together with their water-binding capacity and their inability to dissociate under physiological conditions, traps aggrecan in cartilage and provides the osmotic swelling pressure that is required for weight-bearing and cartilage function. Aggrecan is degraded by proteolysis at multiple sites along the core protein, but the most detrimental cleavage occurs within the rod-like interglobular domain which separates the tissue-anchored G1 domain from the water-binding CS and KS chains. Proteolysis within the CS-rich region is considered less detrimental to cartilage function and likely reflects normal maturation and ageing, since most aggrecan in adult cartilage lacks the G3 domain and varying portions of the CS-rich region (Ilic et al. 1998).

3.1.2 Type II Collagen Structure and Function

Type II collagen is the major collagen in cartilage; its expression is restricted to cartilaginous tissues and the vitreous of the eye (further details, see Chap. 2 in Volume 1 (Grässel 2016)). Type II collagen is a homotrimer of three identical α

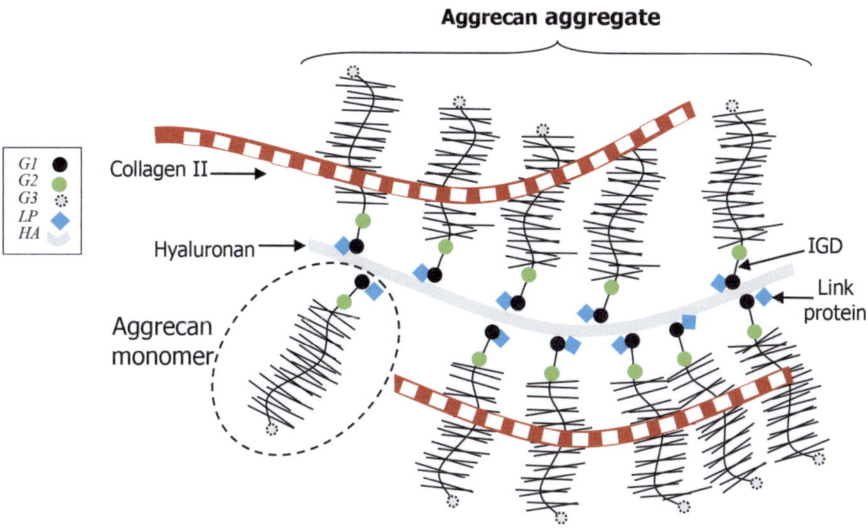

Fig. 3.1 Schematic cartoon showing arrangement of aggrecan and collagen in cartilage. The aggrecan core protein contains three globular domains including a C-terminal G3 domain, a G2 domain (*green*) and an N-terminal G1 domain (*black*). Aggrecan is anchored in the cartilage matrix by binding to link protein (*blue diamond*) and hyaluronan to form large multi-molecular aggregates (Adapted, with publisher's permission, from Fosang and Beier (2011))

chains, each with an uninterrupted helical domain characterised by repeating Gly-X-Y motifs where X and Y are bulky amino acids, often proline and hydroxyproline. The helical domain is flanked by N- and C-terminal propeptide domains that are removed during secretion and N- and C-terminal telopeptides that form inter- and intra-helical cross links, essential for helix stability (Metsaranta et al. 1991; Bateman et al. 1996). Collagen monomers assemble outside the cell into heterofibrils with collagens IX and XI, in a process that is aided by the SLRPs. Type II collagen fibres in the cartilage superficial zone are aligned parallel with the cartilage surface, whereas type II collagen fibres in the midzone have a cross-over pattern; fibres in the deep zone, adjacent to the calcified cartilage, are aligned perpendicular with the cartilage surface. This arrangement of type II collagen provides tensile strength and a structural scaffold for the cartilage. Proteolytic clipping of the collagen network loosens the elastic restraint within the cartilage matrix, causing it to soften, and, at the same time, allows aggrecan to swell and occupy increasingly greater hydrodynamic domains (Maroudas 1976). Only a limited amount of type II collagenolysis is tolerated before the onset of irreversible cartilage damage.

3.1.3 SLRP Structure and Function

The SLRP core proteins are characterised by a central region of contiguous leucine-rich repeat motifs, flanked by cysteine-rich domains (further details, see Chap. 1 in Volume 1 (Aspberg 2016)). The leucine-rich repeats adopt a horseshoe shape that binds to extracellular matrix molecules, in particular, the collagens. The variable N-termini of the SLRPs defines the unique function of family members and may contain CS, KS or dermatan sulphate chains. The SLRPs biglycan, decorin, lumican, fibromodulin and keratocan are common to cartilage, tendon and meniscus (Melrose et al. 2008; Rees et al. 2009a). Epiphycan, mimecan, osteoadherin, chondroadherin, chondroadherin-like, asporin, osteoglycin, opticin and prolargin are also found in cartilage (for structure and classification of SLRPs found in joint tissues, see Chap. 1, Volume 1 (Aspberg 2016)). SLRPs are best known for their role as structural constituents in joint matrices. SLRPs such as biglycan, decorin, fibromodulin and lumican bind to collagen fibrils, likely aiding fibril formation and modulating tissue organisation. SLRPs coat the surface of collagen fibrils, providing a steric barrier that potentially limits the access of collagen-degrading enzymes. Aside from their structural role, SLRPs directly regulate cell function by binding a range of cytokines, growth factors and cell surface receptors. The importance of SLRPs in OA is evident from the phenotypes of the SLRP-deficient mice. Major phenotypes of SLRP single and double knockout mice are tendon and ligament defects leading to joint laxity, ectopic ossification and early-onset OA (reviewed in Ni et al. (2014)).

3.2 Cartilage-Degrading Proteinases in OA

In contrast to the aggressive inflammation that typifies rheumatoid arthritis, the hallmark of OA is radiographic joint space narrowing caused by proteinase-driven cartilage erosion in a relatively mild inflammatory environment. Cartilage erosion in OA can progress for many years in the absence of symptoms. Given that type II collagen in adult human cartilage has a half-life of ~117 years (Verzijl et al. 2000), any significant damage to the collagen network is generally considered irreparable (Stoop et al. 1999). Data from mouse models of arthritis have confirmed that cartilage can resist only a limited attack on type II collagen before the onset of irreversible damage (Stoop et al. 1999). In contrast, in vivo studies have shown that large glycosaminoglycan-bearing aggrecan fragments lost from cartilage following stimulated aggrecanolysis are readily replaced by newly synthesised aggrecan (Thomas 1956). The major protagonists of cartilage degradation in OA are the metal-dependent proteinases of the matrix metalloproteinase (MMP) and a disintegrin and metalloproteinase with thrombospondin motifs (ADAMTS) families. Cathepsins, and cysteine proteinases such as m-calpain, might also have roles in aggrecan catabolism (Struglics and Hansson 2010; Troeberg and Nagase 2012) but are not discussed further.

3.2.1 MMPs

The MMPs are members of the metzincin superfamily. They have a zinc ion at the active site that is essential for catalytic activity and a conserved motif of histidine residues (HExxHxxGxxH) that coordinate the zinc ion. Following this motif is a conserved methionine that forms a characteristic turn important for structure and function. The MMPs are expressed as inactive zymogens, with an N-terminal propeptide that maintains latency, a catalytic domain and, in the majority of MMPs, a flexible hinge region linked to a hemopexin domain. The membrane-bound MMPs have an additional transmembrane domain and a cytoplasmic tail or are GPI anchored. Dysregulated MMP activity against type II collagen and aggrecan is thought to contribute to cartilage pathology in OA (Troeberg and Nagase 2012; Goldring et al. 2011). Specific cleavage sites and the products of MMP-mediated aggrecanolysis and collagenolysis have been well characterised, and fragments of type II collagen have been investigated for their utility as OA biomarkers (Nemirovskiy et al. 2007). SLRPs are also substrates for MMPs in cartilage.

Much attention has focused on MMP-13 (collagenase-3) due to its potent activity against type II collagen. Selective inhibitor studies have shown that MMP-13 is the major cartilage collagenase in vivo (Billinghurst et al. 1997; Dahlberg et al. 2000). MMP-13 expression is increased in cartilage and synovium during late-stage OA (Mitchell et al. 1996; Bau et al. 2002; Kevorkian et al. 2004; Davidson et al. 2006), and animal studies have shown that overexpression of MMP-13 in mouse cartilage induces OA-like pathology in the knee joint (Neuhold et al. 2001). *Mmp13−/−* mice are protected against cartilage erosion in a surgical model of OA (Little et al. 2009). Accordingly, MMP-13 is a prime target for OA therapies. Early attempts to develop MMP inhibitors focused on chelation of the active site zinc ion, using hydroxamic acid derivatives. These inhibitors failed in clinical trials due to poor selectivity and tendinitis-like side effects (reviewed in Jacobsen et al. (2010)). Since then, inhibitor designs have targeted secondary binding sites, known as exosites, with the aim of improving selectivity. Several pharmaceutical companies have reported the development of potent and highly selective inhibitors to MMP-13 (Gao et al. 2010; Nara et al. 2014; Engel et al. 2005; Gege et al. 2012; Jungel et al. 2010; Ruminski et al. 2016; Fischer and Riedl 2016), but to date there are no reports of clinical trials of these new classes of MMP-13 inhibitors (reviewed in Fields (2015)).

3.2.2 ADAMTS Enzymes

Like MMPs, the ADAMTS enzymes are a subset of the metzincin superfamily. There are 19 members of the mammalian ADAMTS family, all of which are secreted and active extracellularly. All ADAMTS enzymes comprise an N-terminal prodomain with a furin-recognition sequence, a catalytic domain with the consensus sequence HEBxHxBGBxH, a disintegrin-like domain, a cysteine-rich domain, a spacer domain and one or more thrombospondin repeats. ADAMTS-4 and ADAMTS-5 degrade aggrecan and they are considered the most important ADAMTS enzymes in OA. These aggrecanases are also active against SLRPs, including biglycan, fibromodulin and keratocan. ADAMTS-4 and ADAMTS-5 have a similar domain structure, with ADAMTS-4 having one and ADAMTS-5 having two thrombospondin motifs. The non-catalytic ancillary domains of ADAMTS-4 and ADAMTS-5 have roles in substrate recognition and matrix localisation (Gendron et al. 2007; Fushimi et al. 2008). Mice with ADAMTS-5 deletion or expressing catalytically inactive ADAMTS-5 have significantly reduced cartilage erosion in models of OA and inflammatory arthritis, respectively (Glasson et al. 2005; Stanton et al. 2005). Monoclonal antibodies targeting the catalytic and disintegrin domains of ADAMTS-5 have shown efficacy in inhibiting cartilage erosion in a mouse model of OA (Larkin et al. 2015).

3.2.3 Regulation of MMP and ADAMTS Enzymes

In healthy cartilage the expression of MMPs and ADAMTS enzymes is tightly regulated to allow low turnover remodelling of the matrix. Evidence from in vitro and in vivo studies has shown that in OA, cells of the joint produce, and respond to, a range of pro-inflammatory mediators, particularly interleukin (IL)-1α, IL-1β and tumour necrosis factor α (TNFα) (reviewed in Goldring and Otero (2011)). IL-1α and β are well known for their ability to upregulate MMP expression in OA cartilage (Mitchell et al. 1996; Kobayashi et al. 2005; Bau et al. 2002). For the aggrecanases, studies with inflammatory mediators have produced inconsistent results, with many studies supporting the constitutive expression of ADAMTS-5 mRNA in joint tissues and cells, but others reporting regulation of ADAMTS-5 mRNA by IL-1α and β, oncostatin M and FGF-2 (reviewed in Fosang et al. (2008); Fosang and Rogerson (2010)). ADAMTS-4 mRNA, in contrast, is reproducibly upregulated by cytokines in many experimental systems. Interestingly, the regulation of ADAMTS-4 and ADAMTS-5 expression differs according to species, cell type, tissue origin (hip or knee) and perhaps disease stage.

The post-transcriptional regulation of MMP and ADAMTS enzymes in OA cartilage is mediated by epigenetic modification to DNA (Tsezou 2014), histone modification (Troeberg and Nagase 2012) and alternative splicing in the case of ADAMTS-4 (Wainwright et al. 2013). One area of post-transcriptional regulation that is rapidly evolving is the contribution that non-coding microRNAs (miRNAs) make to the regulation of gene expression in OA. miRNAs bind to the 3′-untranslated region of target RNAs and silence gene expression. Over 40 miRNAs have now been reported as differentially regulated in OA (Nugent 2016). Compared with wild-type mice, miR-140 $^{-/-}$ mice develop OA-like pathology with increased levels of ADAMTS-5 mRNA and increased aggrecan loss in vitro (Miyaki et al. 2009). Conversely, overexpression of miR-140 in chondrocytes reduced aggrecan loss in IL-1β-treated cells in vitro and also reduced in vivo aggrecan loss in the antigen-induced model of inflammatory arthritis (Miyaki et al. 2010). miR-146a (Li et al. 2011) and miR-148a (Vonk et al. 2014) also negatively regulate ADAMTS-5 mRNA, whereas miR-125b regulates ADAMTS-4 (Matsukawa et al. 2013). MiR-33a (Kostopoulou et al. 2015) and miR-27b (Akhtar et al. 2010) function as positive and negative regulators, respectively, of MMP-13 in human chondrocytes, and miRNA-127-5p suppresses IL-1β-induced MMP-13 production in human chondrocytes (Park et al. 2013). Most recently, miR-30a has been identified as an essential regulator of IL-1β-induced ADAMTS-5 expression and cartilage degradation (Ji et al. 2016).

In normal tissues, the post-translational regulation of the MMP and ADAMTS enzymes is tightly controlled to keep levels of enzyme activity in check. The first checkpoint is activation, which requires proteolytic processing at the N-terminus to remove the propeptide. For most MMPs, this activation occurs via a 'cysteine-switch' mechanism, involving proteolytic disruption of a pairing between a conserved cysteine residue in the propeptide and the zinc ion at the active site. A range of serine proteinases are candidates for MMP propeptide removal in arthritic joints, including plasmin, matriptase-1, activated protein-C, mast cell tryptases and chymases (reviewed in Fosang and Beier (2011)). Active MMPs are also capable of initiating MMP activation cascades; in particular, membrane-type MMP-1 (MT1-MMP) can activate proMMP-2 and proMMP-13 (Murphy et al. 1999). Some MMPs, including MMP-11, MMP-28 and several MT-MMPs, as well as ADAMTS-4 and ADAMTS-5 have a furin-recognition sequence (RXK/RR) in their propeptide domains, which is cleaved by proprotein convertases. There are seven proprotein convertases in mammals, including the archetypal furin, which compartmentalises to various subcellular, cell surface and extracellular locations, providing spatial control of enzyme activation. PACE4 is the most abundant proprotein convertase in extracts of human osteoarthritic cartilage and is likely to be the primary activator of ADAMTS-4 and ADAMTS-5 in cartilage (Malfait et al. 2008).

For ADAMTS-4 and ADAMTS-5, C-terminal truncation by proteolytic (Gao et al. 2004) or autocatalytic mechanisms (Zeng et al. 2006; Flannery et al. 2002) provides a further level of control. Autocatalytic cleavages occur in the spacer domain of ADAMTS-4 and ADAMTS-5 (Flannery et al. 2002; Zeng et al. 2006). The spacer and cysteine-rich domains are important for locating the enzymes to the cell surface or to the pericellular or extracellular matrices. The endogenous tissue inhibitors of metalloproteinases (TIMPs) inhibit MMPs and ADAMTS enzymes. All four members of the TIMP family inhibit MMPs, showing relatively little selectivity, with the exception of some membrane-type MMPs that are poorly inhibited by TIMP-1. ADAMTS-4 and ADAMTS-5, in contrast, are only inhibited by TIMP-3, but it is effective at sub-nanomolar concentrations and is the only TIMP that localises to the cartilage matrix, via binding to heparan sulphate-bearing proteoglycans, including perlecan (Troeberg et al. 2014).

The levels of MMPs and ADAMTS enzymes in cartilage are further controlled by endocytotic clearance via low-density lipoprotein receptor-related protein-1 (LRP-1). ADAMTS-4 and ADAMTS-5 share the same binding site on LRP-1, but ADAMTS-5 has the greater affinity for LRP-1 (Yamamoto et al. 2016). TIMP-3 is also internalised by LRP-1, and, interestingly, TIMP-3 binding to sulphated glycosaminoglycans reduces TIMP-3 clearance rates and increases TIMP-3 affinity for ADAMTS-5 (and likely ADAMTS-4 and several MMPs), making synthetic glycan mimetics potential targets for blocking cartilage destruction in OA (Troeberg et al. 2014). More recently, Yamamoto and colleagues showed that MMP-13 is also internalised by LRP-1 and that it can be co-endocytosed with ADAMTS-4, ADAMTS-5 or TIMP-3 (Yamamoto et al. 2016). A rapid, LRP-1-mediated clearance of

MMP-13, ADAMTS-5 and TIMP-3 might explain why detection of these proteins in chondrocyte-conditioned medium is difficult, despite measurable levels of mRNA expression. Rapid clearance might also provide tight control over proteinase activity in healthy cartilage. But in OA, shedding of LRP-1 protein, along with elevated expression of MMP-13, ADAMTS-4 and ADAMTS-5, is expected to tip the balance in favour of cartilage degradation.

3.2.4 Aggrecan Degradation

Following the detection of aggrecanase fragments in human synovial fluids from patients with OA, joint injury and inflammatory joint disease (Lohmander et al. 1993; Sandy et al. 1992), two aggrecanase enzymes were cloned and identified as ADAMTS-4 (Tortorella et al. 1999) and ADAMTS-5 (Abbaszade et al. 1999). Both enzymes cleaved at the $E_{373}\downarrow_{374}A$ bond in the aggrecan interglobular domain. ADAMTS-5 is the primary aggrecanase in the mouse, but it is not yet clear whether ADAMTS-4 or ADAMTS-5 is the predominant human aggrecanase. Monoclonal antibodies against a conformational epitope spanning both the catalytic and disintegrin-like domains of either ADAMTS-4 or ADAMTS-5 have been raised to further explore the roles of ADAMTS-4 and ADAMTS-5 in vivo (Larkin et al. 2015). The anti-ADAMTS-5 antibodies blocked aggrecan loss from human cartilage explants cultured with or without cytokine stimulation, whereas anti-ADAMTS-4 antibodies only blocked aggrecan loss from unstimulated cartilage explants. These results do not provide a clear-cut verdict on whether ADAMTS-4 might also be a functional aggrecanase in humans; however, they reveal the interesting possibility that in both ADAMTS-4 and ADAMTS-5, the active site spans the catalytic and the disintegrin-like domains. Other studies have also suggested that both catalytic and disintegrin-like domains of ADAMTS-5 are required for maximum aggrecanase activity (Kosasih et al. 2016; Santamaria et al. 2015). It is unclear which ADAMTS is responsible for aggrecan degradation in tendon. Tendon explant cultures secrete aggrecanase activity in the absence of catabolic stimulation suggesting a role for aggrecanases in constitutive turnover of aggrecan (Rees et al. 2009a).

Cleavage at $E_{373}\downarrow_{374}A$ by ADAMTS enzymes is considered the most detrimental to aggrecan function because it releases all the CS chains from the aggrecan aggregate. Although this large and highly glycosylated $_{374}$ARGS fragment is considered a defining product of cartilage erosion, it is not the preferred cleavage site for the ADAMTS enzymes. Instead ADAMTS-4 and ADAMTS-5 cleave bovine aggrecan in the CS-rich region, at $E_{1666}\downarrow_{1667}G$ and $E_{1480}\downarrow_{1481}G$, prior to cleavage at $E_{373}\downarrow_{374}A$ (Tortorella et al. 2000). Subsequent cleavage events at $E_{1819}\downarrow_{1820}A$ and $E_{1919}\downarrow_{1920}L$ also occur in the CS-rich region. These cleavage sites are conserved in human, pig, mouse and rat aggrecan. Several fragments are further catabolised, for example, the G1-TEGE$_{373}$ fragment is cleaved by MMPs to generate a 32-amino-acid fragment with a $_{342}$FFGV N-terminus and a TEGE$_{373}$ C-terminus, which we have named the '32mer' (Lees et al. 2015). The 32mer has pro-inflammatory and pro-catabolic activity, as discussed below.

3.2.5 Aggrecan Neoepitopes as Markers of Cartilage Catabolism

Neoepitope antibodies are valuable and well-documented tools for identifying proteinase families involved in protein degradation. Aggrecan neoepitope antibodies recognise amino acid sequences at the newly created N- or C-terminus of cleaved aggrecan, but critically, they do not recognise the same sequence of amino acids in undegraded aggrecan or aggrecan fragments derived from different enzyme families. Aggrecan $_{374}$ARGS fragments are the most prominent fragments in OA synovial fluids and are widely used as a marker of ADAMTS activity in human OA studies and other experimental systems. In contrast, aggrecan $_{374}$ARGS fragments are poorly represented in children with juvenile arthritis, suggesting that enzymes other than aggrecanases might have a greater role in childhood arthritides (Struglics et al. 2012). MMP-derived aggrecan $_{342}$FFGV fragments are also present in human and animal synovial fluids. Similarly, the corresponding G1 domains with either IPEN$_{341}$ or TEGE$_{373}$ C-termini are retained in cartilage by binding to hyaluronan and are readily detected by immunohistochemistry in sections from human and animal cartilage.

3.2.6 Type II Collagen Degradation

The collagenases are a specialised subfamily of MMPs that cleave type II collagen at a single site in the triple helix, approximately three-quarters of the way from the N-terminus at G$_{775}$↓$_{776}$L (human sequence). Cleavage at this highly conserved site produces a three-fourths length TC^{A} N-terminal fragment and a one-fourth length TC^{B} C-terminal fragment (TC^{B}) and is the initiating event in collagenolysis (Miller et al. 1976), in all species. Thereafter the thermally unstable TC^{A} and TC^{B} fragments denature to become substrates for further proteolysis, for example, TC^{B} fragments are cleaved by MMPs at G$_{778}$↓$_{779}$Q (Mitchell et al. 1996; Billinghurst et al. 1997) and G$_{781}$↓$_{782}$ (Billinghurst et al. 1997). TC^{A} cleavage products have been detected in the urine of OA patients by mass spectrometry (Nemirovskiy et al. 2007). Although MMP-1, MMP-2, MMP-8, MMP-13 and MMP-14 cleave native type II collagen in vitro, MMP-13 is considered the principal cartilage collagenase (Billinghurst et al. 1997; Mitchell et al. 1996; Knäuper et al. 1996).

Considering that most enzyme substrates are single polypeptide sequences, the helix of native collagen presents a considerable challenge for hydrolysis, given that three polypeptide sequences are tightly wound. The mechanism of hydrolysis is intriguing because the catalytic cleft of the collagenases is not wide enough to accommodate a triple helical substrate (Chung et al. 2004), and isolated catalytic domains of the collagenases fail to cleave collagen (Knäuper et al. 1997; Patterson et al. 2001), suggesting that a sophisticated protease-substrate interaction is involved. Manka et al. (2012) used data from biochemical and crystallographic analyses to construct a model of collagenolysis that is unique in the field of protease-substrate interactions. Collagenolysis is made possible by the plasticity of the collagenases and by the relative thermoflexibility of collagen at the collagenase cleavage site, which is sufficient at body temperature to allow localised loosening of the helix. Mapping has shown that catalytic and hemopexin domains of MMP-1 bind cooperatively to two conserved leucines near the cleavage site, to anchor two collagen α chains. Modelling suggests that flexing of MMP-1 at the linker region between the catalytic and hemopexin domains bends the third α chain, facilitating localised collagen unwinding and positioning the collagenase-cleavable bond in the catalytic cleft. Following the initial cleavage, it is possible that the other two strands unravel, making them susceptible to hydrolysis (Stura et al. 2013).

Collagen type II fibrils can be further compromised by cathepsin or MMP cleavage within the N- and C-terminal telopeptide domains. For example, peptides with the sequence EKGPDP, derived from the collagen type II telopeptide region have been detected in the urine of OA patients (Mazzuca et al. 2006). Similarly, peptide fragments derived from cathepsin K cleavage at $G_{61}\downarrow_{62}K$ (human sequence) near the N-terminus of the triple helix have been detected in human and animal cartilage in vitro (Kafienah et al. 1998) and in vivo (Dejica et al. 2008).

3.2.7 The Relative Contributions of Aggrecanolysis and Collagenolysis to Cartilage Erosion

Cartilage erosion in OA can progress for many years in the absence of clinical symptoms, and there is evidence to suggest that aggrecanolysis precedes collagenolysis, for example:

(i) The consistent finding of weak metachromatic staining in sections of degenerate human and animal cartilage, prior to evidence of clinical symptoms.
(ii) The in vivo studies of Lewis Thomas (1956) which showed that papain injected into an upright rabbit ear produced a loose, floppy ear lacking tissue turgor; aggrecan is readily degraded by papain whereas fibrillar collagen is resistant to all proteinases other than collagenases under physiological conditions.
(iii) The data showing that treatment of cartilage explants with IL-α, IL-1ß, TNF-α, oncostatin M or retinoic acid stimulates marked aggrecan loss within 1–2 days, followed by collagen breakdown a week or more later (Little et al. 2002; Kozaci et al. 1997; Beekman et al. 1998; Pratta et al. 2003).

Possibly, the long period of subclinical OA is due to protection conferred on the collagen scaffold by constantly replenishing levels of aggrecan. Furthermore, it is possible that cartilage failure and radiographic cartilage damage emerge when aggrecan synthesis declines, and the collagen scaffold becomes progressively exposed.

Other than the studies above on the timing of aggrecanolysis versus collagenolysis, there are currently no data to indicate which of these processes is most important for driving OA. To explore this aspect, we produced knock-in mice with mutations at either the aggrecanase cleavage site in the aggrecan IGD (Jaffa mice) (Little et al. 2007) or the collagenase cleavage site in type II collagen. To block cleavage of type II collagen, we introduced a mutation at the primary collagenase cleavage site in mouse collagen II that changes amino acids PQG$^{775}\downarrow^{776}$LAG to PPG$^{775}\downarrow^{776}$MPG in the α1(II) chain. This mutation blocks cleavage at the primary collagenase cleavage site and thus blocks production of the TC^{A} and TC^{B} fragments that are the substrates for further collagen processing by gelatinases. This change, which is identical to the changes made previously in the α1(I) chain of type I collagen (Wu et al. 1990; Liu et al. 1995; Hasty et al. 1993; Zhao et al. 1999), completely blocked cleavage at the three-fourths to one-fourth collagenase cleavage site but, importantly, had no impact on collagen fibrillogenesis, which remained normal. Whereas type II collagen is expressed in cartilage and the vitreous of the eye, type I collagen is expressed abundantly in many tissues. The type I collagen knock-in mouse has a mild bone phenotype that resolves with age.

In the experiments described here, Jaffa mice were homozygous for the aggrecanase-resistant mutation, whereas the collagenase-resistant mutation was heterozygous in both the single- and double-mutant mice. We first compared passive and stimulated aggrecan loss in response to short-term IL-1α treatment and found that while the levels of passive aggrecan loss were low, and similar for each genotype (data not shown), the levels of stimulated aggrecan loss were different for each genotype (Fig. 3.2a). The greatest level of aggrecan loss was from wild-type

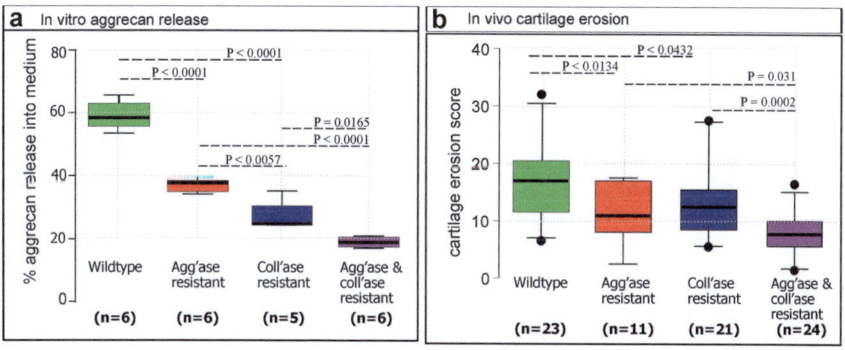

Fig. 3.2 Aggrecan loss and cartilage erosion in wild-type mice or mice resistant to aggrecanase activity, collagenase activity or both activities. (**a**) Femoral head cartilage was harvested from 3-week-old wild-type mice or mice resistant to aggrecanase activity (Agg'ase) (Little et al. 2007), collagenase activity (Coll'ase) or resistant to both aggrecanase and collagenase activity (Agg'ase and Coll'ase). The explants were stimulated with IL-1α for 3 days, and the amount of aggrecan released into the medium was quantitated by the dimethylmethylene blue dye-binding assay. For each genotype the number of biological replicates is shown in brackets. Statistical significance was calculated by the unpaired *t*-test with Welsh's correction. The relevant *P* values are shown. (**b**) Ten-week-old wild-type mice, or mice resistant to aggrecanase activity (Little et al. 2007), collagenase activity (Fosang, 2009 unpublished data) or resistant to both aggrecanase and collagenase activity, received surgery to induce experimental OA by destabilising the medial meniscus (DMM surgery). After 8 weeks, the operated knees were harvested and processed for histology to score for cartilage erosion. For each genotype the number of biological replicates is shown in brackets. The whiskers represent the 5–95 percentiles and outliers are represented by solid circles. Statistical significance was calculated by unpaired *t*-test with Welsh's correction. The relevant *P* values are shown

cartilage, with 59 % of aggrecan lost to the conditioned medium after 3 days in culture. In contrast, the amount of aggrecan released from Jaffa cartilage was reduced to 37 % and reduced to 27 % loss from the collagenase-resistant mice. Aggrecan released from the double aggrecanase- and collagenase-resistant mice was only 18 % (Fig. 3.2a). While these data confirm the key role of ADAMTS activity in stimulated aggrecanolysis, it also suggests that there is a relationship between aggrecanolysis and collagenolysis, with specific inhibition of the latter conferring protection against the loss of aggrecan from cartilage.

To examine the effect of genotype on in vivo cartilage erosion, male 10-week-old mice from each genotype had OA surgically induced by destabilisation of the medial meniscus (DMM), and cartilage erosion was scored for 8 weeks after surgery, as previously described (Little et al. 2007). The results showed that compared with wild-type mice, cartilage erosion was significantly reduced in both the homozygous aggrecanase-resistant ($P < 0.0134$) and heterozygous collagenase-resistant ($P < 0.0432$) mouse lines. Furthermore, it was clear that blocking both catalytic activities together substantially reduced the severity of cartilage erosion ($P < 0.00001$) (Fig. 3.2b). Since the collagenase-resistant mice were heterozygous for the knock-in mutation, the results suggest that blocking collagenase activity is likely to be a more efficacious OA target than blocking aggrecanase activity, and clearly blocking both activities simultaneously had the greatest effect on reducing cartilage erosion.

3.2.8 SLRP Degradation

Unlike the fields of aggrecanolysis and collagenolysis which are relatively mature, the field of SLRP proteolysis is still emerging. Biochemical-style studies in which recombinant enzymes have been used to digest recombinant or purified SLRPs in solution have identified specific cleavage sites in SLRP core proteins (Melching et al. 2006; Monfort et al. 2006; Imai et al. 1997; Gendron et al. 2007; Rees et al. 2009b; Kashiwagi et al. 2004), but such studies have included relatively few SLRPs found in joint tissues and only a small pool of enzymes. Other studies on SLRP degradation processes in tissues ex vivo or cultured in vitro have identified SLRPs that are more susceptible to proteolysis in normal and OA cartilage and allowed the timing of SLRP degradation to be placed in context with timing of degradation of other matrix molecules such as aggrecan and collagen (Monfort et al. 2006; Heathfield et al. 2004; Zhen et al. 2008; Sztrolovics et al. 1999). Analysis of the size and abundance of SLRP fragments in human joint tissues has revealed increased fragmentation of SLRPs in OA joints, compared with age-matched non-arthritic controls (Melrose et al. 2008; Cs-Szabo et al. 1997). However, SLRP cleavage sites defined in vitro using soluble SLRPs do not always correlate in size with SLRP fragments detected in tissues ex vivo (Melrose et al. 2008). This might reflect the degree of glycosylation or the accessibility of cleavage sites in matrix-bound SLRPs, compared with soluble SLRPs, or that enzymes other than those tested in vitro are responsible for cleaving SLRPs in tissues.

One consequence of SLRP degradation in OA tissues is breaching of the steric barrier that SLRPs provide on the surface of collagen fibrils. This activity is predicted to expose collagen fibrils to proteolytic attack and to accelerate cartilage degeneration (reviewed in Ni et al. (2014)). Fibromodulin fragmentation precedes collagen degradation in IL-1α-stimulated cartilage explants, and this fragmentation occurs long after the peak of aggrecan loss (Sztrolovics et al. 1999; Heathfield et al. 2004). Fibromodulin is cleaved by MMP-13, provided it is matrix bound, producing a product that is identical to the fibromodulin fragment that is generated in cartilage explant cultures (Heathfield et al. 2004). Biglycan is also highly susceptible to MMP-13 digestion (Monfort et al. 2006), suggesting that poorly controlled MMP-13 activity could degrade the SLRP surface coat leading to degradation of the collagen fibril and progressive degeneration of the fibrillar network. Other MMPs, including MMP-2, MMP-3, MMP-8, MMP-9 and MMP-12, as well as the aggrecanases, can also release SLRP fragments from human cartilage (Zhen et al. 2008).

In a study mapping fragments of decorin, biglycan, fibromodulin, lumican and keratocan in meniscus, knee and hip cartilage from OA patients (Melrose et al. 2008), the same molecular mass fragments were found in all tissues, suggesting common proteolytic mechanisms at play, despite differences in cell type, matrix composition and vascularity between meniscus and cartilage. Furthermore, although SLRP fragments were less abundant in normal cartilage, the fragment sizes were similar (with the exception of fibromodulin), suggesting that the enzymes responsible for SLRP proteolysis in OA are the same as those regulating homeostasis in normal cartilage.

3.3 Cartilage Degradation Products Promote Inflammatory Mediators

In addition to compromising the structural and mechanical properties of cartilage, aggrecan degradation generates fragments with bioactivity that can compromise tissue function. A synthetic 32-amino acid peptide (^{342}FFG-EGE^{374}), representing the peptide product of MMP cleavage at the amino terminus and aggrecanase cleavage at the carboxy terminus, is pro-inflammatory and pro-catabolic when added to cultures of primary mouse chondrocytes, synovial fibroblasts or peritoneal macrophages (Lees et al. 2015). Furthermore, native, KS-containing 32mer purified from pig articular cartilage, also upregulates the expression of MMP-3 and MMP-13 mRNA and protein in human chondrocytes. These results suggest that the 32mer aggrecan degradation product is a DAMP (*damage* (*or danger*)*-associated molecular pattern*).

DAMPS are host-derived molecules found often (but not always) 'out of context' for the cells. For example, DAMPs are molecules that may be hydrolysed, reduced, oxidised, denatured, fragmented, overexpressed or mislocalised, such as DNA outside the nucleus or mitochondria. DAMPs mediate their actions via interactions with members of the Toll-like receptor (TLR) family, leading to activation of major signalling pathways including the nuclear factor kB and mitogen-activated protein kinase pathways (reviewed in O'Neill (2006) and details in Chap. 1). The TLRs expressed in human cartilage include TLR1, TLR2, TLR3, TLR4 and TLR6 (Zhang et al. 2008). Activation of these TLRs can upregulate collagenase expression and activation, leading to resorption of cartilage collagen. The aggrecan 32mer signals via TLR2 in mouse chondrocytes (Lees et al. 2015). Because the 32mer is derived from the G1-EGE^{373} fragment, which is retained in the cartilage matrix with a half-life of ~20 years (Verzijl et al. 2001), repeated cycles of glycosaminoglycan loss and new aggrecan synthesis produce a readily renewable supply of G1-EGE^{373} substrate, making the aggrecan 32mer an important addition to the list of OA-relevant DAMPs.

Other DAMPs present in degraded, inflamed or OA cartilage include fibronectin fragments (Su et al. 2005), tenascin C (Chevalier et al. 1994; Midwood et al. 2009) and fragments of low MW hyaluronan (Liu-Bryan and Terkeltaub 2010). From the proteoglycan family, soluble biglycan, decorin and versican show capabilities or characteristics of DAMPs (reviewed in Schaefer (2014)), and recently the proteoglycan lubricin was added to the list (Alqurqini et al. 2015). In chondrocytes, biglycan signalling (mostly via TLR-4) upregulates MMP and inflammatory cytokine mRNA and protein expression and downregulates aggrecan and type II collagen mRNA expression (Barreto et al. 2015). Biglycan must be both soluble and intact to act as a DAMP, since matrix-bound biglycan does not signal, nor does deglycosylated biglycan core protein or biglycan-derived glycosaminoglycan chains (Schaefer et al. 2005) (reviewed in Nastase et al. (2012)). Intact, soluble biglycan is present in the circulation (Moreth et al. 2010) and in synovial fluids from patients with OA and is most abundant in joint fluids from patients with advanced OA (Barreto et al. 2015).

Conclusions

Together, aggrecan and collagen II provide the compressive resilience and mechanical strength required for weight bearing in articular cartilage. SLRPs decorating the surface of collagen fibrils further augment collagen resilience by modulating fibril formation and providing a steric barrier against proteinase attack. The proteolytic activities of MMP-13, together with ADAMTS-5 and possibly ADAMTS-4 activities, compromise cartilage function and are therefore valid targets for the development of OA therapies. The modulation of these proteinase activities by cytokines and non-coding RNAs provides a layer of regulation that may also be amenable to pharmacological intervention. Early results from genetic approaches targeting aggrecan and collagen-degrading activity in vivo suggest that simultaneous blockade of both activities provides greater efficacy in limiting cartilage damage, than either activity alone.

Several products of cartilage erosion are recognised as DAMPs. DAMPs are 'danger signals' derived *not* from bacteria or pathogens but from pathogen-free cell or tissue injury, and they drive inflammation. While low levels of inflammation provide benefits to tissues and organisms by eliminating cellular and tissue debris following a physical or biological insult, persistently high levels of harmful stimuli can produce DAMPS that act as renewable danger signals, triggering ongoing and escalating cycles of inflammation, that can become chronic. Specific fragments of degraded cartilage molecules including aggrecan, hyaluronan and fibronectin are now recognised as DAMPs, and the degradative events that generate these DAMPS are also well known. Other DAMPs are likely to emerge as we learn more about the dynamic signalling between healthy and unhealthy cartilage in OA.

References

Abbaszade I, Liu RQ, Yang F, Rosenfeld SA, Ross OH, Link JR, Ellis DM, Tortorella MD, Pratta MA, Hollis JM, Wynn R, Duke JL, George HJ, Hillman MC Jr, Murphy K, Wiswall BH, Copeland RA, Decicco CP, Bruckner R, Nagase H, Itoh Y, Newton RC, Magolda RL, Trzaskos JM, Hollis GF, Arner EC, Burn TC (1999) Cloning and characterization of ADAMTS11, an aggrecanase from the ADAMTS family. J Biol Chem 274(33):23443–23450

Akhtar N, Rasheed Z, Ramamurthy S, Anbazhagan AN, Voss FR, Haqqi TM (2010) MicroRNA-27b regulates the expression of matrix metalloproteinase 13 in human osteoarthritis chondrocytes. Arthritis Rheum 62(5):1361–1371

Alqurqini A, Garguilo S, D'Souza G, Zhang LX, Schmidt TA, Jay GD, Elsai KA (2015) The interaction of lubricin/proteoglycan 4 (PRG4) with toll-like receptors 2 and 4: an anti-inflammatory role for PRG4 in synovial fluid. Arthritis Res Ther 17:353

Aspberg A (2016) Cartilage proteoglycans. In: Grässel S, Aszódi A (eds) Cartilage: volume 1: physiology and development. Springer International Publishing, Cham, pp 1–22

Barreto G, Soininen A, Ylinen P, Sandelin J, Konttinen YT, Nordstrom DC, Eklund KK (2015) Soluble biglycan: a potential mediator of cartilage degradation in osteoarthritis. Arthritis Res Ther 17(1):379

Bateman JF, Lamande SR, Ramshaw JAM (1996) Collagen superfamily. In: Comper WD (ed) Extracellular matrix, vol 2, Molecular components adn interactions. Harwood Academic Publishers, The Netherlands, pp 22–67

Bau B, Gebhard PM, Haag J, Knorr T, Bartnik E, Aigner T (2002) Relative messenger RNA expression profiling of collagenases and aggrecanases in human articular chondrocytes in vivo and in vitro. Arthritis Rheum 46(10):2648–2657

Beekman B, Verzijl N, de Roos JA, TeKoppele JM (1998) Matrix degradation by chondrocytes cultured in alginate: IL-1 beta induces proteoglycan degradation and proMMP synthesis but does not result in collagen degradation. Osteoarthr Cartil 6(5):330–340

Billinghurst RC, Dahlberg L, Ionescu M, Reiner A, Bourne R, Rorabeck C, Mitchell P, Hambor J, Diekmann O, Tschesche H, Chen J, Van Wart H, Poole AR (1997) Enhanced cleavage of type II collagen by collagenases in osteoarthritic articular cartilage. J Clin Invest 99(7):1534–1545

Chevalier X, Groult N, Larget-Piet B, Zardi L, Hornebeck W (1994) Tenascin distribution in articular cartilage from normal subjects and from patients with osteoarthritis and rheumatoid arthritis. Arthritis Rheum 37(7):1013–1022

Chung L, Dinakarpandian D, Yoshida N, Lauer-Fields JL, Fields GB, Visse R, Nagase H (2004) Collagenase unwinds triple-helical collagen prior to peptide bond hydrolysis. Embo J 23(15):3020–3030

Cs-Szabo G, Melching LI, Roughley PJ, Glant TT (1997) Changes in messenger RNA and protein levels of proteoglycans and link protein in human osteoarthritic cartilage samples. Arthritis Rheum 40(6):1037–1045

Dahlberg L, Billinghurst RC, Manner P, Nelson F, Webb G, Ionescu M, Reiner A, Tanzer M, Zukor D, Chen J, Van Wart HE, Poole AR (2000) Selective enhancement of collagenase-mediated cleavage of resident type II collagen in cultured osteoarthritic cartilage and arrest with a synthetic inhibitor that spares collagenase 1 (matrix metalloproteinase 1). Arthritis Rheum 43(3):673–682

Davidson RK, Waters JG, Kevorkian L, Darrah C, Cooper A, Donell ST, Clark IM (2006) Expression profiling of metalloproteinases and their inhibitors in synovium and cartilage. Arthritis Res Ther 8(4):R124

Dejica VM, Mort JS, Laverty S, Percival MD, Antoniou J, Zukor DJ, Poole AR (2008) Cleavage of type II collagen by cathepsin K in human osteoarthritic cartilage. Am J Pathol 173(1):161–169

Engel CK, Pirard B, Schimanski S, Kirsch R, Habermann J, Klingler O, Schlotte V, Weithmann KU, Wendt KU (2005) Structural basis for the highly selective inhibition of MMP-13. Chem Biol 12(2):181–189

Fields GB (2015) New strategies for targeting matrix metalloproteinases. Matrix Biol 44–46:239–246

Fischer T, Riedl R (2016) Molecular recognition of the catalytic zinc (II) Ion in MMP-13: structure-based evolution of an allosteric inhibitor to dual binding mode inhibitors with improved lipophilic ligand efficiencies. Int J Mol Sci 17(3):314

Flannery CR, Zeng W, Corcoran C, Collins-Racie LA, Chockalingam PS, Hebert T, Mackie SA, McDonagh T, Crawford TK, Tomkinson KN, LaVallie ER, Morris EA (2002) Autocatalytic cleavage of ADAMTS-4 (aggrecanase-1) reveals multiple glycosaminoglycan-binding sites. J Biol Chem 277(45):42775–42780

Fosang AJ, Beier F (2011) Emerging frontiers in cartilage and chondrocyte biology. Best Pract Res Clin Rheumatol 25(6):751–766

Fosang AJ, Rogerson FM (2010) Identifying the human aggrecanase. Osteoarthr Cartil 18(9):1109–1116

Fosang AJ, Rogerson FM, East CJ, Stanton H (2008) ADAMTS-5: the story so far. Eur Cell Mater 15:11–26

Fushimi K, Troeberg L, Nakamura H, Lim NH, Nagase H (2008) Functional differences of the catalytic and non-catalytic domains in human ADAMTS-4 and ADAMTS-5 in aggrecanolytic activity. J Biol Chem 283:6706–6716

Gao G, Plaas A, Thompson VP, Jin S, Zuo F, Sandy JD (2004) ADAMTS4 (aggrecanase-1) activation on the cell surface involves C-terminal cleavage by glycosylphosphatidyl inositol-anchored membrane type 4-matrix metalloproteinase and binding of the activated proteinase to chondroitin sulfate and heparan sulfate on syndecan-1. J Biol Chem 279(11):10042–10051

Gao DA, Xiong Z, Heim-Riether A, Amodeo L, August EM, Cao X, Ciccarelli L, Collins BK, Harrington K, Haverty K, Hill-Drzewi M, Li X, Liang S, Margarit SM, Moss N, Nagaraja N, Proudfoot J, Roman R, Schlyer S, Keenan LS, Taylor S, Wellenzohn B, Wiedenmayer D, Li J, Farrow NA (2010) SAR studies of non-zinc-chelating MMP-13 inhibitors: improving selectivity and metabolic stability. Bioorg Med Chem Lett 20(17):5039–5043

Gege C, Bao B, Bluhm H, Boer J, Gallagher BM, Korniski B, Powers TS, Steeneck C, Taveras AG, Baragi VM (2012) Discovery and evaluation of a non-Zn chelating, selective matrix metalloproteinase 13 (MMP-13) inhibitor for potential intra-articular treatment of osteoarthritis. J Med Chem 55(2):709–716

Gendron C, Kashiwagi M, Lim NH, Enghild JJ, Thogersen IB, Hughes C, Caterson B, Nagase H (2007) Proteolytic activities of human ADAMTS-5: comparative studies with ADAMTS-4. J Biol Chem 282(25):18294–18306

Glasson SS, Askew R, Sheppard B, Carito B, Blanchet T, Ma HL, Flannery CR, Peluso D, Kanki K, Yang Z, Majumdar MK, Morris EA (2005) Deletion of active ADAMTS5 prevents cartilage degradation in a murine model of osteoarthritis. Nature 434(7033):644–648

Goldring MB, Otero M (2011) Inflammation in osteoarthritis. Curr Opin Rheumatol 23:471–478

Goldring MB, Otero M, Plumb DA, Dragomir C, Favero M, El Hachem K, Hashimoto K, Roach HI, Olivotto E, Borzi RM, Marcu KB (2011) Roles of inflammatory and anabolic cytokines in cartilage metabolism: signals and multiple effectors converge upon MMP-13 regulation in osteoarthritis. Eur Cell Mater 21:202–220

Grässel S (2016) Collagens in hyaline cartilage. In: Grässel S, Aszódi A (eds) Cartilage: volume 1: physiology and development. Springer International Publishing, Cham, pp 23–53

Hasty KA, Wu H, Byrne M, Goldring MB, Seyer JM, Jaenisch R, Krane SM, Mainardi CL (1993) Susceptibility of type I collagen containing mutated a1(1) chains to cleavage by human neutrophil collagenase. Matrix 13:181–186

Heathfield TF, Onnerfjord P, Dahlberg L, Heinegard D (2004) Cleavage of fibromodulin in cartilage explants involves removal of the N-terminal tyrosine sulfate-rich region by proteolysis at a site that is sensitive to matrix metalloproteinase-13. J Biol Chem 279(8):6286–6295

Ilic MZ, Robinson HC, Handley CJ (1998) Characterization of aggrecan retained and lost from the extracellular matrix of articular cartilage. Involvement of carboxyl-terminal processing in the catabolism of aggrecan. J Biol Chem 273(28):17451–17458

Imai K, Hiramatsu A, Fukushima D, Pierschbacher MD, Okada Y (1997) Degradation of decorin by matrix metalloproteinases: identification of the cleavage sites, kinetic analyses and transforming growth factor-beta1 release. Biochem J 322:809–814

Jacobsen JA, Major Jourden JL, Miller MT, Cohen SM (2010) To bind zinc or not to bind zinc: an examination of innovative approaches to improved metalloproteinase inhibition. Biochim Biophys Acta 1803(1):72–94

Ji Q, Xu X, Zhang Q, Kang L, Xu Y, Zhang K, Li L, Liang Y, Hong T, Ye Q, Wang Y (2016) The IL-1beta/AP-1/miR-30a/ADAMTS-5 axis regulates cartilage matrix degradation in human osteoarthritis. J Mol Med (Berl)

Jungel A, Ospelt C, Lesch M, Thiel M, Sunyer T, Schorr O, Michel BA, Gay RE, Kolling C, Flory C, Gay S, Neidhart M (2010) Effect of the oral application of a highly selective MMP-13 inhibitor in three different animal models of rheumatoid arthritis. Ann Rheum Dis 69(5):898–902

Kafienah W, Bromme D, Buttle DJ, Croucher LJ, Hollander AP (1998) Human cathepsin K cleaves native type I and II collagens at the N-terminal end of the triple helix. Biochem J 331(Pt 3):727–732

Kashiwagi M, Enghild JJ, Gendron C, Hughes C, Caterson B, Itoh Y, Nagase H (2004) Altered proteolytic activities of ADAMTS-4 expressed by C-terminal processing. J Biol Chem 279(11):10109–10119

Kevorkian L, Young DA, Darrah C, Donell ST, Shepstone L, Porter S, Brockbank SM, Edwards DR, Parker AE, Clark IM (2004) Expression profiling of metalloproteinases and their inhibitors in cartilage. Arthritis Rheum 50(1):131–141

Knäuper V, Lopez-Otin C, Smith B, Knight G, Murphy G (1996) Biochemical characterization of human collagenase-3. J Biol Chem 271:1544–1550

Knäuper V, Cowell S, Smitj B, Lopez-Otin C, O'Shea M, Morris H, Zardi L, Murphy G (1997) The role of the C-terminal domain of human collagenase-3 (MMP-13) in the activation of procollagenase-3, substrate specificity, and tissue inhibitor of metalloproteinase interaction. J Biol Chem 272:7608–7616

Kobayashi M, Squires GR, Mousa A, Tanzer M, Zukor DJ, Antoniou J, Feige U, Poole AR (2005) Role of interleukin-1 and tumor necrosis factor alpha in matrix degradation of human osteoarthritic cartilage. Arthritis Rheum 52(1):128–135

Kosasih HJ, Last K, Rogerson FM, Golub SB, Gauci SJ, Russo VC, Stanton H, Wilson R, Lamande SR, Holden P, Fosang AJ (2016) A disintegrin and metalloproteinase with thrombospondin motifs-5 (ADAMTS-5) forms catalytically active oligomers. J Biol Chem 291(7):3197–3208

Kostopoulou F, Malizos KN, Papathanasiou I, Tsezou A (2015) MicroRNA-33a regulates cholesterol synthesis and cholesterol efflux-related genes in osteoarthritic chondrocytes. Arthritis Res Ther 17:42

Kozaci LD, Buttle DJ, Hollander AP (1997) Degradation of type II collagen, but not proteoglycan correlates with matrix metalloproteinase activity in cartilage explant cultures. Arthritis Rheum 40:164–174

Larkin J, Lohr TA, Elefante L, Shearin J, Matico R, Su JL, Xue Y, Liu F, Genell C, Miller RE, Tran PB, Malfait AM, Maier CC, Matheny CJ (2015) Translational development of an ADAMTS-5 antibody for osteoarthritis disease modification. Osteoarthr Cartil 23:1254–1266

Lees SJ, Golub SB, Last K, Zeng W, Jackson DC, Sutton P, Fosang AJ (2015) Bioactivity in an aggrecan 32mer fragment is mediated via toll-like receptor 2. Arthritis Rheum 67(5):1240–1249

Li X, Gibson G, Kim JS, Kroin J, Xu S, van Wijnen AJ, Im HJ (2011) MicroRNA-146a is linked to pain-related pathophysiology of osteoarthritis. Gene 480(1–2):34–41

Little CB, Hughes CE, Curtis CL, Janusz MJ, Bohne R, Wang-Weigand S, Taiwo YO, Mitchell PG, Otterness IG, Flannery CR, Caterson B (2002) Matrix metalloproteinases are involved in C-terminal and interglobular domain processing of cartilage aggrecan in late stage cartilage degradation. Matrix Biol 21(3):271–288

Little CB, Meeker CT, Golub SB, Lawlor KE, Farmer PJ, Smith SM, Fosang AJ (2007) Blocking aggrecanase cleavage in the aggrecan interglobular domain abrogates cartilage erosion and promotes cartilage repair. J Clin Invest 117(6):1627–1636

Little CB, Barai A, Burkhardt D, Smith SM, Fosang AJ, Werb Z, Shah M, Thompson EW (2009) Matrix metalloproteinase 13-deficient mice are resistant to osteoarthritic cartilage erosion but not chondrocyte hypertrophy or osteophyte development. Arthritis Rheum 60(12):3723–3733

Liu X, Wu H, Byrne M, Jeffrey J, Krane S, Jaenisch R (1995) A targeted mutation at the known collagenase cleavage site in mouse type I collagen impairs tissue remodeling. J Cell Biol 130(1):227–237

Liu-Bryan R, Terkeltaub R (2010) Chondrocyte innate immune myeloid differentiation factor 88-dependent signaling drives procatabolic effects of the endogenous Toll-like receptor 2/Toll-like receptor 4 ligands low molecular weight hyaluronan and high mobility group box chromosomal protein 1 in mice. Arthritis Rheum 62(7):2004–2012

Lohmander LS, Neame PJ, Sandy JD (1993) The structure of aggrecan fragments in human synovial fluid: evidence that aggrecanase mediates cartilage degradation in inflammatory joint disease, joint injury, and osteoarthritis. Arthritis Rheum 36:1214–1222

Malfait AM, Arner EC, Song RH, Alston JT, Markosyan S, Staten N, Yang Z, Griggs DW, Tortorella MD (2008) Proprotein convertase activation of aggrecanases in cartilage in situ. Arch Biochem Biophys 478(1):43–51

Manka SW, Carafoli F, Visse R, Bihan D, Raynal N, Farndale RW, Murphy G, Enghild JJ, Hohenester E, Nagase H (2012) Structural insights into triple-helical collagen cleavage by matrix metalloproteinase 1. Proc Natl Acad Sci U S A 109(31):12461–12466

Maroudas AI (1976) Balance between swelling pressure and collagen tension in normal and degenerate cartilage. Nature 260(5554):808–809

Matsukawa T, Sakai T, Yonezawa T, Hiraiwa H, Hamada T, Nakashima M, Ono Y, Ishizuka S, Nakahara H, Lotz MK, Asahara H, Ishiguro N (2013) MicroRNA-125b regulates the expression of aggrecanase-1 (ADAMTS-4) in human osteoarthritic chondrocytes. Arthritis Res Ther 15(1):R28

Mazzuca SA, Brandt KD, Eyre DR, Katz BP, Askew J, Lane KA (2006) Urinary levels of type II collagen C-telopeptide crosslink are unrelated to joint space narrowing in patients with knee osteoarthritis. Ann Rheum Dis 65(8):1055–1059

Melching LI, Fisher WD, Lee ER, Mort JS, Roughley PJ (2006) The cleavage of biglycan by aggrecanases. Osteoarthr Cartil 14(11):1147–1154

Melrose J, Fuller ES, Roughley PJ, Smith MM, Kerr B, Hughes CE, Caterson B, Little CB (2008) Fragmentation of decorin, biglycan, lumican and keratocan is elevated in degenerate human meniscus, knee and hip articular cartilages compared with age-matched macroscopically normal and control tissues. Arthritis Res Ther 10(4):R79

Metsaranta M, Toman D, De Crombrugghe B, Vuorio E (1991) Mouse type II collagen gene. Complete nucleotide sequence, exon structure, and alternative splicing. J Biol Chem 266(25):16862–16869

Midwood K, Sacre S, Piccinini AM, Inglis J, Trebaul A, Chan E, Drexler S, Sofat N, Kashiwagi M, Orend G, Brennan F, Foxwell B (2009) Tenascin-C is an endogenous activator of Toll-like receptor 4 that is essential for maintaining inflammation in arthritic joint disease. Nat Med 15(7):774–780

Miller EJ, Harris ED, Chung E, Finch JE, McCroskery PA, Butler WT (1976) Cleavage of type II and III collagens with mammalian collagenase: site of cleavage and primary structure at the NH2-terminal portion of the smaller fragment released from both collagens. Biochemistry 15(4):787–792

Mitchell PG, Magna HA, Reeves LM, Lopresti-Morrow LL, Yocum SA, Rosner PJ, Geoghegan KF, Hambor JE (1996) Cloning, expression, and type II collagenolytic activity of matrix metalloproteinase-13 from human osteoarthritic cartilage. J Clin Invest 97:761–768

Miyaki S, Nakasa T, Otsuki S, Grogan SP, Higashiyama R, Inoue A, Kato Y, Sato T, Lotz MK, Asahara H (2009) MicroRNA-140 is expressed in differentiated human articular chondrocytes and modulates interleukin-1 responses. Arthritis Rheum 60(9):2723–2730

Miyaki S, Sato T, Inoue A, Otsuki S, Ito Y, Yokoyama S, Kato Y, Takemoto F, Nakasa T, Yamashita S, Takada S, Lotz MK, Ueno-Kudo H, Asahara H (2010) MicroRNA-140 plays dual roles in both cartilage development and homeostasis. Genes Dev 24(11):1173–1185

Monfort J, Tardif G, Reboul P, Mineau F, Roughley P, Pelletier JP, Martel-Pelletier J (2006) Degradation of small leucine-rich repeat proteoglycans by matrix metalloprotease-13: identification of a new biglycan cleavage site. Arthritis Res Ther 8(1):R26

Moreth K, Brodbeck R, Babelova A, Gretz N, Spieker T, Zeng-Brouwers J, Pfeilschifter J, Young MF, Schaefer RM, Schaefer L (2010) The proteoglycan biglycan regulates expression of the B cell chemoattractant CXCL13 and aggravates murine lupus nephritis. J Clin Invest 120(12):4251–4272

Murphy G, Stanton H, Cowell S, Butler G, Knauper V, Atkinson S, Gavrilovic J (1999) Mechanisms for pro matrix metalloproteinase activation. Apmis 107(1):38–44

Nara H, Sato K, Naito T, Mototani H, Oki H, Yamamoto Y, Kuno H, Santou T, Kanzaki N, Terauchi J, Uchikawa O, Kori M (2014) Discovery of novel, highly potent, and selective quinazoline-2-carboxamide-based matrix metalloproteinase (MMP)-13 inhibitors without a zinc binding group using a structure-based design approach. J Med Chem 57(21):8886–8902

Nastase MV, Young MF, Schaefer L (2012) Biglycan: a multivalent proteoglycan providing structure and signals. J Histochem Cytochem 60(12):963–975

Nemirovskiy OV, Dufield DR, Sunyer T, Aggarwal P, Welsch DJ, Mathews WR (2007) Discovery and development of a type II collagen neoepitope (TIINE) biomarker for matrix metalloproteinase activity: from in vitro to in vivo. Anal Biochem 361(1):93–101

Neuhold LA, Killar L, Zhao W, Sung ML, Warner L, Kulik J, Turner J, Wu W, Billinghurst C, Meijers T, Poole AR, Babij P, DeGennaro LJ (2001) Postnatal expression in hyaline cartilage of constitutively active human collagenase-3 (MMP-13) induces osteoarthritis in mice. J Clin Invest 107(1):35–44

Ni GX, Li Z, Zhou YZ (2014) The role of small leucine-rich proteoglycans in osteoarthritis pathogenesis. Osteoarthr Cartil 22(7):896–903

Nugent M (2016) MicroRNAs: exploring new horizons in osteoarthritis. Osteoarthr Cartil 24(4):573–580

O'Neill LA (2006) How toll-like receptors signal: what we know and what we don't know. Curr Opin Immunol 18(1):3–9

Park SJ, Cheon EJ, Lee MH, Kim HA (2013) MicroRNA-127-5p regulates matrix metalloproteinase 13 expression and interleukin-1beta-induced catabolic effects in human chondrocytes. Arthritis Rheum 65(12):3141–3152

Patterson ML, Atkinson SJ, Knauper V, Murphy G (2001) Specific collagenolysis by gelatinase A, MMP-2, is determined by the hemopexin domain and not the fibronectin-like domain. FEBS Lett 503(2–3):158–162

Pratta MA, Yao W, Decicco C, Tortorella MD, Liu RQ, Copeland RA, Magolda R, Newton RC, Trzaskos JM, Arner EC (2003) Aggrecan protects cartilage collagen from proteolytic cleavage. J Biol Chem 30:30

Rees SG, Dent CM, Caterson B (2009a) Metabolism of proteoglycans in tendon. Scand J Med Sci Sports 19(4):470–478

Rees SG, Waggett AD, Kerr BC, Probert J, Gealy EC, Dent CM, Caterson B, Hughes CE (2009b) Immunolocalisation and expression of keratocan in tendon. Osteoarthr Cartil 17(2):276–279

Ruminski PG, Massa M, Strohbach J, Hanau CE, Schmidt M, Scholten JA, Fletcher TR, Hamper BC, Carroll JN, Shieh HS, Caspers N, Collins B, Grapperhaus M, Palmquist KE, Collins J, Baldus JE, Hitchcock J, Kleine HP, Rogers MD, McDonald J, Munie GE, Messing DM, Portolan S, Whiteley LO, Sunyer T, Schnute ME (2016) Discovery of N-(4-fluoro-3-

methoxybenzyl)-6-(2-(((2S,5R)-5-(hydroxymethyl)-1,4-dioxan-2-yl) met hyl)-2H-tetrazol-5-yl)-2-methylpyrimidine-4-carboxamide. A highly selective and orally bioavailable matrix metalloproteinase-13 inhibitor for the potential treatment of osteoarthritis. J Med Chem 59(1):313–327

Sandy JD, Flannery CR, Neame PJ, Lohmander LS (1992) The structure of aggrecan fragments in human synovial fluid. Evidence for the involvement in osteoarthritis of a novel proteinase which cleaves the Glu 373 - Ala 374 bond of the interglobular domain. J Clin Invest 89:1512–1516

Santamaria S, Yamamoto K, Botkjaer K, Tape C, Dyson MR, McCafferty J, Murphy G, Nagase H (2015) Antibody-based exosite inhibitors of ADAMTS-5 (aggrecanase-2). Biochem J 471(3):391–401

Schaefer L (2014) Complexity of danger: the diverse nature of damagae-associated molecular patterns. J Biol Chem 289:35237–35245

Schaefer L, Babelova A, Kiss E, Hausser HJ, Baliova M, Krzyzankova M, Marsche G, Young MF, Mihalik D, Gotte M, Malle E, Schaefer RM, Grone HJ (2005) The matrix component biglycan is proinflammatory and signals through toll-like receptors 4 and 2 in macrophages. J Clin Invest 115(8):2223–2233

Stanton H, Rogerson FM, East CJ, Golub SB, Lawlor KE, Meeker CT, Little CB, Last K, Farmer PJ, Campbell IK, Fourie AM, Fosang AJ (2005) ADAMTS5 is the major aggrecanase in mouse cartilage in vivo and in vitro. Nature 434(7033):648–652

Stoop R, van der Kraan PM, Buma P, Hollander AP, Poole AR, Van Den Berg WB (1999) Denaturation of type II collagen in articular cartilage in experimental murine arthritis. Evidence for collagen degradation in both reversible and irreversible cartilage damage. J Pathol 188(3):329–337

Struglics A, Hansson M (2010) Calpain is involved in C-terminal truncation of human aggrecan. Biochem J 430(3):531–538

Struglics A, Lohmander LS, Last K, Akikusa J, Allen R, Fosang AJ (2012) Aggrecanase cleavage in juvenile idiopathic arthritis patients is minimally detected in the aggrecan interglobular domain but robust at the aggrecan C-terminus. Arthritis Rheum 64(12):4151–4161

Stura EA, Visse R, Cuniasse P, Dive V, Nagase H (2013) Crystal structure of full-length human collagenase 3 (MMP-13) with peptides in the active site defines exosites in the catalytic domain. FASEB J 27(11):4395–4405

Su SL, Tsai CD, Lee CH, Salter DM, Lee HS (2005) Expression and regulation of toll-like receptor 2 by IL-1beta and fibronectin fragments in human articular chondrocytes. Osteoarthr Cartil 13(10):879–886

Sztrolovics R, White RJ, Poole AR, Mort JS, Roughley PJ (1999) Resistance of small leucine-rich repeat proteoglycans to proteolytic degradation during interleukin-1-stimulated cartilage catabolism. Biochem J 339(Pt:3):3–577

Thomas L (1956) Reversible collapse of rabbit ears after intravenous papain, and prevention of recovery by cortisone. J Exp Med 104:245–261

Tortorella MD, Burn TC, Pratta MA, Abbaszade I, Hollis JM, Liu R, Rosenfeld SA, Copeland RA, Decicco CP, Wynn R, Rockwell A, Yang F, Duke JL, Solomon K, George H, Bruckner R, Nagase H, Itoh Y, Ellis DM, Ross H, Wiswall BH, Murphy K, Hillman MC Jr, Hollis GF, Newton RC, Magolda RL, Trzaskos JM, Arner EC (1999) Purification and cloning of aggrecanase-1: a member of the ADAMTS family of proteins. Science 284(5420):1664–1666

Tortorella MD, Pratta M, Liu RQ, Austin J, Ross OH, Abbaszade I, Burn T, Arner E (2000) Sites of aggrecan cleavage by recombinant human aggrecanase-1 (ADAMTS-4). J Biol Chem 275(24):18566–18573

Troeberg L, Nagase H (2012) Proteases involved in cartilage matrix degradation in osteoarthritis. Biochim Biophys Acta 1824(1):133–145

Troeberg L, Lazenbatt C, Anower EKMF, Freeman C, Federov O, Habuchi H, Habuchi O, Kimata K, Nagase H (2014) Sulfated glycosaminoglycans control the extracellular trafficking and the activity of the metalloprotease inhibitor TIMP-3. Chem Biol 21(10):1300–1309

Tsezou A (2014) Osteoarthritis year in review 2014: genetics and genomics. Osteoarthr Cartil 22(12):2017–2024

Verzijl N, DeGroot J, Thorpe SR, Bank RA, Shaw JN, Lyons TJ, Bijlsma JW, Lafeber FP, Baynes JW, TeKoppele JM (2000) Effect of collagen turnover on the accumulation of advanced glycation end products. J Biol Chem 275(50):39027–39031

Verzijl N, DeGroot J, Bank RA, Bayliss MT, Bijlsma JW, Lafeber FP, Maroudas A, TeKoppele JM (2001) Age-related accumulation of the advanced glycation endproduct pentosidine in human articular cartilage aggrecan: the use of pentosidine levels as a quantitative measure of protein turnover. Matrix Biol 20(7):409–417

Vonk LA, Kragten AH, Dhert WJ, Saris DB, Creemers LB (2014) Overexpression of hsa-miR-148a promotes cartilage production and inhibits cartilage degradation by osteoarthritic chondrocytes. Osteoarthr Cartil 22(1):145–153

Wainwright SD, Bondeson J, Caterson B, Hughes CE (2013) ADAMTS-4_v1 is a splice variant of ADAMTS-4 that is expressed as a protein in human synovium and cleaves aggrecan at the interglobular domain. Arthritis Rheum 65(11):2866–2875

Wu H, Byrne MH, Stacey A, Goldring MB, Birkhead JR, Jaenisch R, Krane SM (1990) Generation of collagenase-resistant collagen by site-directed mutagenesis of murine pro alpha 1(I) collagen gene. Proc Natl Acad Sci U S A 87(15):5888–5892

Yamamoto K, Okano H, Miyagawa W, Visse R, Shitomi Y, Santamaria S, Dudhia J, Troeberg L, Strickland DK, Hirohata S, Nagase H (2016) MMP-13 is constitutively produced in human chondrocytes and co-endocytosed with ADAMTS-5 and TIMP-3 by the endocytic receptor LRP1. Matrix Biol

Zeng W, Corcoran C, Collins-Racie LA, Lavallie ER, Morris EA, Flannery CR (2006) Glycosaminoglycan-binding properties and aggrecanase activities of truncated ADAMTSs: Comparative analyses with ADAMTS-5, -9, -16 and -18. Biochim Biophys Acta 1760(3):517–524

Zhang Q, Hui W, Litherland GJ, Barter MJ, Davidson R, Darrah C, Donell ST, Clark IM, Cawston TE, Robinson JH, Rowan AD, Young DA (2008) Differential Toll-like receptor-dependent collagenase expression in chondrocytes. Ann Rheum Dis 67(11):1633–1641

Zhao W, Byrne MH, Boyce BF, Krane SM (1999) Bone resorption induced by parathyroid hormone is strikingly diminished in collagenase-resistant mutant mice. J Clin Invest 103(4):517–524

Zhen EY, Brittain IJ, Laska DA, Mitchell PG, Sumer EU, Karsdal MA, Duffin KL (2008) Characterization of metalloprotease cleavage products of human articular cartilage. Arthritis Rheum 58(8):2420–2431

Role of Proteoglycans in Osteoarthritis

4

Jessica Bertrand and Annelena Held

Abstract

The articular cartilage covers the end of the long bones and ensures frictionless motion of the joints. Chondrocytes produce and maintain in a homeostatic equilibrium a large amount of extracellular matrix (ECM), thereby ensuring the cartilage function in locomotion and the absorption of biomechanical stress throughout life. To achieve this, chondrocytes are extremely sensitive to mechanical stress and injury signals and can rapidly adapt their metabolic activity ultimately ensuring that remodeling and repair of the tissue follow injury.

Proteoglycans as one of the main components of the ECM have diverse functions in the cartilage. They bind water and provide the basis for absorbing high compressive loads. Additionally, they bind cytokines, chemokines, growth factors, and morphogens, thereby protecting these factors against proteolysis and/or acting as a depot of regulatory factors when matrix degradation occurs. They also modulate signaling pathways and create morphogen gradients by immobilization of ligands in the ECM and regulation of the turnover of ligands. The extracellular domains of proteoglycans can be shed from the cell surface, generating soluble decoy receptors, and they can act as co-receptors for various tyrosine-kinase-type growth factor receptors. Furthermore, they can act as endocytic receptors and contribute to the clearance of bound ligands and cooperate with integrins and other cell adhesion receptors to facilitate cell attachment, cell–cell interactions, and cell motility.

Given these important roles of proteoglycans in regulating cell functions, it is well understandable that loss of ECM and degradation of proteoglycans during OA induce severe changes in cartilage homeostasis. The following review will discuss the role of different proteoglycans in the aforementioned processes with special emphasis on the role in osteoarthritis.

J. Bertrand, PhD (✉) • A. Held, M Sc
Department of Orthopaedic Surgery, Otto-von-Guericke University Magdeburg,
Leipziger Straße 44, 39120 Magdeburg, Germany
e-mail: jessica.bertrand@med.ovgu.de; Annelena.held@med.ovgu.de

S. Grässel, A. Aszódi (eds.), *Cartilage*, DOI 10.1007/978-3-319-45803-8_4

4.1 Introduction

The articular cartilage is an avascular, aneural tissue that covers the end of the long bones and ensures frictionless motion of the joints. Its only cell type, the chondrocyte, produces and maintains in a homeostatic equilibrium a large amount of extracellular matrix (ECM). The ECM is rich in highly sulfated glycosaminoglycans (GAGs), which provide capacity for shock absorption. Because of the cartilage function in locomotion, the absorption of biomechanical stress and the maintenance of homeostatic integrity of this tissue are important throughout life. To achieve this, chondrocytes are extremely sensitive to mechanical stress and injury signals and can rapidly adapt their metabolic activity ultimately ensuring that remodeling and repair of the tissue follow injury. If the stress factors and damage to the cartilage, however, are too extreme for repair by chondrocytes, the degeneration of cartilage, called osteoarthritis (OA), is the ultimate result. OA is widely considered a degenerative disease that is characterized by progressive structural changes in joint tissues, principally in articular cartilage, but also in the subchondral bone, the synovial membrane, and the synovial fluid. Articular cartilage fibrillation and erosions accompanied by chondrocyte dedifferentiation and loss of ECM components such as proteoglycans are key features associated with OA.

The ECM is principally composed of water, different collagens, various proteoglycans, and other non-collagenous proteins and glycoproteins present in lesser amounts (Buckwalter and Rosenberg 1988; Buckwalter and Lane 1997). This composition helps to retain water in the cartilage ECM and thereby provides the basis for the biomechanical properties of the cartilage. In healthy articular cartilage, tissue fluid represents between 65 and 80 % of the total weight (Mow et al. 1992), proteoglycans account for 25–35 %, and collagens account for about 60 % of the cartilage dry weight (Buckwalter et al. 2005).

Proteoglycans are the second-largest group of macromolecules in the ECM of the articular cartilage (Aspberg 2016). They consist of a core protein with one or more covalently attached linear glycosaminoglycan (GAG) chains. These side chains can be very long as they might be composed of more than 100 monosaccharides. Proteoglycans can be separated into different subgroups according to the nature of GAG side chains attached to the core protein or the size of the core protein. The side chains can be composed of heparan, chondroitin, dermatan, and keratan sulfate. According to their size, aggrecan, versican, agrin, and perlecan belong to the large-sized proteoglycans, and syndecan, glypican, decorin, biglycan, asporin, fibromodulin, and lumican belong to the group of small-sized proteoglycans, whereby the last five proteoglycans form the subgroup of small leucine-rich proteoglycans (SLRPs).

The most important function of interstitial proteoglycans is their capacity to bind water and provide the basis for absorbing high compressive loads by water desorption and resorption. Furthermore, proteoglycans can bind cytokines, chemokines, growth factors, and morphogens such as bone morphogenetic proteins (BMPs) (Ruppert et al. 1996), fibroblast growth factors (FGFs) (Rapraeger et al. 1991), and Wnt ligands (Wnts) (Reichsman et al. 1996). These bound factors are protected against proteolysis and act as a depot of regulatory factors that can be liberated either by selective degradation of the matrix or in disease when matrix degradation occurs. It has been shown that heparan sulfates modulate signaling pathways, e.g., Wnts, through several mechanisms (Whitelock and Iozzo 2005). First, Wnts avidly bind to heparan sulfate proteoglycans (HSPGs), thereby creating morphogen gradients (Han et al. 2005) (for details, see (Hartmann 2016)). Second, HSPGs immobilize and regulate the turnover of ligands that act at the cell surface (Filion and Popel 2004). Third, the extracellular domains of these proteoglycans can be shed from the cell surface, generating soluble heparan sulfate proteoglycans that can inhibit interactions at the cell surface (Bernfield et al. 1999). Biochemical studies have implicated HSPGs not only in the binding of growth factors but also in Wnt signal transduction (Reichsman et al. 1996; Uren et al. 2000). Cell surface or transmembrane proteoglycans can act as co-receptors for various tyrosine-kinase-type growth factor receptors, lowering the threshold for activation of signaling cascades or changing the duration of signaling reactions. Furthermore, they can be internalized together with the respective receptor and ligand(s) and thereby act as endocytic receptors and contribute to the clearance of bound ligands. They can also cooperate with integrins and other cell adhesion receptors to facilitate cell attachment, cell–cell interactions, and cell motility (Esko and Linhardt 2009). Finally, HSPG binding to Wnt ligands is strictly required to maintain these hydrophobic molecules in solution, thereby allowing their biological activity (Fuerer et al. 2010) (Fig. 4.1).

Given these important roles of proteoglycans in regulating cell functions, it is well understandable that loss of ECM and degradation of proteoglycans during OA induce severe changes in cartilage homeostasis. The following paragraphs will discuss the role of different proteoglycans in the aforementioned processes with special emphasis on the role in osteoarthritis.

4.2 Large Proteoglycans

Prominent large proteoglycans in the cartilage are aggrecan, versican, agrin, and perlecan (Aspberg 2016).

4.2.1 Aggrecan

Aggrecan belongs to the family of chondroitin sulfate proteoglycans and is the largest in size and the most abundant proteoglycan in the cartilage. It possesses more than 100 GAG chains and is composed of three globular domains (G1, G2, and G3). The attachment sites for the GAG chains are located between G2 and G3. The G1 domain interacts with hyaluronan (HA) and the link protein, mediating the formation of large proteoglycan aggregates (Watanabe et al. 1998; Schwartz et al. 1999) (for details, see Chap. 3). Aggrecan occupies the interfibrillar space of the cartilage ECM and provides the cartilage with osmotic properties, which are critical to its ability to resist compressive loads. In former times it has been suggested

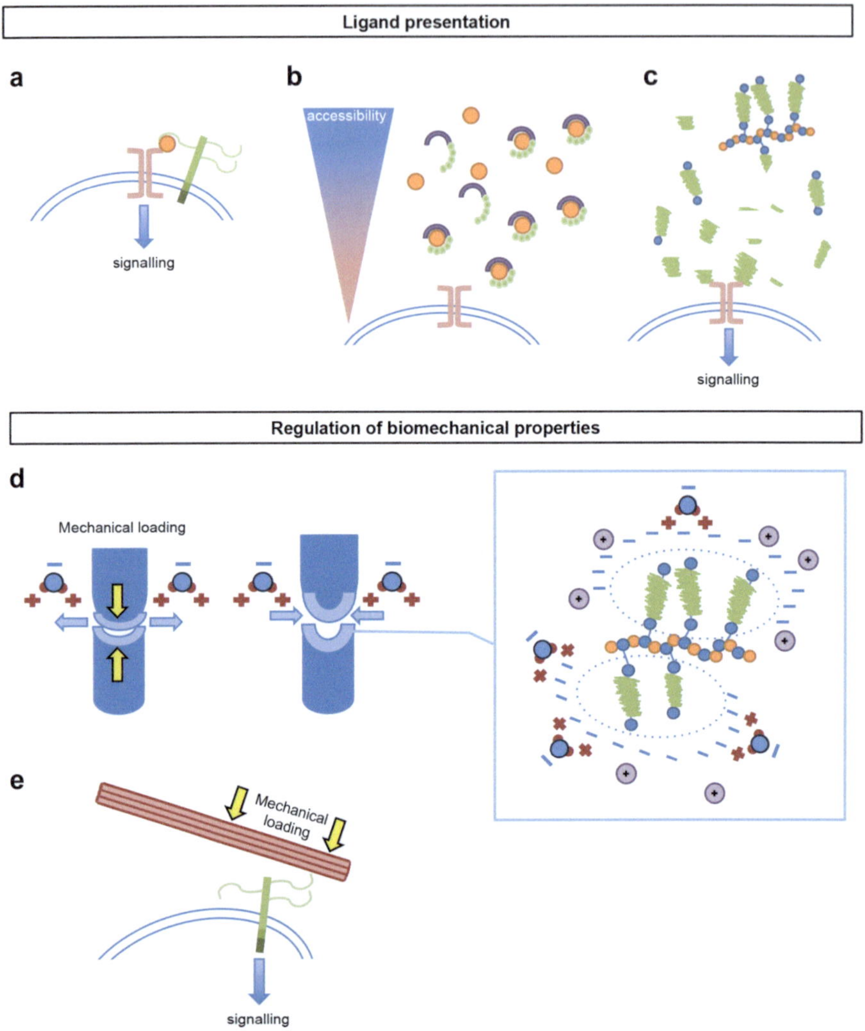

Fig. 4.1 Proteoglycans have diverse functions in modulating signal transduction and biomechanical properties in the cartilage. Induction of some signal pathways can be influenced by proteoglycans in various ways. Proteoglycans can act as co-receptor (**a**) or create morphogen/ligand gradients (**b**). Under inflammatory/pathological conditions, fragments of some proteoglycans can induce signaling cascades (**c**). Proteoglycans are also important for the biomechanical properties of the cartilage. Due to their predominantly negative charge, given by their side chains, proteoglycans comprise an inevitable osmotic function in the cartilage. Furthermore, proteoglycans are able to sense mechanical loading (**d**) and biomechanical stress (**e**), thereby initiating downstream signaling pathways

that aggrecan is not immediately incorporated in aggregates and that this incorporation is age dependent (Oegema 1980). Newer studies, however, suggest that newly synthesized aggrecan is processed in pools with different capacities for aggregation, which might be defined by factors like extracellular pH, age, and disease state (Bayliss et al. 2000). The C-terminal G3 domain of aggrecan includes different modules such as an epithelial growth factor (EGF) module, a C-type lectin module, and a C-reactive protein (CRP) module. Interestingly, it has been shown that with aging, there is an increase in the population of aggrecan monomers lacking the G3 domain, which is most likely due to extracellular proteolytic degradation. This cleavage results in the production of one fragment that remains bound to HA and one free fragment. The HA bound fragment is retained within the cartilage tissue, whereas the latter fragment is lost by diffusion into the synovial fluid. This cleavage of aggrecan results in a net decrease in the anionic charge of the aggrecan molecule and therefore a decreased ability of the cartilage to resist compression due to the reduction in osmotic force (Roughley and Mort 2014). The major contributors to aggrecan degradation in vivo are members of the matrix metalloproteinase (MMP) family, in particular stromelysin (MMP3) and MMP13, which are produced under inflammatory conditions and during OA (Troeberg and Nagase 2012). Furthermore, the two members of the ADAMTS (a disintegrin and metalloprotease with thrombospondin motifs) family of metalloproteases, ADAMTS-4 and ADAMTS-5, are also potent aggrecan-degrading enzymes (Nagase and Kashiwagi 2003), which are also highly expressed during OA (for further details, see Chap. 3).

A very recent study investigated the extent of aggrecan cleavage between individuals and found a high variability in the aggrecan cleavage, which was greatest in areas adjacent to sites of cartilage erosion compared to sites more remote within the same joint. The proportion of aggrecan cleavage attributable to MMPs or aggrecanases was also variable, but aggrecanase action was usually predominant in late-stage OA (Mort et al. 2016). Therefore, the proteolytic cleavage of aggrecan due to the increased expression of MMPs and aggrecanases under disease conditions is directly linked to a reduction and deterioration of biomechanical properties of the cartilage.

Lees and coworkers have investigated another aspect of aggrecan cleavage in the perpetuation of OA. They analyzed the bioactivity of an aggrecan fragment that is generated by ADAMTS-4/ADAMTS-5 cleavage and subsequent cleavage of the fragment by MMPs, resulting in a 32-amino-acid fragment (Lees et al. 2015). The originated 32-mer peptide caused a pro-catabolic, anti-anabolic, and pro-inflammatory response in murine and human chondrocytes in vitro. The incubation of chondrocytes with this 32-mer aggrecan fragment increased mRNA expression for several proteases, including MMP-13 and ADAMTS-5, and decreased mRNA expression for matrix molecules, including type II collagen alpha 1 (COL2A1) and aggrecan (ACAN). The authors could show that this effect was mediated through binding of the aggrecan fragment to the Toll-like receptor (TLR)-2. It has long been known that some matrix molecules and fragments of these proteins, including fibronectin, tenascin C, and hyaluronan fragments, can act as damage-associated molecular patterns (DAMPs) and activate Toll-like receptors (TLRs) that are locally expressed in the joint, initiating a cascade of inflammatory cytokine production (Liu-Bryan and Terkeltaub 2015).

4.2.2 Versican

Versican, like aggrecan, consists of globular structures at the N-terminal (G1 domain) and C-terminal (G3 domain). Five splice variants of versican are known, which are expressed under specific conditions or in different tissues. The V3 variant, however, is generally expressed at low levels in adult tissues (Cattaruzza et al. 2002). The G1 domain of versican has the capability to bind hyaluronan, and the G3 domain consists of various modules. Between G1 and G3, the chondroitin sulfate GAG side chains attach to the core protein (Zimmermann and Ruoslahti 1989). The G3 domain promotes cell adhesion (Wu et al. 2002), and the G1 promotes cell proliferation, migration and adhesion, and inflammation (Yang et al. 1999; Wight et al. 2014). The first description of versican in association with OA was by Nishida et al. (Nishida et al. 1994). The authors showed that the cytoplasm of inflammatory cells invading the OA cartilage matrix was stained for a protein with an extremely high homology to versican (Nishida et al. 1994). Some other studies also found an upregulation of versican expression in OA cartilage, mainly linking this phenomenon to an attempt of chondrocytes to repair the cartilage damage (Heard et al. 2013; Martin et al. 2001; Cs-Szabo et al. 1997). Versican can be degraded by the aggrecanases ADAMTS-4/ADAMTS-5 (Westling et al. 2004; McCulloch et al. 2009), but the degradation or the relevance of degradation in cartilage has not been investigated. That cleavage of versican also leads to a release of a free fragment like it has been shown for aggrecan degradation, has also not been proven for cartilage, but is very likely. However, it has been shown that versican also acts as an endogenous ligand of TLRs, generating a rapid inflammatory response (Frey et al. 2013), although this has not been shown for the cartilage yet.

4.2.3 Agrin

Agrin is a large HS proteoglycan that is abundantly distributed at most basement membranes. It is predominantly known to exhibit its function in neurotransmission (Iozzo and San Antonio 2001) and in mechanotransduction, connecting the cytoskeleton to other members of the basement membrane (Bezakova and Ruegg 2003; Denzer et al. 1997, 1998). Microarray screening of cartilage samples after injury identified agrin gene expression to be differentially regulated (Dell'accio et al. 2008). Besides, a transgenic mouse model showed that agrin deficiency leads to changes in the growth plate cartilage, specifically affecting the viability of hypertrophic chondrocytes (Hausser et al. 2007). In the context of OA, Eldridge et al. observed a progressive loss of agrin in human OA cartilage and in a surgical mouse model of this disease (destabilization of the medial meniscus (DMM)) compared to preserved cartilage or sham-operated joints (Eldridge et al. 2015). According to this, in vitro loss-of-function experiments indicated decreased glycosaminoglycan content and a downregulation of the cartilage-specific transcription factor SOX9 and of COL2A1 and ACAN as marker genes for the chondrocyte homeostasis (Eldridge et al. 2015). The regulation of SOX9 expression by agrin seems to depend on the availability of α-dystroglycan (DAG1) and the low-density lipoprotein receptor-related protein 4 (LRP4) (Eldridge et al. 2015). LRP4 is a receptor for the Wnt and BMP inhibitors Sost (Choi et al. 2009) and Sostdc1 (Wise) (Ohazama et al. 2008), as well as for the Wnt inhibitor Dickkopf1 (Dkk1) (Choi et al. 2009) in osteoblasts. Therefore, the LRP4 receptor constitutes an interesting link to BMP and Wnt signaling pathways (Eldridge et al. 2015), which are proposed to affect chondrocyte differentiation and may interfere in the progression of OA (rev. in (van der Kraan et al. 2010; Luyten et al. 2009; Corr 2008)). This study emphasizes the role of a pericellular HS proteoglycan in indirect modulation of morphogen gradients, which are essential for cartilage homeostasis.

4.2.4 Perlecan

Perlecan is a large multi-domain proteoglycan to which three glycosaminoglycan chains – often heparan sulfate but also chondroitin sulfate – are attached. It is present in all basement membranes, cartilage, and bone marrow stromal cells (Hassell et al. 1980; SundarRaj et al. 1995). Perlecan is associated with bone formation during endochondral ossification. Consequently, functional null mutations of the perlecan gene in humans and mice are associated with dyssegmental dysplasia of the Silverman-Handmaker type, Schwartz-Jampel syndrome (SJS) (Giedion et al. 1997), and perlecan-null mice develop severe chondrodysplasia (Arikawa-Hirasawa et al. 1999, 2001; Costell et al. 1999). Perlecan protein and mRNA levels are upregulated in OA cartilage adjacent to the main cartilage defect. This might be seen as an attempt on the part of the cartilage tissue to stabilize the extracellular matrix (Tesche and Miosge 2004). Another study shows that perlecan is required for the chondrogenic and adipogenic differentiation from synovial mesenchymal cells (SMCs) via its regulation of SOX9 and PPARγ gene expression but not for osteogenic differentiation via RUNX2 (Sadatsuki et al. 2016). Another study also describes a role for synovial perlecan, which is required for osteophyte formation in knee osteoarthritis in mice (Kaneko et al. 2013). Although these findings seem to be at least in part contradictionary, the role of perlecan in chondrocyte differentiation and bone formation is consistent.

Perlecan is enriched mainly in the pericellular environment of chondrocytes (Melrose et al. 2002) and has been identified as an HS proteoglycan that sequesters FGF-2, which is a regulator of chondrocyte proliferation (Sahni et al. 1999; Takahashi et al. 2005). In experiments using chondrocytes embedded in alginate, pericellular perlecan and FGF-2 were accumulated and delivered an FGF-dependent activation of the extracellular signal-regulated kinase (ERK) when loaded (Vincent et al. 2007), giving rise to the assumption that perlecan might have a function in biomechanical-induced OA. This finding was corroborated by another study showing that the HS side chains of perlecan play a significant role in directing the development of posttraumatic OA. The authors demonstrated that via the alteration of FGF/HS/FGFR signaling, perlecan exhibits a chondroprotective function, at least in part, via the preservation of FGFR-3 and increased FGF signaling (Shu et al. 2016).

4.3 Small Proteoglycans

Small proteoglycans in the cartilage comprise glypican, syndecan, decorin, biglycan, asporin, fibromodulin, and lumican.

4.3.1 Glypicans

Glypicans are glycosylphosphatidylinositol (GPI) anchored and can bind extracellular ligands via heparan sulfate chains (David et al. 1990; Bernfield et al. 1999). Using glypicans as an example, the role of proteoglycans and their side chains as modulators of signal transduction and facilitators of tissue homeostasis by affecting the local formation of morphogen gradients can be highlighted. Glypican-3 (GPC3) knockout mice showed an inhibited noncanonical Wnt/c-Jun N-terminal kinase (JNK) and an activated canonical Wnt/β-catenin signaling (Song et al. 2005). The formation of glypican–Wnt complexes has been reported (Ohkawara et al. 2003), but there are conflicting reports on the involvement of HS chains in this process (Song et al. 2005; Ohkawara et al. 2003) (De Cat et al. 2003). De Cat et al. clearly indicated that both the protein and the HS chains seem to be required for the proteoglycan function in signal transduction (De Cat et al. 2003).

Grover and Roughley showed low expression of glypican in human articular cartilage (Grover and Roughley 1995). Although there is a skeletal phenotype as mutations in the glypican-3 (GPC3) gene have been linked to the Simpson–Golabi–Behmel syndrome (SGBS), characterizing pre- and postnatal overgrowth (gigantism) with visceral and skeletal anomalies (Pilia et al. 1996), no reports for the function of glypican in OA are published until now.

4.3.2 Syndecans

Syndecans represent a family of type I transmembrane HS proteoglycans, which comprises four members in vertebrates (Couchman 2003). They interact via their GAG chains with a variety of extracellular matrix molecules and can act as co-receptors for integrins and for a growing number of growth factor receptors like FGF receptors (Tkachenko et al. 2005). Due to these interactions, syndecans are involved in growth control, cell spreading, cellular recognition, cellular adhesion, and signaling (Choi et al. 2011; Couchman 2010). Each syndecan has a tissue-specific and developmentally regulated expression pattern. While syndecan-1 is mainly expressed in epithelial cells, syndecan-2 is detectable in mesenchymal cells. Syndecan-3 is found during bone development and in the nervous system, whereas syndecan-4 is ubiquitously expressed (Kim et al. 1994). Of the four mammalian syndecan family members, syndecan-1 and syndecan-3 and syndecan-2 and syndecan-4 are closely related and can be considered to form subfamilies (Couchman 2010).

Syndecan-4 is expressed ubiquitously and is highly distributed during development (David et al. 1992); furthermore, it is upregulated under inflammatory conditions (Wang et al. 2011). Syndecan-2, however, is mainly expressed during embryogenesis in the bone, where it can compensate for the loss of syndecan-4 (Bertrand et al. 2013). Under inflammatory conditions syndecan-2 is downregulated, so it cannot compensate the loss of function of syndecan-4 in knockout mice (Bertrand et al. 2013). Syndecans can interact with several ligands such as growth factors, cytokines, proteinases, adhesion receptors, ECM components, and morphogens most likely through their HS chains (Pap and Bertrand 2013). Thus, they are able to modulate tissue homeostatic processes and tissue remodeling upon injury. Syndecan-4 expression is specifically induced in hypertrophic chondrocytes both in human OA and in murine models of this disease (Echtermeyer et al. 2009). Syndecan-4 controls the activation of ADAMTS-5 through direct interaction via its HS chains, as it immobilizes ADAMTS-5 at the cell surface. Furthermore, it regulates the activation of ADAMTS-5 by regulating the IL-1β-induced expression of MMP-3, via reduced activation of mitogen-activated protein kinase (MAPK) upon IL-1β stimulation (Echtermeyer et al. 2009). The role of syndecan-4 in matrix remodeling during OA was confirmed using wild-type mice, which were treated with a blocking syndecan-4 antibody, as well as syndecan-4-deficient mice, which are protected from OA-induced cartilage degradation, most likely because of decreased MAPK-initiated expression of MMP-3, resulting in reduced ADAMTS-5 activation (Echtermeyer et al. 2009). In line with this, ADAMTS-5-deficient mice are also protected from cartilage degradation in a murine model of OA (Stanton et al. 2005; Glasson et al. 2005) (for further details, see Chap. 3).

Another mechanism of syndecan-4 in cartilage protection during OA might be its function in regulating the dynamic recycling of receptors, e.g., α5β1 and αVβ3 integrins, by disrupting the dynamin- and caveolin-dependent endocytosis (Bass et al. 2011). It was shown that cells, expressing a mutagenized syndecan 4 in one of the signaling domains, exhibited a severe reduction in syndecan-4/integrin receptor internalization. These data suggest that syndecan-4-dependent endocytic pathways may be intimately associated with receptor internalization and therefore regulating signal transduction (Bass et al. 2011; Humphries et al. 2005). Both integrins have been associated with OA and cartilage homeostasis (Garciadiego-Cazares et al. 2015).

4.4 Small Leucine-Rich Repeat Proteoglycans (SLRPs)

The small leucine-rich repeat proteoglycan (SLRP) family is represented by decorin, biglycan, asporin (Lorenzo et al. 2001), fibromodulin (Hedlund et al. 1994), and lumican (Sztrolovics et al. 1999) in the articular cartilage (Martel-Pelletier et al. 2008). Decorin and biglycan contain one or two chondroitin sulfate GAG chains (Iozzo 1999), while fibromodulin exhibits keratin sulfate GAG chains (Jepsen et al. 2002). Although asporin is highly identical with decorin and biglycan, by strict definition, it is no proteoglycan, as the consensus sequence for glycosaminoglycan attachment between the propeptide and the amino-terminal cysteine motif is missing (Lorenzo et al. 2001; Henry et al. 2001; Nakajima et al. 2007). The SLRP members asporin and decorin consist of collagen binding sites (Kalamajski et al. 2007,

2009; Svensson et al. 1995) and can interact with soluble factors such as members of the transforming growth factor β superfamily (TGFβ) (Martel-Pelletier et al. 2008; Embree et al. 2010).

Several studies have proposed that SLRPs play a role in cartilage remodeling during OA (Young et al. 2005; Kizawa et al. 2005; Bock et al. 2001), as increased mRNA expression of these SLRPs has been described in OA cartilage (Poole et al. 1996; Cs-Szabo et al. 1997). Young et al. reported also an enhanced expression of biglycan and lumican, although he found a decreased synthesis of decorin in the articular cartilage in a murine model of OA (Young et al. 2005). In total mRNA expression and protein synthesis of most SLRPs seem to be increased in late stages of OA (Bock et al. 2001).

4.4.1 Asporin

Asporin has been shown to bind TGFβ1 and inhibit TGFβ1-induced expression of COL2A1 and ACAN in ATDC5 cells in vitro. Presumably, asporin thereby is able to interfere with chondrocyte differentiation and thus cartilage homeostasis (Kizawa et al. 2005). Kizawa et al. showed an increased expression of asporin in human OA. Furthermore, the study proposed that a functional polymorphism in the aspartic acid repeat of the asporin gene might result in a higher susceptibility to OA (Kizawa et al. 2005). Interestingly, they could show that the ability of asporin to inhibit TGFβ signaling was significantly different in the polymorphic variants of asporin (Kizawa et al. 2005), thereby providing a possible mechanism for the role of asporin in OA.

4.4.2 Decorin

Decorin also has high binding affinities not only to TGFβ1 but also to TGFβ2 (Iozzo and Sanderson 2011; Goldoni and Iozzo 2008). It has been reported that bone-derived decorin can enhance the ligand-receptor binding of TGFβ and thereby augment cell responses (Takeuchi et al. 1994), which might alter the balance in bone formation and degeneration toward bone resorption. Interestingly, Gronau et al. demonstrated that decorin-deficient mice are less susceptible to develop OA in a forced exercise mouse model. The study shows higher TGFβ activity and enhanced compressive stiffness of the articular cartilage accompanied by increased sulfation of GAG chains (Gronau 2016). They propose that decorin regulates the TGFβ-dependent expression of the sulfotransferase CHST11, which has been linked to OA by a genome-wide association study (GWAS) of OA cartilage samples (Reynard and Loughlin 2013).

4.4.3 Biglycan

Biglycan can interact (and sequester) with type I, II, III, and VI collagen, tumor necrosis factor α (TNFα) (Tufvesson and Westergren-Thorsson 2002), BMP-2, BMP-4, and BMP-6 (Chen et al. 2004; Mochida et al. 2006), and TGFβ (Hildebrand et al. 1994) via its tissue-specific chondroitin or dermatan sulfate GAG chains. Consequently, biglycan-deficient mice show an osteoporosis-like phenotype and abnormalities of collagen fibrils (Xu et al. 1998).

The study of Schaefer et al. showed for the first time that an SLRP, in this case biglycan, can act as ligand of TLR-2 and TLR-4. They found that upon binding of biglycan, the mitogen-activated protein kinase p38, ERK, and nuclear factor kappa-light-chain enhancer of activated B cells (NF-kB) got activated, leading to the secretion of TNFα, thereby triggering an inflammatory response (Schaefer et al. 2005). Furthermore, biglycan is capable of clustering several types of receptors and thereby orchestrating the signal transduction of these receptors. Babelova et al. showed that soluble, unsequestered biglycan can stimulate receptor interaction of TLR-2/TLR-4 and purinergic P2X (4)/P2X (7) receptors, resulting in the activation of the NLRP3 inflammasome. Although these data were generated in macrophages, the signaling mechanism most likely will be the same in the cartilage, giving an explanation of how tissue injury is sensed by innate immune receptors by binding of ECM components and translating this signal into an inflammatory response (Babelova et al. 2009). The expression of TLRs (TLR-2 and TLR-4, but also others) has been proven for chondrocytes (Sillat et al. 2013; Radwan et al. 2013). Interestingly, TLR-2 gene expression gets upregulated upon stimulation with IL-1β and fibronectin proteolytic fragments (Su et al. 2005).

4.5 Perspectives

Given the described functions of proteoglycans in the cartilage, the importance of these molecules in OA is not surprising. Due to their electrostatic loading, they retain water and largely determine the biomechanical properties of the cartilage. Upon degradation of proteoglycans, biomechanical properties of the cartilage change due to the loss of negative charges from the GAG chains. Furthermore, fragments of various proteoglycans can act as ligands for TLR receptors, inducing inflammatory cascades. Proteoglycans also bind various ligands, acting as co-receptor and control the sensitivity of cells for extracellular stimuli. One interesting function of proteoglycans has not been mentioned before, as the responsible proteoglycan has not been identified yet. Sherwood et al. have shown that the chemokine CXCL6 is expressed in healthy cartilage and retained through binding to heparan sulfate proteoglycans. The study shows that CXCL6 signaling via CXCR1/CXCR2 is important for maintaining cartilage homeostasis and that the loss of ELR+CXC chemokines during cartilage breakdown in osteoarthritis contributes to the characteristic loss of chondrocyte phenotypic stability and hypertrophic differentiation (Sherwood et al. 2015). This study highlights the role of proteoglycans in keeping a reservoir of different factors bound in the ECM. Whether all these ligands are needed to keep the cartilage homeostasis, or whether they might act as morphogens or cytokines upon release during cartilage degradation, is unknown. The understanding of proteoglycan biochemistry, however, under OA conditions is of great importance in order to understand the pathomechanisms of this disease (Fig. 4.1).

References

Arikawa-Hirasawa E, Watanabe H, Takami H, Hassell JR, Yamada Y (1999) Perlecan is essential for cartilage and cephalic development. Nat Genet 23(3):354–358. doi:10.1038/15537

Arikawa-Hirasawa E, Wilcox WR, Le AH, Silverman N, Govindraj P, Hassell JR, Yamada Y (2001) Dyssegmental dysplasia, Silverman-Handmaker type, is caused by functional null mutations of the perlecan gene. Nat Genet 27(4):431–434. doi:10.1038/86941

Aspberg A (2016) Cartilage proteoglycans. In: Grässel S, Aszódi A (eds) Cartilage, vol 1, Physiology and development. Springer International Publishing, Cham, pp 1–22. doi:10.1007/978-3-319-29568-8_1

Babelova A, Moreth K, Tsalastra-Greul W, Zeng-Brouwers J, Eickelberg O, Young MF, Bruckner P, Pfeilschifter J, Schaefer RM, Grone HJ, Schaefer L (2009) Biglycan, a danger signal that activates the NLRP3 inflammasome via toll-like and P2X receptors. J Biol Chem 284(36):24035–24048. doi:10.1074/jbc.M109.014266

Bass MD, Williamson RC, Nunan RD, Humphries JD, Byron A, Morgan MR, Martin P, Humphries MJ (2011) A syndecan-4 hair trigger initiates wound healing through caveolin- and RhoG-regulated integrin endocytosis. Dev Cell 21(4):681–693. doi:10.1016/j.devcel.2011.08.007

Bayliss MT, Howat S, Davidson C, Dudhia J (2000) The organization of aggrecan in human articular cartilage. Evidence for age-related changes in the rate of aggregation of newly synthesized molecules. J Biol Chem 275(9):6321–6327

Bernfield M, Gotte M, Park PW, Reizes O, Fitzgerald ML, Lincecum J, Zako M (1999) Functions of cell surface heparan sulfate proteoglycans. Annu Rev Biochem 68:729–777. doi:10.1146/annurev.biochem.68.1.729

Bertrand J, Stange R, Hidding H, Echtermeyer F, Nalesso G, Godmann L, Timmen M, Bruckner P, Dell'Accio F, Raschke MJ, Pap T, Dreier R (2013) Syndecan 4 supports bone fracture repair, but not fetal skeletal development, in mice. Arthritis Rheum 65(3):743–752. doi:10.1002/art.37817

Bezakova G, Ruegg MA (2003) New insights into the roles of agrin. Nat Rev Mol Cell Biol 4(4):295–308. doi:10.1038/nrm1074

Bock HC, Michaeli P, Bode C, Schultz W, Kresse H, Herken R, Miosge N (2001) The small proteoglycans decorin and biglycan in human articular cartilage of late-stage osteoarthritis. Osteoarthr Cartil/OARS Osteoarthr Res Soc 9(7):654–663. doi:10.1053/joca.2001.0420

Buckwalter JA, Lane NE (1997) Athletics and osteoarthritis. Am J Sports Med 25(6):873–881

Buckwalter JA, Rosenberg LC (1988) Electron microscopic studies of cartilage proteoglycans. Electron Microsc Rev 1(1):87–112

Buckwalter JA, Mankin HJ, Grodzinsky AJ (2005) Articular cartilage and osteoarthritis. Instr Course Lect 54:465–480

Cattaruzza S, Schiappacassi M, Ljungberg-Rose A, Spessotto P, Perissinotto D, Morgelin M, Mucignat MT, Colombatti A, Perris R (2002) Distribution of PG-M/versican variants in human tissues and de novo expression of isoform V3 upon endothelial cell activation, migration, and neoangiogenesis in vitro. J Biol Chem 277(49):47626–47635. doi:10.1074/jbc.M206521200

Chen XD, Fisher LW, Robey PG, Young MF (2004) The small leucine-rich proteoglycan biglycan modulates BMP-4-induced osteoblast differentiation. FASEB J: Off Publ Fed Am Soc Exp Biol 18(9):948–958. doi:10.1096/fj.03-0899com

Choi HY, Dieckmann M, Herz J, Niemeier A (2009) Lrp4, a novel receptor for Dickkopf 1 and sclerostin, is expressed by osteoblasts and regulates bone growth and turnover in vivo. PLoS One 4(11):e7930. doi:10.1371/journal.pone.0007930

Choi Y, Chung H, Jung H, Couchman JR, Oh ES (2011) Syndecans as cell surface receptors: unique structure equates with functional diversity. Matrix Biol: J Int Soc Matrix Biol 30(2):93–99. doi:10.1016/j.matbio.2010.10.006

Corr M (2008) Wnt-beta-catenin signaling in the pathogenesis of osteoarthritis. Nat Clin Pract Rheumatol 4(10):550–556. doi:10.1038/ncprheum0904

Costell M, Gustafsson E, Aszodi A, Morgelin M, Bloch W, Hunziker E, Addicks K, Timpl R, Fassler R (1999) Perlecan maintains the integrity of cartilage and some basement membranes. J Cell Biol 147(5):1109–1122

Couchman JR (2003) Syndecans: proteoglycan regulators of cell-surface microdomains? Nat Rev Mol Cell Biol 4(12):926–937. doi:10.1038/nrm1257

Couchman JR (2010) Transmembrane signaling proteoglycans. Annu Rev Cell Dev Biol 26:89–114. doi:10.1146/annurev-cellbio-100109-104126

Cs-Szabo G, Melching LI, Roughley PJ, Glant TT (1997) Changes in messenger RNA and protein levels of proteoglycans and link protein in human osteoarthritic cartilage samples. Arthritis Rheum 40(6):1037–1045. doi:10.1002/1529-0131(199706)40:6<1037::AID-ART6>3.0.CO;2-A

David G, Lories V, Decock B, Marynen P, Cassiman JJ, Van den Berghe H (1990) Molecular cloning of a phosphatidylinositol-anchored membrane heparan sulfate proteoglycan from human lung fibroblasts. J Cell Biol 111(6 Pt 2):3165–3176

David G, van der Schueren B, Marynen P, Cassiman JJ, van den Berghe H (1992) Molecular cloning of amphiglycan, a novel integral membrane heparan sulfate proteoglycan expressed by epithelial and fibroblastic cells. J Cell Biol 118(4):961–969

De Cat B, Muyldermans SY, Coomans C, Degeest G, Vanderschueren B, Creemers J, Biemar F, Peers B, David G (2003) Processing by proprotein convertases is required for glypican-3 modulation of cell survival, Wnt signaling, and gastrulation movements. J Cell Biol 163(3):625–635. doi:10.1083/jcb.200302152

Dell'accio F, De Bari C, Eltawil NM, Vanhummelen P, Pitzalis C (2008) Identification of the molecular response of articular cartilage to injury, by microarray screening: Wnt-16 expression and signaling after injury and in osteoarthritis. Arthritis Rheum 58(5):1410–1421. doi:10.1002/art.23444

Denzer AJ, Brandenberger R, Gesemann M, Chiquet M, Ruegg MA (1997) Agrin binds to the nerve-muscle basal lamina via laminin. J Cell Biol 137(3):671–683

Denzer AJ, Schulthess T, Fauser C, Schumacher B, Kammerer RA, Engel J, Ruegg MA (1998) Electron microscopic structure of agrin and mapping of its binding site in laminin-1. EMBO J 17(2):335–343. doi:10.1093/emboj/17.2.335

Echtermeyer F, Bertrand J, Dreier R, Meinecke I, Neugebauer K, Fuerst M, Lee YJ, Song YW, Herzog C, Theilmeier G, Pap T (2009) Syndecan-4 regulates ADAMTS-5 activation and cartilage breakdown in osteoarthritis. Nat Med 15(9):1072–1076. doi:10.1038/nm.1998

Eldridge S, Nalesso G, Ismail H, Vicente-Greco K, Kabouridis P, Ramachandran M, Niemeier A, Herz J, Pitzalis C, Perretti M, Dell'Accio F (2015) Agrin mediates chondrocyte homeostasis and requires both LRP4 and alpha-dystroglycan to enhance cartilage formation in vitro and in vivo. Ann Rheum Dis 75:1228–1235. doi:10.1136/annrheumdis-2015-207316

Embree MC, Kilts TM, Ono M, Inkson CA, Syed-Picard F, Karsdal MA, Oldberg A, Bi Y, Young MF (2010) Biglycan and fibromodulin have essential roles in regulating chondrogenesis and extracellular matrix turnover in temporomandibular joint osteoarthritis. Am J Pathol 176(2):812–826. doi:10.2353/ajpath.2010.090450

Esko JD, Linhardt RJ (2009) Proteins that bind sulfated glycosaminoglycans. In: Varki A, Cummings RD, Esko JD et al (eds) Essentials of glycobiology, 2nd edn. Cold Spring Harbor Laboratory Press, Cold Spring Harbor

Filion RJ, Popel AS (2004) A reaction-diffusion model of basic fibroblast growth factor interactions with cell surface receptors. Ann Biomed Eng 32(5):645–663

Frey H, Schroeder N, Manon-Jensen T, Iozzo RV, Schaefer L (2013) Biological interplay between proteoglycans and their innate immune receptors in inflammation. FEBS J 280(10):2165–2179. doi:10.1111/febs.12145

Fuerer C, Habib SJ, Nusse R (2010) A study on the interactions between heparan sulfate proteoglycans and Wnt proteins. Dev Dyn: Off Publ Am Assoc Anatomists 239(1):184–190. doi:10.1002/dvdy.22067

Garciadiego-Cazares D, Aguirre-Sanchez HI, Abarca-Buis RF, Kouri JB, Velasquillo C, Ibarra C (2015) Regulation of alpha5 and alphaV integrin expression by GDF-5 and BMP-7 in chondrocyte differentiation and osteoarthritis. PLoS One 10(5):e0127166. doi:10.1371/journal.pone.0127166

Giedion A, Boltshauser E, Briner J, Eich G, Exner G, Fendel H, Kaufmann L, Steinmann B, Spranger J, Superti-Furga A (1997) Heterogeneity in Schwartz-Jampel chondrodystrophic myotonia. Eur J Pediatr 156(3):214–223

Glasson SS, Askew R, Sheppard B, Carito B, Blanchet T, Ma HL, Flannery CR, Peluso D, Kanki K, Yang Z, Majumdar MK, Morris EA (2005) Deletion of active ADAMTS5 prevents cartilage degradation in a murine model of osteoarthritis. Nature 434(7033):644–648. doi:10.1038/nature03369

Goldoni S, Iozzo RV (2008) Tumor microenvironment: modulation by decorin and related molecules harboring leucine-rich tandem motifs. Int J of Cancer J Int Cancer 123(11):2473–2479. doi:10.1002/ijc.23930

Gronau T, Kruger K, Prein C, Aszodi A, Gronau I, Iozzo RV, Mooren FC, Clausen-Schaumann H, Bertrand J, Pap T, Bruckner P, Dreier R (2016) Forced exercise-induced osteoarthritis is attenuated in mice lacking the small leucine-rich proteoglycan decorin. Ann Rheum Dis. 2016 Jul 4. pii: annrheumdis-2016-209319. doi: 10.1136/annrheumdis-2016-209319.

Grover J, Roughley PJ (1995) Expression of cell-surface proteoglycan mRNA by human articular chondrocytes. Biochem J 309(Pt 3):963–968

Han B, Liu J, Ma MJ, Zhao L (2005) Clinicopathological significance of heparanase and basic fibroblast growth factor expression in human esophageal cancer. World J Gastroenterol: WJG 11(14):2188–2192

Hartmann C (2016) Wnt signaling in cartilage development. In: Grässel S, Aszódi A (eds) Cartilage, vol 1, Physiology and development. Springer International Publishing, Cham, pp 229–252. doi:10.1007/978-3-319-29568-8_10

Hassell JR, Robey PG, Barrach HJ, Wilczek J, Rennard SI, Martin GR (1980) Isolation of a heparan sulfate-containing proteoglycan from basement membrane. Proc Natl Acad Sci U S A 77(8):4494–4498

Hausser HJ, Ruegg MA, Brenner RE, Ksiazek I (2007) Agrin is highly expressed by chondrocytes and is required for normal growth. Histochem Cell Biol 127(4):363–374. doi:10.1007/s00418-006-0258-2

Heard BJ, Solbak NM, Achari Y, Chung M, Hart DA, Shrive NG, Frank CB (2013) Changes of early post-traumatic osteoarthritis in an ovine model of simulated ACL reconstruction are associated with transient acute post-injury synovial inflammation and tissue catabolism. Osteoarthr Cartil/OARS Osteoarthr Res Soc 21(12):1942–1949. doi:10.1016/j.joca.2013.08.019

Hedlund H, Mengarelli-Widholm S, Heinegard D, Reinholt FP, Svensson O (1994) Fibromodulin distribution and association with collagen. Matrix Biol: J Int Soc Matrix Biol 14(3):227–232

Henry SP, Takanosu M, Boyd TC, Mayne PM, Eberspaecher H, Zhou W, de Crombrugghe B, Hook M, Mayne R (2001) Expression pattern and gene characterization of asporin. a newly discovered member of the leucine-rich repeat protein family. J Biol Chem 276(15):12212–12221. doi:10.1074/jbc.M011290200

Hildebrand A, Romaris M, Rasmussen LM, Heinegard D, Twardzik DR, Border WA, Ruoslahti E (1994) Interaction of the small interstitial proteoglycans biglycan, decorin and fibromodulin with transforming growth factor beta. Biochem J 302(Pt 2):527–534

Humphries MJ, Mostafavi-Pour Z, Morgan MR, Deakin NO, Messent AJ, Bass MD (2005) Integrin-syndecan cooperation governs the assembly of signalling complexes during cell spreading. Novartis Found Symp 269:178–188; discussion 188–192, 223–130

Iozzo RV (1999) The biology of the small leucine-rich proteoglycans. Functional network of interactive proteins. J Biol Chem 274(27):18843–18846

Iozzo RV, San Antonio JD (2001) Heparan sulfate proteoglycans: heavy hitters in the angiogenesis arena. J Clin Invest 108(3):349–355. doi:10.1172/JCI13738

Iozzo RV, Sanderson RD (2011) Proteoglycans in cancer biology, tumour microenvironment and angiogenesis. J Cell Mol Med 15(5):1013–1031. doi:10.1111/j.1582-4934.2010.01236.x

Jepsen KJ, Wu F, Peragallo JH, Paul J, Roberts L, Ezura Y, Oldberg A, Birk DE, Chakravarti S (2002) A syndrome of joint laxity and impaired tendon integrity in lumican- and fibromodulin-deficient mice. J Biol Chem 277(38):35532–35540. doi:10.1074/jbc.M205398200

Kalamajski S, Aspberg A, Oldberg A (2007) The decorin sequence SYIRIADTNIT binds collagen type I. J Biol Chem 282(22):16062–16067. doi:10.1074/jbc.M700073200

Kalamajski S, Aspberg A, Lindblom K, Heinegard D, Oldberg A (2009) Asporin competes with decorin for collagen binding, binds calcium and promotes osteoblast collagen mineralization. Biochem J 423(1):53–59. doi:10.1042/BJ20090542

Kaneko H, Ishijima M, Futami I, Tomikawa-Ichikawa N, Kosaki K, Sadatsuki R, Yamada Y, Kurosawa H, Kaneko K, Arikawa-Hirasawa E (2013) Synovial perlecan is required for osteophyte formation in knee osteoarthritis. Matrix Biol: J Int Soc Matrix Biol 32(3-4):178–187. doi:10.1016/j.matbio.2013.01.004

Kim CW, Goldberger OA, Gallo RL, Bernfield M (1994) Members of the syndecan family of heparan sulfate proteoglycans are expressed in distinct cell-, tissue-, and development-specific patterns. Mol Biol Cell 5(7):797–805

Kizawa H, Kou I, Iida A, Sudo A, Miyamoto Y, Fukuda A, Mabuchi A, Kotani A, Kawakami A, Yamamoto S, Uchida A, Nakamura K, Notoya K, Nakamura Y, Ikegawa S (2005) An aspartic acid repeat polymorphism in asporin inhibits chondrogenesis and increases susceptibility to osteoarthritis. Nat Genet 37(2):138–144. doi:10.1038/ng1496

Lees S, Golub SB, Last K, Zeng W, Jackson DC, Sutton P, Fosang AJ (2015) Bioactivity in an aggrecan 32-mer fragment is mediated via toll-like receptor 2. Arthritis Rheumatol 67(5):1240–1249. doi:10.1002/art.39063

Liu-Bryan R, Terkeltaub R (2015) Emerging regulators of the inflammatory process in osteoarthritis. Nat Rev Rheumatol 11(1):35–44. doi:10.1038/nrrheum.2014.162

Lorenzo P, Aspberg A, Onnerfjord P, Bayliss MT, Neame PJ, Heinegard D (2001) Identification and characterization of asporin. a novel member of the leucine-rich repeat protein family closely related to decorin and biglycan. J Biol Chem 276(15):12201–12211. doi:10.1074/jbc.M010932200

Luyten FP, Tylzanowski P, Lories RJ (2009) Wnt signaling and osteoarthritis. Bone 44(4):522–527. doi:10.1016/j.bone.2008.12.006

Martel-Pelletier J, Boileau C, Pelletier JP, Roughley PJ (2008) Cartilage in normal and osteoarthritis conditions. Best Pract Res Clin Rheumatol 22(2):351–384. doi:10.1016/j.berh.2008.02.001

Martin I, Jakob M, Schafer D, Dick W, Spagnoli G, Heberer M (2001) Quantitative analysis of gene expression in human articular cartilage from normal and osteoarthritic joints. Osteoarthr Cartil/OARS Osteoarthr Res Soc 9(2):112–118. doi:10.1053/joca.2000.0366

McCulloch DR, Nelson CM, Dixon LJ, Silver DL, Wylie JD, Lindner V, Sasaki T, Cooley MA, Argraves WS, Apte SS (2009) ADAMTS metalloproteases generate active versican fragments that regulate interdigital web regression. Dev Cell 17(5):687–698. doi:10.1016/j.devcel.2009.09.008

Melrose J, Tammi M, Smith S (2002) Visualisation of hyaluronan and hyaluronan-binding proteins within ovine vertebral cartilages using biotinylated aggrecan G1-link complex and biotinylated hyaluronan oligosaccharides. Histochem Cell Biol 117(4):327–333. doi:10.1007/s00418-002-0392-4

Mochida Y, Parisuthiman D, Yamauchi M (2006) Biglycan is a positive modulator of BMP-2 induced osteoblast differentiation. Adv Exp Med Biol 585:101–113

Mort JS, Geng Y, Fisher WD, Roughley PJ (2016) Aggrecan heterogeneity in articular cartilage from patients with osteoarthritis. BMC Musculoskelet Disord 17:89. doi:10.1186/s12891-016-0944-8

Mow VC, Ratcliffe A, Poole AR (1992) Cartilage and diarthrodial joints as paradigms for hierarchical materials and structures. Biomaterials 13(2):67–97

Nagase H, Kashiwagi M (2003) Aggrecanases and cartilage matrix degradation. Arthritis Res Ther 5(2):94–103

Nakajima M, Kizawa H, Saitoh M, Kou I, Miyazono K, Ikegawa S (2007) Mechanisms for asporin function and regulation in articular cartilage. J Biol Chem 282(44):32185–32192. doi:10.1074/jbc.M700522200

Nishida Y, Shinomura T, Iwata H, Miura T, Kimata K (1994) Abnormal occurrence of a large chondroitin sulfate proteoglycan, PG-M/versican in osteoarthritic cartilage. Osteoarthr Cartil/OARS Osteoarthr Res Soc 2(1):43–49

Oegema TR Jr (1980) Delayed formation of proteoglycan aggregate structures in human articular cartilage disease states. Nature 288(5791):583–585

Ohazama A, Johnson EB, Ota MS, Choi HY, Porntaveetus T, Oommen S, Itoh N, Eto K, Gritli-Linde A, Herz J, Sharpe PT (2008) Lrp4 modulates extracellular integration of cell signaling pathways in development. PLoS One 3(12):e4092. doi:10.1371/journal.pone.0004092

Ohkawara B, Yamamoto TS, Tada M, Ueno N (2003) Role of glypican 4 in the regulation of convergent extension movements during gastrulation in xenopus laevis. Development 130(10):2129–2138

Pap T, Bertrand J (2013) Syndecans in cartilage breakdown and synovial inflammation. Nat Rev Rheumatol 9(1):43–55. doi:10.1038/nrrheum.2012.178

Pilia G, Hughes-Benzie RM, MacKenzie A, Baybayan P, Chen EY, Huber R, Neri G, Cao A, Forabosco A, Schlessinger D (1996) Mutations in GPC3, a glypican gene, cause the Simpson-Golabi-Behmel overgrowth syndrome. Nat Genet 12(3):241–247. doi:10.1038/ng0396-241

Poole AR, Rosenberg LC, Reiner A, Ionescu M, Bogoch E, Roughley PJ (1996) Contents and distributions of the proteoglycans decorin and biglycan in normal and osteoarthritic human articular cartilage. J Orthop Res 14(5):681–689. doi:10.1002/jor.1100140502

Radwan M, Gavriilidis C, Robinson JH, Davidson R, Clark IM, Rowan AD, Young DA (2013) Matrix metalloproteinase 13 expression in response to double-stranded RNA in human chondrocytes. Arthritis Rheum 65(5):1290–1301. doi:10.1002/art.37868

Rapraeger AC, Krufka A, Olwin BB (1991) Requirement of heparan sulfate for bFGF-mediated fibroblast growth and myoblast differentiation. Science 252(5013):1705–1708

Reichsman F, Smith L, Cumberledge S (1996) Glycosaminoglycans can modulate extracellular localization of the wingless protein and promote signal transduction. J Cell Biol 135(3):819–827

Reynard LN, Loughlin J (2013) The genetics and functional analysis of primary osteoarthritis susceptibility. Expert Rev Mol Med 15:e2. doi:10.1017/erm.2013.4

Roughley PJ, Mort JS (2014) The role of aggrecan in normal and osteoarthritic cartilage. J Exp Orthop 1(1):8. doi:10.1186/s40634-014-0008-7

Ruppert R, Hoffmann E, Sebald W (1996) Human bone morphogenetic protein 2 contains a heparin-binding site which modifies its biological activity. Eur J Biochem/FEBS 237(1):295–302

Sadatsuki R, Kaneko H, Kinoshita M, Futami I, Nonaka R, Culley KL, Otero M, Hada S, Goldring MB, Yamada Y, Kaneko K, Arikawa-Hirasawa E, Ishijima M (2016) Perlecan is required for the chondrogenic differentiation of synovial mesenchymal cells through regulation of Sox9 gene expression. J Orthop Res: Off Publ Orthop Res Soc. doi:10.1002/jor.23318

Sahni M, Ambrosetti DC, Mansukhani A, Gertner R, Levy D, Basilico C (1999) FGF signaling inhibits chondrocyte proliferation and regulates bone development through the STAT-1 pathway. Genes Dev 13(11):1361–1366

Schaefer L, Babelova A, Kiss E, Hausser HJ, Baliova M, Krzyzankova M, Marsche G, Young MF, Mihalik D, Gotte M, Malle E, Schaefer RM, Grone HJ (2005) The matrix component biglycan is proinflammatory and signals through Toll-like receptors 4 and 2 in macrophages. J Clin Invest 115(8):2223–2233. doi:10.1172/JCI23755

Schwartz NB, Pirok EW 3rd, Mensch JR Jr, Domowicz MS (1999) Domain organization, genomic structure, evolution, and regulation of expression of the aggrecan gene family. Prog Nucleic Acid Res Mol Biol 62:177–225

Sherwood J, Bertrand J, Nalesso G, Poulet B, Pitsillides A, Brandolini L, Karystinou A, De Bari C, Luyten FP, Pitzalis C, Pap T, Dell'Accio F (2015) A homeostatic function of CXCR2 signalling in articular cartilage. Ann Rheum Dis 74(12):2207–2215. doi:10.1136/annrheumdis-2014-205546

Shu CC, Jackson MT, Smith MM, Smith SM, Penm S, Lord MS, Whitelock JM, Little CB, Melrose J (2016) Ablation of perlecan domain 1 heparan sulfate reduces progressive cartilage degradation, synovitis, and osteophyte size in a preclinical model of posttraumatic osteoarthritis. Arthritis Rheumatol 68(4):868–879. doi:10.1002/art.39529

Sillat T, Barreto G, Clarijs P, Soininen A, Ainola M, Pajarinen J, Korhonen M, Konttinen YT, Sakalyte R, Hukkanen M, Ylinen P, Nordstrom DC (2013) Toll-like receptors in human chondrocytes and osteoarthritic cartilage. Acta Orthop 84(6):585–592. doi:10.3109/17453674.2013.854666

Song HH, Shi W, Xiang YY, Filmus J (2005) The loss of glypican-3 induces alterations in Wnt signaling. J Biol Chem 280(3):2116–2125. doi:10.1074/jbc.M410090200
Stanton H, Rogerson FM, East CJ, Golub SB, Lawlor KE, Meeker CT, Little CB, Last K, Farmer PJ, Campbell IK, Fourie AM, Fosang AJ (2005) ADAMTS5 is the major aggrecanase in mouse cartilage in vivo and in vitro. Nature 434(7033):648–652. doi:10.1038/nature03417
Su SL, Tsai CD, Lee CH, Salter DM, Lee HS (2005) Expression and regulation of Toll-like receptor 2 by IL-1beta and fibronectin fragments in human articular chondrocytes. Osteoarthr Cartil/OARS Osteoarthr Res Soc 13(10):879–886. doi:10.1016/j.joca.2005.04.017
SundarRaj N, Fite D, Ledbetter S, Chakravarti S, Hassell JR (1995) Perlecan is a component of cartilage matrix and promotes chondrocyte attachment. J Cell Sci 108(Pt 7):2663–2672
Svensson L, Heinegard D, Oldberg A (1995) Decorin-binding sites for collagen type I are mainly located in leucine-rich repeats 4-5. J Biol Chem 270(35):20712–20716
Sztrolovics R, White RJ, Poole AR, Mort JS, Roughley PJ (1999) Resistance of small leucine-rich repeat proteoglycans to proteolytic degradation during interleukin-1-stimulated cartilage catabolism. Biochem J 339(Pt 3):571–577
Takahashi T, Ogasawara T, Kishimoto J, Liu G, Asato H, Nakatsuka T, Uchinuma E, Nakamura K, Kawaguchi H, Chung UI, Takato T, Hoshi K (2005) Synergistic effects of FGF-2 with insulin or IGF-I on the proliferation of human auricular chondrocytes. Cell Transplant 14(9):683–693
Takeuchi Y, Kodama Y, Matsumoto T (1994) Bone matrix decorin binds transforming growth factor-beta and enhances its bioactivity. J Biol Chem 269(51):32634–32638
Tesche F, Miosge N (2004) Perlecan in late stages of osteoarthritis of the human knee joint. Osteoarthr Cartil/OARS Osteoarthr Res Soc 12(11):852–862. doi:10.1016/j.joca.2004.07.004
Tkachenko E, Rhodes JM, Simons M (2005) Syndecans: new kids on the signaling block. Circ Res 96(5):488–500. doi:10.1161/01.RES.0000159708.71142.c8
Troeberg L, Nagase H (2012) Proteases involved in cartilage matrix degradation in osteoarthritis. Biochim Biophys Acta 1824(1):133–145. doi:10.1016/j.bbapap.2011.06.020
Tufvesson E, Westergren-Thorsson G (2002) Tumour necrosis factor-alpha interacts with biglycan and decorin. FEBS Lett 530(1-3):124–128
Uren A, Reichsman F, Anest V, Taylor WG, Muraiso K, Bottaro DP, Cumberledge S, Rubin JS (2000) Secreted frizzled-related protein-1 binds directly to Wingless and is a biphasic modulator of Wnt signaling. J Biol Chem 275(6):4374–4382
van der Kraan PM, Blaney Davidson EN, van den Berg WB (2010) Bone morphogenetic proteins and articular cartilage: to serve and protect or a wolf in sheep clothing's? osteoarthritis and cartilage / OARS. Osteoarthr Res Soc 18(6):735–741. doi:10.1016/j.joca.2010.03.001
Vincent TL, McLean CJ, Full LE, Peston D, Saklatvala J (2007) FGF-2 is bound to perlecan in the pericellular matrix of articular cartilage, where it acts as a chondrocyte mechanotransducer. Osteoarthr Cartil/OARS Osteoarthr Res Soc 15(7):752–763. doi:10.1016/j.joca.2007.01.021
Wang J, Markova D, Anderson DG, Zheng Z, Shapiro IM, Risbud MV (2011) TNF-alpha and IL-1beta promote a disintegrin-like and metalloprotease with thrombospondin type I motif-5-mediated aggrecan degradation through syndecan-4 in intervertebral disc. J Biol Chem 286(46):39738–39749. doi:10.1074/jbc.M111.264549
Watanabe H, Yamada Y, Kimata K (1998) Roles of aggrecan, a large chondroitin sulfate proteoglycan, in cartilage structure and function. J Biochem 124(4):687–693
Westling J, Gottschall PE, Thompson VP, Cockburn A, Perides G, Zimmermann DR, Sandy JD (2004) ADAMTS4 (aggrecanase-1) cleaves human brain versican V2 at Glu405-Gln406 to generate glial hyaluronate binding protein. Biochem J 377(Pt 3):787–795. doi:10.1042/BJ20030896
Whitelock JM, Iozzo RV (2005) Heparan sulfate: a complex polymer charged with biological activity. Chem Rev 105(7):2745–2764. doi:10.1021/cr010213m
Wight TN, Kang I, Merrilees MJ (2014) Versican and the control of inflammation. Matrix Biol: J Int Soc Matrix Biol 35:152–161. doi:10.1016/j.matbio.2014.01.015
Wu Y, Chen L, Zheng PS, Yang BB (2002) Beta 1-Integrin-mediated glioma cell adhesion and free radical-induced apoptosis are regulated by binding to a C-terminal domain of PG-M/versican. J Biol Chem 277(14):12294–12301. doi:10.1074/jbc.M110748200

Xu T, Bianco P, Fisher LW, Longenecker G, Smith E, Goldstein S, Bonadio J, Boskey A, Heegaard AM, Sommer B, Satomura K, Dominguez P, Zhao C, Kulkarni AB, Robey PG, Young MF (1998) Targeted disruption of the biglycan gene leads to an osteoporosis-like phenotype in mice. Nat Genet 20(1):78–82. doi:10.1038/1746

Yang BL, Zhang Y, Cao L, Yang BB (1999) Cell adhesion and proliferation mediated through the G1 domain of versican. J Cell Biochem 72(2):210–220

Young AA, Smith MM, Smith SM, Cake MA, Ghosh P, Read RA, Melrose J, Sonnabend DH, Roughley PJ, Little CB (2005) Regional assessment of articular cartilage gene expression and small proteoglycan metabolism in an animal model of osteoarthritis. Arthritis Res Ther 7(4):R852–R861. doi:10.1186/ar1756

Zimmermann DR, Ruoslahti E (1989) Multiple domains of the large fibroblast proteoglycan, versican. EMBO J 8(10):2975–2981

5 Pro- and Anti-inflammatory Cytokine Profiles in Osteoarthritis

Yvonne Bastiaansen-Jenniskens, Daniel Saris, and Laura B. Creemers

Abstract

The presence and production of soluble factors in the osteoarthritic (OA) joint have always been a focus of research, as they are assumed to play a role in initiation and/or progression of disease. Many tissue and cell types in the joint are capable of their production, with the synovial fluid serving as a reservoir into which they can be secreted. Although an increasing interest is directed towards chemokines, growth factors and adipokines, traditionally, a subset of inflammatory, anti-inflammatory and modulatory cytokines has been studied. Differential profiles compared to healthy joints were found in the knee and other OA joints, whereby also joint damage induces a specific change in secretory pattern. However, for the cytokines commonly assumed to play a role in OA, such as IL-1 and TNFα, their consistently low levels, frequent lack of association with disease and the presence of natural inhibitors suggest that other soluble factors may be more promising as possible targets.

Y. Bastiaansen-Jenniskens, PhD
Erasmus MC, University Medical Center Rotterdam, Department of Orthopaedics, Gravendijkwal 230, 3015 CE Rotterdam, The Netherlands
e-mail: y.bastiaansen@erasmusmc.nl

D. Saris, PhD
University Medical Center Utrecht, Department of Orthopaedics, HP G05.228, PO Box 85090, 3508 GA Utrecht, The Netherlands

MIRA Institute, University Twente, ME125, PO Box 217, 7500 AE Enschede, The Netherlands
e-mail: D.saris@umcutrecht.nl

L.B. Creemers, PhD (✉)
University Medical Center Utrecht, Department of Orthopaedics, HP G05.228, PO Box 85090, 3508 GA Utrecht, The Netherlands
e-mail: L.B.Creemers@umcutrecht.nl

S. Grässel, A. Aszódi (eds.), *Cartilage*, DOI 10.1007/978-3-319-45803-8_5

5.1 Cytokines in Osteoarthritis

Cytokines are commonly soluble, but sometimes cell surface-associated proteins capable of autocrine, paracrine and endocrine signalling (Liles and Van Voorhis 1995; Turner et al. 2014). In osteoarthritis (OA), pro-inflammatory cytokines allegedly play a crucial role in initiation and/or progression of disease (Goldring and Otero 2011; Goldring and Goldring 2004; Rahmati et al. 2016). Whether inflammation is the trigger leading to OA, or biomechanical overloading is the initiating event, cytokine production at some point is the converging element in these two theories (Berenbaum 2013; Estes et al. 2004; Guilak 2011; Houard et al. 2013). Pro-inflammatory cytokines can mediate cartilage destruction by enhancing the production and activity of several matrices degrading matrix metalloproteinases (MMPs), in particular the type II collagen degrading MMP-1 and MMP-13 and the aggrecan degrading ADAMTS-4 and ADAMTS-5 (Durigova et al. 2008; Flannery et al. 2000; Koshy et al. 2002a, b; Long et al. 2008; Stove et al. 2000). Although initially assumed to mainly be produced by immune cells, it has become clear that many cell types in the body are capable of their production (Turner et al. 2014). In the joint space, many of these factors will end up in the synovial fluid, which connects all tissues and thereby can mediate solute-drive communication between the joint tissues. Here, combinations of pro- and anti-inflammatory cytokines as part of so-called cytokine networks are likely to act in concert and may synergistically enhance or attenuate cartilage matrix degradation (Schett et al. 2013). Together with their possible role as target in OA, the use of cytokine analysis as biomarker or even prognostic tool incited many studies evaluating their presence inside and outside the joint (Table 5.1).

5.2 Origin (Synovial Membrane, Cartilage, Fat Pad, Bone/Cell Types)

An articular joint typically contains the cartilage, bone and synovium and is also surrounded by different adipose tissue bodies, such as Hoffa's fat pad (also called the infrapatellar fat pad) in the knee joint. Next to fibroblast-like synoviocytes (FLSs), the synovium contains many other cells (de Lange-Brokaar et al. 2015) such as macrophages (Beekhuizen et al. 2011; Deligne et al. 2015; Fahy et al. 2014), mast cells (de Lange-Brokaar et al. 2015), T-cells (Beekhuizen et al. 2011; Deligne et al. 2015), B-cells and neutrophils (Deligne et al. 2015) (Fig. 5.1). The latter three lymphocytes are however mainly present in the sub-intimal layer and not detectable in healthy synovium (Deligne et al. 2015). The presence of mast cells is also correlated with the degree of synovitis (de Lange-Brokaar et al. 2015). With all of these different types of immune cells, the synovium is a usual suspect in contributing to the cytokine levels in synovial fluid. Indeed, synovium explants in culture secrete many different cytokines, adipokines, chemokines and growth factors, including IL-1β, IL-1RA, IL-6, IL-8, IL-10, IL-17, IL-18, IL-22, IL-23, TGFβ1, RANTES/CCL5, CCL18, osteoprotegerin (OPG) and visfatin

Table 5.1 Comparisons of cytokine levels in synovial fluid between healthy and diseased joints

Cytokine	OA vs healthy			(Acute) damage vs healthy			OA vs damage			Late vs early			Concentration range in OA (pg/ml)
	↑	↓	=	↑	↓	=	↑	↓	=	↑	↓	=	
TNFα			Sauerschnig et al. (2014), Teunis et al. (2013), and Tsuchida et al. (2014)	Bigoni et al. (2013) and Cameron et al. (1997)		Cuellar et al. (2010), Cuéllar et al. (2015), Higuchi et al. (2006), Marks and Donaldson (2005), and Teunis et al. (2013)		Irie et al. (2003)	Ribbens et al. (2000) and Scanzello et al. (2009)				UD Heard et al. (2013) -105 Hussein et al. (2008)
IL-1α			Heard et al. (2013) and Tsuchida et al. (2014)			Cameron et al. (1997)			Tsuchida et al. (2014)	Heard et al. (2013)			UD Beekhuizen et al. (2013), Kahle et al. (1992)-15 Tsuchida et al. (2014)

(continued)

Table 5.1 (continued)

Cytokine	OA vs healthy			(Acute) damage vs healthy			OA vs damage			Late vs early			Concentration range in OA (pg/ml)
	↑	↓	=	↑	↓	=	↑	↓	=	↑	↓	=	
IL-1β	Sauerschnig et al. (2014)		Beekhuizen et al. (2013), Heard et al. (2013), and Tsuchida et al. (2014)	Bigoni et al. (2013) and Marks and Donaldson (2005)		Cameron et al. (1997), Cuellar et al. (2010), and Higuchi et al. (2006)		Irie et al. (2003)		Heard et al. (2013)			UD Teunis et al. (2013)-21 Kahle et al. (1992)
IL-6	Beekhuizen et al. (2013), Heard et al. (2013), Teunis et al. (2013), and Tsuchida et al. (2014)			Bigoni et al. (2013), Cameron et al. (1997), Cuellar et al. (2010), Higuchi et al. (2006), Tsuchida et al. (2014), andCuéllar et al. (2015)		Teunis et al. (2013)		Irie et al. (2003)	Ribbens et al. (2000), Scanzello et al. (2009), and Tsuchida et al. (2014)	Heard et al. (2013)			120 Beekhuizen et al. (2013)-4747 Heard et al. (2013)
IL-8	Beekhuizen et al. (2013), Heard et al. (2013)		Teunis et al. (2013), and Tsuchida et al. (2014)	Bigoni et al. (2013) and Cameron et al. (1997)		Cuellar et al. (2010), Teunis et al. (2013), and Tsuchida et al. (2014)		Irie et al. (2003)	Teunis et al. (2013) and Tsuchida et al. (2014)				52 Tsuchida et al. (2014)-714 Heard et al. (2013)

IL-10		Beekhuizen et al. (2013)	Tsuchida et al. (2014)			Cuellar et al. (2010), Cuéllar et al. (2015), and Tsuchida et al. (2014)	Teunis et al. (2013)	Irie et al. (2003)	Tsuchida et al. (2014)	Heard et al. (2013)			2 Beekhuizen et al. (2013)-58 Heard et al. (2013)
IL-13			Tsuchida et al. (2014)	Tsuchida et al. (2014)		Cuellar et al. (2010)			Tsuchida et al. (2014)	Heard et al. (2013)			6 Heard et al. (2013)-18 Tsuchida et al. (2014)
IL-15								Scanzello et al. (2009)					4 Scanzello et al. (2009)
IL-17						Cuellar et al. (2010)							UD (Heard et al. (2013)-0.5 Hussein et al. (2008)
IL-1RA			Teunis et al. (2013)	Cameron et al. (1997), and Marks and Donaldson (2005)		Teunis et al. (2013)		Irie et al. (2003)	Teunis et al. (2013)	Heard et al. (2013)			UD Beekhuizen et al. (2013)-811 Irie et al. (2003)
IFNγ	Teunis et al. (2013) and Tsuchida et al. (2014)		Beekhuizen et al. (2013)	Cuellar et al. (2010) and Tsuchida et al. (2014)		Teunis et al. (2013)	Teunis et al. (2013)		Tsuchida et al. (2014)				UD Heard et al. (2013)-51 Tsuchida et al. (2014)

(continued)

Table 5.1 (continued)

Cytokine	OA vs healthy			(Acute) damage vs healthy			OA vs damage			Late vs early			Concentration range in OA (pg/ml)
	↑	↓	=	↑	↓	=	↑	↓	=	↑	↓	=	
OSM	Tsuchida et al. (2014)		Teunis et al. (2013)	Tsuchida et al. (2014)		Teunis et al. (2013)	Tsuchida et al. (2014)		Teunis et al. (2013)				UD Beekhuizen et al. (2013)-38 Tsuchida et al. (2014)
Leptin	Beekhuizen et al. (2013)												1637 Beekhuizen et al. (2013)-2959 Mabey et al. (2014)
VEGF	Heard et al. (2013)					Cuéllar et al. (2015)	Rübenhagen et al. (2012)			Heard et al. (2013) and Rübenhagen et al. (2012)			307 Heard et al. (2013)-1,000 Rübenhagen et al. (2012)

An arrow down means decreased concentrations in OA compared to healthy (first column), joints with damage versus healthy joints, etc. An arrow up vice versa means increased in OA compared to healthy joints, etc. UD means undetectable, so the cytokine was analysed but the values fell below detection limits. Numbers in superscript denote the accompanying references. Cytokines which were not detectable in a particular study were not included in the overview. For the cytokine concentrations in OA joints, only studies using ELISA for detection were included, quoting the average levels per patient group

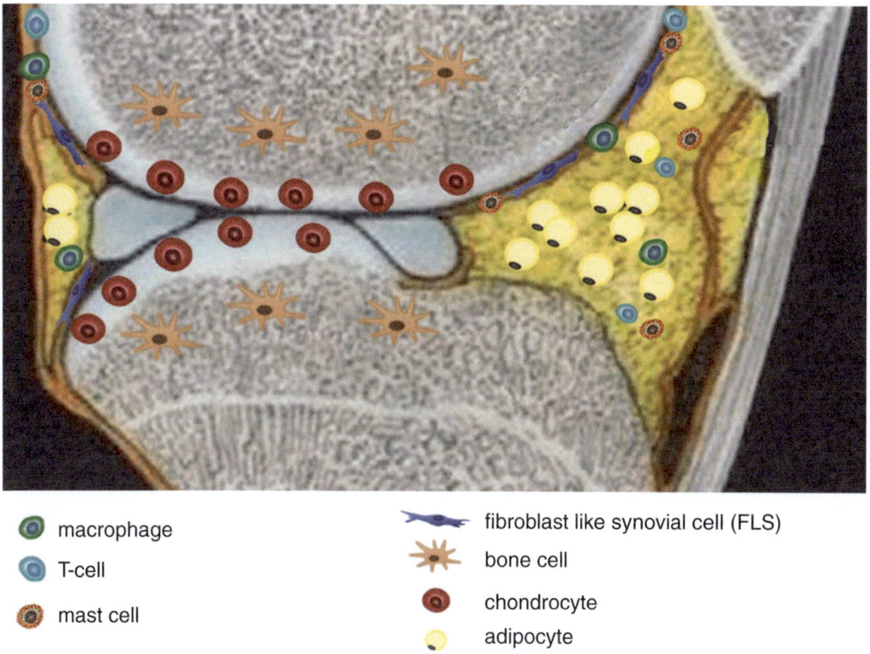

Fig. 5.1 Overview of the cells in the various joint structures capable of cytokine production

(Beekhuizen et al. 2011; Deligne et al. 2015; Fahy et al. 2014; Laiguillon et al. 2014). TNFα and nerve growth factor (NGF) immunopositive cells are also seen in the osteoarthritic synovium (Kc et al. 2016). As might be expected, IL-1β, IL-6, IL-8, and IL-18 but also TGFβ1 release are increased with increased inflammation of the synovium, characterised by hypervascularised areas containing hypertrophic and hyperemic villi (Deligne et al. 2015). However, IL-23 and RANTES secretion was not dependent on the inflammatory status of synovium explants (Deligne et al. 2015). When the OA synovial tissue was digested and single cells were cultured, the cells spontaneously secreted the factors mentioned above, including TNFα, IL-1β and large amounts of IL-6, IL-8 and MCP1. However, when macrophages were depleted from the digest, IL-1β, TNFα, IL-6 and IL-8 production by the remaining FLSs was strongly decreased. As IL-6 and IL-8 are known to be produced by FLSs, this indicates that the synovial macrophages are required for this secretory activity. No effect was seen after T-cell depletion (Bondeson et al. 2006). As shown by immunohistochemistry, the synovium of OA patients also contains IL-35. However, these levels are relatively low when compared to the synovium of rheumatoid arthritis (RA) patients (Filková et al. 2015).

Next to the synovium, Hoffa's fat pad contains many immune cells such as macrophages (Bastiaansen-Jenniskens et al. 2012; Wei et al. 2015), T-cells, mast cells and a few B-cells (Klein-Wieringa et al. 2011). Hoffa's fat pad produces IL-1β,

IL-6, IL-8, interferon (IFN)γ, MCP1, IL-4, IL-10, IL-17, TNFα, the adipokines adipsin, adiponectin, visfatin, resistin and leptin and the growth factors fibroblast growth factor (FGF)2 and vascular endothelial growth factor (VEGF) (Clockaerts et al. 2012; Distel et al. 2009; Klein-Wieringa et al. 2011; Ushiyama et al. 2003). Inflammation in the fat pad explants induced in vitro with IL-1β results in an increase of TNFα release, indicating that protein release might also be altered in the inflamed knee joint (Clockaerts et al. 2012). The CD4+ and CD8+ T-cells and macrophages in Hoffa's fat pad produce TNFα, IL-6 and IL-10, and the T-cells also produce IFNγ (Klein-Wieringa et al. 2011).

Many cytokines, growth factors and adipokines are expressed by the osteoarthritic cartilage, as shown in tissue culture or ex vivo tissue extracts, namely, IL-1α, IL-1β, IL-1RA, IL-4, IL-6, IL-7, IL-8 (actually a chemokine), IL-10, IL-13, TNFα, oncostatin M (OSM), leukaemia inhibitory factor (LIF), adiponectin, leptin, visfatin monocyte chemotactic protein 1 (MCP1), RANTES, FGF2, hepatocyte growth factor (HGF) and VEGF (Beekhuizen et al. 2011; Kc et al. 2016; Laiguillon et al. 2014; Tsuchida et al. 2014). The severity of OA also seems to influence the production of IL-1β, TNFα and most likely other cytokines, since, with more severe OA, more chondrocytes positive for IL-1β and TNFα were observed (Kc et al. 2016; Tetlow et al. 2001). Surprisingly, most of the aforementioned factors, except for TNFα, adiponectin, FGF2 and HGF, were produced at a higher level by chondrocytes that were isolated from the cartilage compared to their production levels inside the tissue (Tsuchida et al. 2014). This indicates that the surrounding matrix influences the production of these factors and cell cultures may not be optimal tools to study OA-related processes.

Finally, the (subchondral) bone is also a source of soluble factors detected in the synovial fluid; however, little is known about this source. In cell culture, the subchondral bone releases IL-1β, IL-6, IL-8, FGF2 and visfatin (Laiguillon et al. 2014; Leyh et al. 2014). Although it is not clear from which bone cells these factors arise, at least IL-6 and IL-1 can be produced by osteoblasts and osteocytes (Bakker et al. 2014; Gordeladze et al. 2002; Maeda et al. 2015).

5.3 Cytokine Profiling in Synovial Fluid of Osteoarthritic Joints

Many patient-based comparative studies have aimed to shed light on synovial fluid cytokine composition in OA (Teunis et al. 2013). Amongst all soluble factors mentioned above, mostly the pro-inflammatory, anti-inflammatory and modulatory cytokines have been studied.

In general, pro-inflammatory cytokine release into the synovial fluid in OA comes nowhere near the high secretory activity seen in RA joints (Hussein et al. 2008; Kahle et al. 1992; Neidel et al. 1997; Ribbens et al. 2000; Richette et al. 2008). However, if compared to healthy subjects, patients with OA do appear to also show cytokine upregulation and differential profiles. Amongst those, the cytokines postulated by many to be instrumental in initiating and/or advancing OA degenerative processes, IL-1β and TNFα, have been subject to close scrutiny. However, where TNFα is clearly upregulated in RA (Vervoordeldonk and Tak 2002), its upregulation in OA is far from convincing. Except for two studies on temporomandibular joint (TMJ) OA (Kaneyama et al. 2002, 2005), its synovial fluid levels do not seem higher in OA than in healthy joints (Sauerschnig et al. 2014; Teunis et al. 2013; Tsuchida et al. 2014; Kaneyama et al. 2003), if not entirely undetectable (Heard et al. 2013; Beekhuizen et al. 2013). IL-1β is either upregulated in the knee and TMJ OA (Sauerschnig et al. 2014; Kubota et al. 1998) or present at similar levels (Beekhuizen et al. 2013; Heard et al. 2013; Tsuchida et al. 2014) while undetectable in wrist OA (Teunis et al. 2013). IL-1α did not show any differential expression at all in the relatively limited numbers of studies describing this cytokine (Beekhuizen et al. 2013; Heard et al. 2013; Tsuchida et al. 2014). Increased but also similar levels compared to healthy joints were described for IL-8, which is actually a chemokine (Beekhuizen et al. 2013; Heard et al. 2013; Kaneyama et al. 2002; Teunis et al. 2013; Tsuchida et al. 2014). Other pro-inflammatory cytokines found to be upregulated in OA in several studies are oncostatin M (OSM) (Beekhuizen et al. 2013; Tsuchida et al. 2014) and INFγ (Teunis et al. 2013; Tsuchida et al. 2014), of which the latter was even suggested to be inversely correlated with acute inflammation (Kahle et al. 1992). However, as for each of these two factors in which also unchanged levels were demonstrated (Beekhuizen et al. 2013; Teunis et al. 2013), more research is required to give proper insight on their association with OA. This also goes for IL-15 (Scanzello et al. 2009) and leptin (Beekhuizen et al. 2013), for which even less data are currently available. For the anti-inflammatory cytokines, information is relatively limited too. Despite the wealth of preclinical studies using exogenously applied IL-1RA to interfere with the degenerative process in OA, very little is known of its concentrations in the osteoarthritic joint compared to healthy joints, and it may be upregulated (Heard et al. 2013) or relatively unchanged (Teunis et al. 2013). More but also contradictory data are available for the anti-inflammatory IL-10, with higher (Heard et al. 2013; Teunis et al. 2013; Fang et al. 1999), similar (Tsuchida et al. 2014) or even lower (Beekhuizen et al. 2013) levels in the OA joint.

Surprisingly, the only cytokine that does show an almost unequivocal elevation in OA synovial fluid compared to healthy controls is the modulatory cytokine IL-6, shown for the knee (Beekhuizen et al. 2013; Tsuchida et al. 2012, 2014; Heard et al. 2013), wrist (Teunis et al. 2013) and temporomandibular osteoarthritis (Kaneyama et al. 2002, 2005; Kubota et al. 1998).

5.4 Profiling in Trauma and Early OA

Apart from characterisation of established OA, synovial fluid cytokine content also changes during progression from joint trauma via early radiographic to end-stage OA. As the acute inflammatory response occurring after injury may trigger degeneration and hence timely interference may prevent this, many efforts were devoted to characterising this response. After anterior cruciate ligament (ACL) injury, either or not combined with meniscal damage, indeed, several cytokines including TNFα, IL-6, IL-8, IL-10 and IL-1RA are strongly upregulated. This upregulation occurs already within the first 2 days, thereafter subsiding immediately again over a period of a few weeks to a few months, when injury is defined as chronic (Bigoni et al. 2013; Cameron et al. 1997; Cuellar et al. 2010; Irie et al. 2003). Changes in IL-1β levels here again were less unequivocal, while for TNFα levels, these were very moderate. Joints with isolated cartilage defects were shown to contain upregulated levels of IL-6, IFNγ, OSM and IL-13 (Tsuchida et al. 2012, 2014), although also unchanged IL-6 concentrations were reported (Cuéllar et al. 2015).

Interestingly, it appears that after entering the chronic phase and several years post-ACL and meniscus injury, in particular IL-6 (Cuellar et al. 2010; Cuéllar et al. 2015; Higuchi et al. 2006) and possibly IL-1RA (Marks and Donaldson 2005) become elevated in the joint again. To what extent this correlates with progression towards OA is not clear. IL-6 and TNFα were shown to be associated with osteophyte formation (Larsson et al. 2015), and IL-6, IL-10, IL-13 and IL-1RA levels were higher in late compared to early radiographic OA (Heard et al. 2013). In addition, also the angiogenic factor VEGF has been shown to be associated with progression of OA (Heard et al. 2013; Larsson et al. 2015; Mabey et al. 2014; Rübenhagen et al. 2012).

5.5 Synovial Fluid Cytokines as Therapeutic Targets in OA?

Although many of the pro-inflammatory cytokines mentioned above have been postulated to play a crucial role in initiation or progression of OA, their true involvement remains unclear. The role of some of the factors traditionally assumed to play a role in OA, in particular IL-1 and TNFα, has been disputed by a paucity of effects seen in clinical trials with OA patients using biological inhibitors of these cytokines (Chevalier et al. 2009, 2014; Cohen et al. 2011; Grunke and Schulze-Koops 2006).

The discrepancy between these clinical results and the plethora of in vitro and in vivo studies suggesting a causative role of these and other inflammatory cytokines may find its origin in several phenomena. First of all, in vitro studies showing their effect on cartilage degradation usually depart from concentrations that are much higher than those found in the osteoarthritic joints. TNFα, IL-1β and OSM are typically used at concentrations in the 1–50 ng per ml range, whereas their levels in the joint synovial fluid commonly range from undetectable to up to tens of picograms per ml (Beekhuizen et al. 2013b, b; Heard et al. 2013; Neidel et al. 1997; Ribbens et al. 2000; Richette et al. 2008; Tsuchida et al. 2014). Though arguably it is possible that due to the synergy shown between several of these factors (Durigova et al. 2008; Henderson and Pettipher 1989; Henrotin et al. 2000; Koshy et al. 2002a, b; Rowan et al. 2001), much lower concentrations are required to achieve cartilage degrading effects, also this synergy was shown using supraphysiologically high concentrations only, thus not providing much support for these mechanisms in vivo.

Apart from the fact that most pro-inflammatory cytokines are usually present at much lower levels in OA than in RA (Fraser et al. 2003; Hussein et al. 2008; Neidel et al. 1997; Ribbens et al. 2000; Richette et al. 2008), the presence of natural antagonists in the osteoarthritic joint may have provided neutralisation of their action. The IL-1RA/IL-1 ratio was found to be higher in OA than in RA (Richette et al. 2008), with IL-1RA often present at over 100 times higher levels than IL-1 (Heard et al. 2013; Irie et al. 2003; Rutgers et al. 2010; Richette et al. 2008), which is the ratio required to inhibit the actions of IL-1 (Arend et al. 1990; Granowitz et al. 1991). Similarly, the presence of endogenously produced soluble TNF receptors at concentrations around 1,000-fold of those of TNFα in the OA joint (Ribbens et al. 2000; Simao et al. 2014) may preclude biological activity of anti-TNFα treatment in OA patients. Finally, the levels of cytokines in the synovial fluid may deviate from the concentrations in the articular cartilage itself, possibly related to fast clearance from the joint space (Rutgers et al. 2010) and diffusion properties of the cartilage.

One approach to more accurately define the role of cytokines in this complex environment is by interfering with specific cytokines in the synovial fluid. Using this approach in vitro, OSM in OA synovial fluid was shown to play a role by inhibiting matrix production in the cartilage, but without affecting degradation. Exposure to OSM alone at the levels found in vivo did not have any effect, while the commonly employed high concentrations did reproduce the enhanced degradation shown before (Beekhuizen et al. 2013). Interference with IL-6 in the synovial fluid did not affect proteoglycan metabolism in the osteoarthritic cartilage, while adding this cytokine enhanced regeneration by isolated chondrocytes even slightly (Tsuchida et al. 2012).

Ideally, the role of cytokines is investigated in vivo. However, although genetically modified mouse models can provide some information, the similarity of OA induced in animals to human disease remains a matter of debate (Teeple et al. 2013). Moreover, accumulating evidence suggests that cartilage degradation in OA is partially related to the phenotype of the underlying bone, via altered biomechanical properties and secretion of several soluble factors (Favero et al. 2015; Hardcastle et al. 2015), so any generalised genetic modification of inflammatory cytokine production may indirectly rather than directly affect cartilage integrity by targeting the underlying bone. Using tissue-specific and/or conditional knockout models may therefore be more informative. In the end, more studies using medium to high throughput approaches based on large patient cohorts and extensive proteomic profiling will allow to identify factors that consistently correlate with disease up and may point towards targets as reliable and proven markers that are sufficiently elevated to distinguish with certainty between patient populations.

5.6 Cytokines as Biomarkers

Biomarkers are useful if they can non-invasively predict or diagnose disease progression. Factors present in the synovial fluid may therefore be less useful as biomarkers; synovial fluid aspiration is relatively invasive, induces a risk for infection and is not always possible from relatively healthy joints which typically contain less synovial fluid (Pascual and Doherty 2009). However, until now, only limited information is available on cytokines as serum markers for disease. Only one study suggested several pro- and anti-inflammatory factors to be altered in moderate to severe OA serum compared to normal, with a decrease for IL-1β, IL-2 and IL-4 and an increase for IL-4, IL-5, IL-6, IL-8, IL-9, IL-10, TNFα and VEGF (Heard et al. 2013). Although not supported here, serum VEGF and IL-8 have been suggested to correlate with radiographic severity (Mabey et al. 2014). Also here extensive profiling studies should be carried out with more certainty pinpoint markers that can reliably distinguish subsets of patients with joint disease.

Acknowledgements The authors wish to thank Prof. G. van Osch, Department of Orthopaedics, Erasmus Medical Centre, for his critical review of the manuscript.

References

Arend WP, Welgus HG, Thompson RC, Eisenberg SP (1990) Biological properties of recombinant human monocyte-derived interleukin 1 receptor antagonist. J Clin Invest 85(5):1694–1697. doi:10.1172/JCI114622

Bakker AD, Kulkarni RN, Klein-Nulend J, Lems WF (2014) IL-6 alters osteocyte signaling toward osteoblasts but not osteoclasts. J Dent Res 93(4):394–399. doi:10.1177/0022034514522485

Bastiaansen-Jenniskens YM, Clockaerts S, Feijt C, Zuurmond A-M, Stojanovic-Susulic V, Bridts C, de Clerck L, DeGroot J, Verhaar JA, Kloppenburg M, van Osch GJVM (2012) Infrapatellar fat pad of patients with end-stage osteoarthritis inhibits catabolic mediators in cartilage. Ann Rheum Dis 71(2):288–294. doi:10.1136/ard.2011.153858

Beekhuizen M, Bastiaansen-Jenniskens YM, Koevoet W, Saris DBF, Dhert WJ, Creemers LB, van Osch GJVM (2011) Osteoarthritic synovial tissue inhibition of proteoglycan production in human osteoarthritic knee cartilage: establishment and characterization of a long-term cartilage-synovium coculture. Arthritis Rheum 63(7):1918–1927. doi:10.1002/art.30364

Beekhuizen M, Gierman LM, van Spil WE, Van Osch GJVM, Huizinga TWJ, Saris DBF, Creemers LB, Zuurmond A-M (2013a) An explorative study comparing levels of soluble mediators in control and osteoarthritic synovial fluid. Osteoarthritis Cartilage/OARS Osteoarthritis Res Soc 21(7):918–922. doi:10.1016/j.joca.2013.04.002

Beekhuizen M, van Osch GJVM, Bot AGJ, Hoekstra MCL, Saris DBF, Dhert WJ, Creemers LB (2013a) Inhibition of oncostatin M in osteoarthritic synovial fluid enhances GAG production in osteoarthritic cartilage repair. Eur Cell Mater 26:80–90; discussion 90. Retrieved from http://www.ncbi.nlm.nih.gov/pubmed/24027021

Berenbaum F (2013) Osteoarthritis as an inflammatory disease (osteoarthritis is not osteoarthrosis!). Osteoarthritis Cartilage 21(1):16–21. doi:10.1016/j.joca.2012.11.012

Bigoni M, Sacerdote P, Turati M, Franchi S, Gandolla M, Gaddi D, Moretti S, Munegato D, Augusti CA, Bresciani E, Omeljaniuk RJ, Locatelli V, Torsello A (2013) Acute and late changes in intraarticular cytokine levels following anterior cruciate ligament injury. J Orthop Res : Off Publ Orthop Res Soc 31(2):315–321. doi:10.1002/jor.22208

Bondeson J, Wainwright SD, Lauder S, Amos N, Hughes CE (2006) The role of synovial macrophages and macrophage-produced cytokines in driving aggrecanases, matrix metalloproteinases, and other destructive and inflammatory responses in osteoarthritis. Arthritis Res Ther 8(6):R187. doi:10.1186/ar2099

Cameron M, Buchgraber A, Passler H, Vogt M, Thonar E, Fu F, Evans CH (1997) The natural history of the anterior cruciate ligament-deficient knee. Changes in synovial fluid cytokine and keratan sulfate concentrations. Am J Sports Med 25(6):751–754. doi:10.1177/036354659702500605

Chevalier X, Goupille P, Beaulieu a D, Burch FX, Bensen WG, Conrozier T, Loeuille D, Kivitz AJ, Silver D, Appleton BE (2009). Intraarticular injection of anakinra in osteoarthritis of the knee: a multicenter, randomized, double-blind, placebo-controlled study. Arthritis Rheum 61(3):344–352. doi:10.1002/art.24096

Chevalier X, Ravaud P, Maheu E, Baron G, Rialland a, Vergnaud P, Roux C, Maugars Y, Mulleman D, Lukas C, Wendling D, Lafforgue P, Loeuille D, Foltz V, Richette P (2014). Adalimumab in patients with hand osteoarthritis refractory to analgesics and NSAIDs: a randomised, multicentre, double-blind, placebo-controlled trial. Ann Rheum Dis. doi:10.1136/annrheumdis-2014-205348

Clockaerts S, Bastiaansen-Jenniskens YM, Feijt C, De Clerck L, Verhaar JaN, Zuurmond a.-M, Stojanovic-Susulic V, Somville J, Kloppenburg M, van Osch GJVM (2012) Cytokine production by infrapatellar fat pad can be stimulated by interleukin 1 and inhibited by peroxisome proliferator activated receptor agonist. Ann Rheum Dis 71(6):1012–1018. doi:10.1136/annrheumdis-2011-200688

Cohen SB, Proudman S, Kivitz AJ, Burch FX, Donohue JP, Burstein D, Sun YN, Banfield C, Vincent MS, Ni L, Zack DJ (2011) A randomized, double-blind study of AMG 108 (a fully human monoclonal antibody to IL-1R1) in patients with osteoarthritis of the knee. Arthritis Res Ther 13(4):R125. doi:10.1186/ar3430

Cuellar VG, Cuellar JM, Golish SR, Yeomans DC, Scuderi GJ (2010) Cytokine profiling in acute anterior cruciate ligament injury. Arthroscopy 26(10):1296–1301. doi:10.1016/j.arthro.2010.02.011, S0749-8063 (10)00180-5 [pii]

Cuéllar VG, Cuéllar JM, Kirsch T, Strauss EJ (2015) Correlation of synovial fluid biomarkers with cartilage pathology and associated outcomes in knee arthroscopy. Artrosc: J Arthrosc Relat Surg 1–11. doi:10.1016/j.arthro.2015.08.033

de Lange-Brokaar BJE, Kloppenburg M, Andersen SN, Dorjée AL, Yusuf E, Herb-van Toorn L, Kroon HM, Zuurmond AM, Stojanovic-Susulic V, Bloem JL, Nelissen RG, Toes RE, Ioan-Facsinay A (2015) Characterization of synovial mast cells in knee osteoarthritis: association with clinical parameters. Osteoarthritis Cartilage 1–8. doi:10.1016/j.joca.2015.11.011

Deligne C, Casulli S, Pigenet A, Bougault C, Campillo-Gimenez L, Nourissat G, Berenbaum F, Elbim C, Houard X (2015) Differential expression of interleukin-17 and interleukin-22 in inflamed and non-inflamed synovium from osteoarthritis patients. Osteoarthritis Cartilage 23(11):1843–1852. doi:10.1016/j.joca.2014.12.007

Distel E, Cadoudal T, Durant S, Poignard A, Chevalier X, Benelli C (2009) The infrapatellar fat pad in knee osteoarthritis: an important source of interleukin-6 and its soluble receptor. Arthritis Rheum 60(11):3374–3377. doi:10.1002/art.24881

Durigova M, Roughley PJ, Mort JS (2008) Mechanism of proteoglycan aggregate degradation in cartilage stimulated with oncostatin M. Osteoarthr Cart/OARS Osteoarthritis Res Soc 16(1):98–104. doi:10.1016/j.joca.2007.05.002

Estes BT, Gimble JM, Guilak F (2004) Mechanical signals as regulators of stem cell fate. Curr Top Dev Biol 60:91–126

Fahy N, de Vries-van Melle ML, Lehmann J, Wei W, Grotenhuis N, Farrell E, van der Kraan PM, Murphy JM, Bastiaansen-Jenniskens YM, van Osch GJVM (2014) Human osteoarthritic synovium impacts chondrogenic differentiation of mesenchymal stem cells via macrophage polarisation state. Osteoarthritis Cartilage 22(8):1167–1175. doi:10.1016/j.joca.2014.05.021

Fang PK, Ma XC, Ma DL, Fu KY (1999) Determination of interleukin-1 receptor antagonist, interleukin-10, and transforming growth factor-beta1 in synovial fluid aspirates of patients with temporomandibular disorders. J Oral Maxillofac Surg 57(8):922–928, doi:S0278239199001056 [pii]

Favero M, Giusti A, Geusens P, Goldring SR, Lems W, Schett G, Bianchi G (2015) OsteoRheumatology: a new discipline? RMD Open 1(Suppl 1):e000083. doi:10.1136/rmdopen-2015-000083

Filková M, Vernerová Z, Hulejová H, Prajzlerová K, Veigl D, Pavelka K, Vencovský J, Šenolt L (2015) Pro-inflammatory effects of interleukin-35 in rheumatoid arthritis. Cytokine 73(1):36–43. doi:10.1016/j.cyto.2015.01.019

Flannery CR, Little CB, Hughes CE, Curtis CL, Caterson B, Jones SA (2000) IL-6 and its soluble receptor augment aggrecanase-mediated proteoglycan catabolism in articular cartilage. Matrix Biol 19(6):549–553. doi:10.1016/S0945-053X(00)00111-6

Fraser A, Fearon U, Billinghurst RC, Ionescu M, Reece R, Barwick T, Emery P, Poole AR, Veale DJ (2003) Turnover of type II collagen and aggrecan in cartilage matrix at the onset of inflammatory arthritis in humans: relationship to mediators of systemic and local inflammation. Arthritis Rheum 48(11):3085–3095. doi:10.1002/art.11331

Goldring SR, Goldring MB (2004) The role of cytokines in cartilage matrix degeneration in osteoarthritis. Clin Orthop Relat Res 427(427):S27–S36. doi:10.1097/01.blo.0000144854.66565.8f

Goldring MB, Otero M (2011) Inflammation in osteoarthritis. Curr Opin Rheumatol 23(5):471–478. doi:10.1097/BOR.0b013e328349c2b1

Gordeladze JO, Drevon CA, Syversen U, Reseland JE (2002) Leptin stimulates human osteoblastic cell proliferation, de novo collagen synthesis, and mineralization: impact on differentiation markers, apoptosis, and osteoclastic signaling. J Cell Biochem 85(4):825–836. doi:10.1002/jcb.10156

Granowitz EV, Clark BD, Mancilla J, Dinarello CA (1991) Interleukin-1 receptor antagonist competitively inhibits the binding of interleukin-1 to the type II interleukin-1 receptor. J Biol Chem 266(22):14147–14150

Grunke M, Schulze-Koops H (2006) Successful treatment of inflammatory knee osteoarthritis with tumour necrosis factor blockade. Ann Rheum Dis 65(4):555–556. doi:10.1136/ard.2006.053272

Guilak F (2011) Biomechanical factors in osteoarthritis. Best Pract Res: Clin Rheumatol 25(6):815–823. doi:10.1016/j.berh.2011.11.013

Hardcastle SA, Dieppe P, Gregson CL, Davey Smith G, Tobias JH (2015) Osteoarthritis and bone mineral density: are strong bones bad for joints? BoneKEy Rep 4:624. doi:10.1038/bonekey.2014.119

Heard BJ, Fritzler MJ, Preston Wiley J, McAllister J, Martin L, El-Gabalawy H, Hart DA, Frank CB, Krawetz R (2013) Intraarticular and systemic inflammatory profiles may identify patients with osteoarthritis. J Rheumatol 40(8):1379–1387. doi:10.3899/jrheum.121204

Henderson B, Pettipher ER (1989) Arthritogenic actions of recombinant IL-1 and tumour necrosis factor alpha in the rabbit: evidence for synergistic interactions between cytokines in vivo. Clin Exp Immunol 75:306–310

Henrotin YE, Zheng SX, Labasse AH, Deby GP, Crielaard JMR, Reginster JYL (2000) Modulation of human chondrocyte metabolism by recombinant human interferon. Osteoarthritis Cartilage 8(6):474–482. doi:10.1053/joca.1999.0323

Higuchi H, Shirakura K, Kimura M, Terauchi M, Shinozaki T, Watanabe H, Takagishi K (2006) Changes in biochemical parameters after anterior cruciate ligament injury. Int Orthop 30(1):43–47. doi:10.1007/s00264-005-0023-5

Houard X, Goldring MB, Berenbaum F (2013) Homeostatic mechanisms in articular cartilage and role of inflammation in osteoarthritis. Curr Rheumatol Rep 15(11):375. doi:10.1007/s11926-013-0375-6

Hussein MR, Fathi NA, El-Din AME, Hassan HI, Abdullah F, Al-Hakeem E, Backer EA (2008) Alterations of the CD4 (+), CD8 (+) T cell subsets, interleukins-1beta, IL-10, IL-17, tumor necrosis factor-alpha and soluble intercellular adhesion molecule-1 in rheumatoid arthritis and osteoarthritis: preliminary observations. Pathol Oncol Res: POR 14(3):321–328. doi:10.1007/s12253-008-9016-1

Irie K, Uchiyama E, Iwaso H (2003) Intraarticular inflammatory cytokines in acute anterior cruciate ligament injured knee. Knee 10(1):93–96. doi:10.1016/S0968-0160(02)00083-2

Kahle P, Saal JG, Schaudt K, Zacher J, Fritz P, Pawelec G (1992) Determination of cytokines in synovial fluids: correlation with diagnosis and histomorphological characteristics of synovial tissue. Ann Rheum Dis 51(6):731–734. doi:10.1136/ard.51.6.731

Kaneyama K, Segami N, Nishimura M, Suzuki T, Sato J (2002) Importance of proinflammatory cytokines in synovial fluid from 121 joints with temporomandibular disorders. Br J Oral Maxillofac Surg 40(5):418–423. doi:10.1016/S0266-4356(02)00215-2

Kaneyama K, Segami N, Nishimura M, Sato J, Suzuki T, Fujimura K (2003) Osteoclastogenesis inhibitory factor/osteoprotegerin in synovial fluid from patients with temporomandibular disorders. Int J Oral Maxillofac Surg 32(4):404–407. doi:10.1054/ijom.2002.0363

Kaneyama K, Segami N, Sun W, Sato J, Fujimura K (2005) Analysis of tumor necrosis factor-α, interleukin-6, interleukin-1β, soluble tumor necrosis factor receptors I and II, interleukin-6 soluble receptor, interleukin-1 soluble receptor type II, interleukin-1 receptor antagonist, and protein in the synovial fl. Oral Surg Oral Med Oral Pathol Oral Radiol Endodontol 99(3):276–284. doi:10.1016/j.tripleo.2004.06.074

Kc R, Li X, Kroin JS, Liu Z, Chen D, Xiao G, Levine B, Li J, Hamilton JL, van Wijnen AJ, Piel M, Shelly DA, Brass D, Kolb E, Im H (2016) PKC δ null mutations in a mouse model of osteoarthritis alter osteoarthritic pain independently of joint pathology by augmenting NGF/TrkA-induced axonal outgrowth. Ann Rheum Dis 1–9. doi:10.1136/annrheumdis-2015-208444

Klein-Wieringa IR, Kloppenburg M, Bastiaansen-Jenniskens YM, Yusuf E, Kwekkeboom JC, El-Bannoudi H, Nelissen RG, Zuurmond A, Stojanovic-Susulic V, Van Osch GJ, Toes RE, Ioan-Facsinay a (2011) The infrapatellar fat pad of patients with osteoarthritis has an inflammatory phenotype. Ann Rheum Dis 7(5):851–857. doi:10.1136/ard.2010.140046

Koshy PJ, Henderson N, Logan C, Life PF, Cawston TE, Rowan AD (2002a) Interleukin 17 induces cartilage collagen breakdown: novel synergistic effects in combination with proinflammatory cytokines. Ann Rheum Dis 61(8):704–713. doi:10.1136/ard.61.8.704

Koshy PJT, Lundy CJ, Rowan AD, Porter S, Edwards DR, Hogan A, Clark IM, Cawston TE (2002b) The modulation of matrix metalloproteinase and ADAM gene expression in human chondrocytes by interleukin-1 and oncostatin M: a time-course study using real-time quantitative reverse transcription-polymerase chain reaction. Arthritis Rheum 46(4):961–967. doi:10.1002/art.10212

Kubota E, Kubota T, Matsumoto J, Shibata T, Murakami K-II (1998) Synovial fluid cytokines and proteinases as markers of temporomandibular joint disease. J Oral Maxillofac Surg 56(2):192–198. doi:10.1016/S0278-2391(98)90868-0

Laiguillon M.-C, Houard X, Bougault C, Gosset M, Nourissat G, Sautet A, Jacques C, Berenbaum F, Sellam J (2014) Expression and function of visfatin (Nampt), an adipokine-enzyme involved in inflammatory pathways of osteoarthritis. Arthritis Res Ther 16(1):R38. doi:10.1186/ar4467

Larsson S, Englund M, Struglics A, Lohmander LS (2015) Interleukin-6 and tumor necrosis factor alpha in synovial fluid are associated with progression of radiographic knee osteoarthritis in subjects with previous meniscectomy. Osteoarthritis Cartilage 23(11):1906–1914. doi:10.1016/j.joca.2015.05.035

Leyh M, Seitz A, Dürselen L, Schaumburger J, Ignatius A, Grifka J, Grässel S (2014) Subchondral bone influences chondrogenic differentiation and collagen production of human bone marrow-derived mesenchymal stem cells and articular chondrocytes. Arthritis Res Ther 16(5):453. doi:10.1186/s13075-014-0453-9

Liles W, Van Voorhis W (1995) Review : nomenclature and biologic significance of cytokines involved in inflammation and the host immune response. Author(s): W. Conrad Liles and Wesley C, Van Voorhis Published by : Oxford University Press Stable URL : http://www.jstor.org/stable/30. J Infect Dis Infect Dis 172(6):1573–1580

Long D, Blake S, Song X-Y, Lark M, Loeser RF (2008) Human articular chondrocytes produce IL-7 and respond to IL-7 with increased production of matrix metalloproteinase-13. Arthritis Res Ther 10(1):R23. doi:10.1186/ar2376

Mabey T, Honsawek S, Saetan N, Poovorawan Y, Tanavalee A, Yuktanandana P (2014) Angiogenic cytokine expression profiles in plasma and synovial fluid of primary knee osteoarthritis. Int Orthop 38(9):1885–1892. doi:10.1007/s00264-014-2406-y

Maeda A, Bandow K, Kusuyama J, Kakimoto K, Ohnishi T, Miyawaki S, Matsuguchi T (2015) Induction of CXCL2 and CCL2 by pressure force requires IL-1β-MyD88 axis in osteoblasts. Bone 74:76–82. doi:10.1016/j.bone.2015.01.007

Marks PH, Donaldson MLC (2005) Inflammatory cytokine profiles associated with chondral damage in the anterior cruciate ligament-deficient knee. Artrosc: J Arthrosc Relat Surg: Off Publ Arthrosc Assoc N Am Int Arthrosc Assoc 21(11):1342–1347. doi:10.1016/j.arthro.2005.08.034

Neidel J, Blum WF, Schaeffer HJ, Schulze M, Schönau E, Lindschau J, Föll J (1997) Elevated levels of insulin-like growth factor (IGF) binding protein-3 in rheumatoid arthritis synovial fluid inhibit stimulation by IGF-I of articular chondrocyte proteoglycan synthesis. Rheumatol Int 17(1):29–37, Retrieved from http://www.ncbi.nlm.nih.gov/pubmed/9194211

Pascual E, Doherty M (2009) Aspiration of normal or asymptomatic pathological joints for diagnosis and research: indications, technique and success rate. Ann Rheum Dis 68(1):3–7. doi:10.1136/ard.2008.088328

Rahmati M, Mobasheri A, Mozafari M (2016) Inflammatory mediators in osteoarthritis: a critical review of the state-of-the art, prospects, and future challenges. Bone. doi:10.1016/j.bone.2016.01.019

Ribbens C, Andre B, Kaye O, Kaiser M, Bonnet V, Jaspar J, de Groote D, Franchimont N, Malaise MG (2000) Synovial fluid matrix metalloproteinase-3 levels are increased in inflammatory arthritides whether erosive or not.. Rheumatology 39:1357–1365.

Richette P, François M, Vicaut E, Fitting C, Bardin T, Corvol M, Savouret J (2008) A high interleukin 1 receptor antagonist/IL-1 ß ratio occurs naturally in knee osteoarthritis. J Rheumatol 35(8):1650–1654

Rowan AD, Koshy PJT, Shingleton WD, Degnan BA, Heath JK, Vernallis AB, Spaull JR, Life PF, Hudson K, Cawston TE (2001) Synergistic effects of glycoprotein 130 binding cytokines in combination with interleukin-1 on cartilage collagen breakdown. Arthritis Rheum 44(7):1620–1632. doi:10.1002/1529-0131. (200107)44:7<1620::AID-ART285>3.0.CO;2-B

Rübenhagen R, Schüttrumpf JP, Stürmer KM, Frosch K-H (2012) Interleukin-7 levels in synovial fluid increase with age and MMP-1 levels decrease with progression of osteoarthritis. Acta Orthop 83(1):59–64. doi:10.3109/17453674.2011.645195

Rutgers M, Saris DB, Dhert WJ, Creemers LB (2010) Cytokine profile of autologous conditioned serum for treatment of osteoarthritis, in vitro effects on cartilage metabolism and intra-articular levels after injection. Arthritis Res Ther 12(6):R114. doi:10.1186/ar3192

Sauerschnig M, Stolberg-Stolberg J, Schulze A, Salzmann G, Perka C, Dynybil C (2014) Diverse expression of selected cytokines and proteinases in synovial fluid obtained from osteoarthritic and healthy human knee joints. Eur J Med Res 19(1):65. doi:10.1186/s40001-014-0065-5

Scanzello CR, Umoh E, Pessler F, Diaz-Torne C, Miles T, DiCarlo E, Potter HG, Mandl L, Marx R, Rodeo S, Goldring SR, Crow MK (2009). Local cytokine profiles in knee osteoarthritis: elevated synovial fluid interleukin-15 differentiates early from end-stage disease. Osteoarthritis Cartilage 17(8):1040–1048. doi:10.1016/j.joca.2009.02.011

Schett G, Elewaut D, Mcinnes IB, Dayer J, Neurath MF (2013) How cytokine networks fuel inflammation. Nat Med 19(7):822–826. doi:10.1038/nm.3260

Simao AP, Almeida TMDO, Mendona VA, Santos SA, Gomes WF, Coimbra CC, Lacerda ACR (2014) Soluble TNF receptors are produced at sites of inflammation and are inversely associated with self-reported symptoms (WOMAC) in knee osteoarthritis. Rheumatol Int 34(12):1759–1763. doi:10.1007/s00296-014-3016-0

Stove J, Huch K, Günther KP, Scharf HP (2000) Interleukin-1b induces different gene expression of stromelysin, aggrecan and tumor-necrosis-factor-stimulated gene 6 in human osteoarthritic chondrocytes in vitro. Pathobiology 68(3):144–149. doi:10.1159/000055915

Teeple E, Jay GD, Elsaid KA, Fleming BC (2013) Animal models of osteoarthritis: challenges of model selection and analysis. AAPS J 15(2):438–446. doi:10.1208/s12248-013-9454-x

Tetlow LC, Adlam DJ, Woolley DE (2001) Matrix metalloproteinase and proinflammatory cytokine production by chondrocytes of human osteoarthritic cartilage: associations with degenerative changes. Arthritis Rheum 44(3):585–594. doi:10.1002/1529-0131,(200103)44:3<585::AID-ANR107>3.0.CO;2-C

Teunis T, Beekhuizen M, Kon M, Creemers LB, Schuurman AH, van Minnen LP (2013) Inflammatory mediators in posttraumatic radiocarpal osteoarthritis. J Hand Surg 38(9):1735–1740. doi:10.1016/j.jhsa.2013.06.023

Tsuchida AI, Beekhuizen M, Rutgers M, van Osch GJ, Bekkers JE, Bot AG, Geurts B, Dhert WJ, Saris DB, Creemers LB (2012). Interleukin-6 is elevated in synovial fluid of patients with focal cartilage defects and stimulates cartilage matrix production in an in vitro regeneration model. Arthritis Res Ther 14(6):R262. doi:10.1186/ar4107

Tsuchida AI, Beekhuizen M, Hart MC, Radstake TRDJ, Dhert WJA, Saris DBF, van Osch GJ, Creemers LB (2014) Cytokine profiles in the joint depend on pathology, but are different between synovial fluid, cartilage tissue and cultured chondrocytes. Arthritis Res Ther 1–15. doi:10.1186/s13075-014-0441-0

Turner MD, Nedjai B, Hurst T, Pennington DJ (2014) Cytokines and Chemokines: at the crossroads of cell signalling and inflammatory disease. Biochim Biophys Acta Mol Cell Res 1843(11):2563–2582. doi:10.1016/j.bbamcr.2014.05.014

Ushiyama T, Chano T, Inoue K, Matsusue Y (2003) Cytokine production in the infrapatellar fat pad: another source of cytokines in knee synovial fluids. Ann Rheum Dis 62(2):108–112. doi:10.1136/ard.62.2.108

Vervoordeldonk M, Tak P (2002) Cytokines in rheumatoid arthritis. Curr Rheumatol Rep 4:208–217. doi:10.1007/BF02215851

Wei W, Rudjito R, Fahy N, Verhaar JAN, Clockaerts S, Bastiaansen-Jenniskens YM, van Osch GJVM (2015) The infrapatellar fat pad from diseased joints inhibits chondrogenesis of mesenchymal stem cells. Eur Cells Mater 30:303–314. doi:10.1016/j.joca.2015.02.889

Molecular Genetics of the Cartilage Collagenopathies

6

Shireen R. Lamandé, Trevor L. Cameron, Ravi Savarirayan, and John F. Bateman

Abstract

The cartilage extracellular matrix is rich in fibrillar- and fibril-associated collagens, and mutations in these collagen genes cause a wide range of chondrodysplasias, ranging from premature arthritis through to severe early lethal disorders. Collagen mutations can interfere with matrix organisation and cause cartilage dysfunction by reducing synthesis of structurally normal protein, or through protein misfolding which leads to intracellular retention and degradation, and consequent secretion of reduced amounts of structurally abnormal protein. In addition, collagen misfolding mutations can induce a cellular unfolded protein response which can alter chondrocyte differentiation and ultimately promote

S.R. Lamandé, PhD (✉)
Murdoch Childrens Research Institute, Royal Children's Hospital, Parkville 3052, Australia

Department of Paediatrics, University of Melbourne, Parkville 3052, Australia
e-mail: shireen.lamande@mcri.edu.au

T.L. Cameron, PhD
Murdoch Childrens Research Institute, Royal Children's Hospital, Parkville 3052, Australia
e-mail: trevor.cameron@mcri.edu.au

R. Savarirayan, MD
Murdoch Childrens Research Institute, Royal Children's Hospital, Parkville 3052, Australia

Department of Paediatrics, University of Melbourne, Parkville 3052, Australia

Victorian Clinical Genetics Service, Melbourne, Australia
e-mail: ravi.savarirayan@mcri.edu.au

J.F. Bateman, PhD
Murdoch Childrens Research Institute, Royal Children's Hospital, Parkville 3052, Australia

Department of Biochemistry and Molecular Biology, University of Melbourne, Melbourne, Parkville 3052, Australia
e-mail: john.bateman@mcri.edu.au

S. Grässel, A. Aszódi (eds.), *Cartilage*, DOI 10.1007/978-3-319-45803-8_6

apoptosis and thus contribute to the pathology. The relative impact of intracellular and extracellular consequences is likely mutation and collagen type specific. Manipulating protein degradation pathways and unfolded protein response signalling offers hope for treating genetic chondrodysplasias.

6.1 Introduction

Cartilage is a heterogeneous tissue comprised of chondrocytes embedded within a structurally complex extracellular matrix rich in heterotypic fibrillar collagens and proteoglycans. In the adult, three types of cartilage occur – elastic cartilage such as that of the pinna, fibrocartilage as occurs in intervertebral discs and hyaline cartilage such as the articular cartilage lining diarthrodial joints. Though permanent and specialised for its role in facilitating joint movement and resisting compressive load, articular hyaline cartilage in the adult is developmentally continuous with the temporary hyaline cartilage that first appears in embryogenesis during endochondral ossification, but which is ultimately degraded and replaced by mineralized bone by the end of adolescence. For brevity, we will limit our discussion here to hyaline cartilage since the symptoms of cartilage collagenopathies are most profoundly manifested in these tissues. This article will review our current knowledge of the genetic basis of human chondrodysplasias, disorders that affect cartilage development, differentiation and function, caused by collagen mutations.

6.2 Cartilage Development and Endochondral Ossification

Since chondrodysplasia mutations affect most aspects of the complex pathway of cartilage and endochondral bone development, as well as the structural integrity of mature cartilage, we will provide a brief overview of cartilage development to provide context for understanding pathological pathways. More detailed descriptions of cartilage development are found in chapters of volume 1. Long bone formation in the developing limb constitutes a classic model of endochondral ossification. The first evidence of endochondral ossification in the developing limb is the appearance of mesenchymal condensations. These result from mesenchymal cells migrating into the embryonic limb bud from the lateral plate mesoderm and then aggregating at sites of future bone growth (Eames et al. 2003; Hall and Miyake 1995). These processes are precisely regulated by spatiotemporally dependent gene expression in response to a combination of inputs that include cell matrix interactions, growth factors and non-coding RNAs. Cells within the condensations express cell-cell adhesion molecules such as N-cadherin and NCAM (Oberlender and Tuan 1994; Widelitz et al. 1993), extracellular matrix molecules such as versican (Kamiya et al. 2006), transcription factors including the master regulator of chondrogenesis, SOX9 (Akiyama et al. 2002; Bi et al. 1999) and growth factor receptors including FGFR2 (Ornitz and Marie 2015) and BMPR1B (Lehmann et al. 2003; Baur et al. 2000).

Accordingly, mesenchymal condensations are responsive to the complex interplay between several signalling pathways, including the FGF (Ornitz and Marie 2015), GDF5 (Buxton et al. 2001; Storm et al. 1994) and BMP-SMAD (Wang et al. 2016; Lim et al. 2015) pathways, the latter of which may act in a SOX9-dependent or SOX9-independent manner, as well as non-coding RNAs such as LncRNA-HIT, which appears to promote the expression of BMPR1B (Carlson et al. 2015).

The onset of chondrogenic differentiation within mesenchymal condensations is marked by changes in the morphology and arrangement of cells within the developing anlage. Whereas condensed mesenchymal cells are relatively small, rounded and tightly packed, chondrocytes are comparatively larger and characterised by a classic cobblestone appearance with separation between adjacent cells. This separation results from a second major hallmark of chondrogenic differentiation, namely, the dramatic upregulation of genes encoding critical components of the cartilage extracellular matrix, including aggrecan, type II collagen, COMP, link protein and matrilin within the early anlage (Cameron et al. 2009). Thus chondrocytes are responsible for secreting the extracellular matrix within which they become embedded and with which they interact during subsequent differentiation steps. Surrounding the cartilaginous anlage is a dense perimeter of connective tissue known as the perichondrium, which is comprised of flattened cells that derive from the same mesenchymal population as the chondrocytes of the anlage itself, but which subsequently differentiate into the osteoblasts of the periosteum under the control of canonical WNT signalling (Day et al. 2005). With further maturation, chondrocytes at the centre of the developing anlage undergo hypertrophy, expressing markers of hypertrophic cartilage such as type X collagen (Kielty et al. 1985); RUNX2, which drives expression of COL10A1 (Zheng et al. 2003); matrix metalloproteinases (MMPs) – including MMP9 and MMP13 that facilitate remodelling of the matrix (Stickens et al. 2004); and VEGF, which promotes vascular invasion of the anlage at the site of hypertrophy (Gerber et al. 1999). Ultimately, hypertrophic chondrocytes undergo either programmed cell death or transdifferentiation to form osteoblasts (Tsang et al. 2015), as more osteoblasts enter the area from the perichondrium (Colnot et al. 2004) to continue the replacement of the hypertrophic cartilage matrix by mineralized bone, thus forming the primary centre of ossification.

This process of chondrocyte differentiation, hypertrophy, cartilage matrix remodelling, degradation and replacement by bone continues longitudinally in a bidirectional manner at the growth plates, which separate the cartilaginous epiphyses of growing long bones from the bony diaphysis. Accordingly, growth plate chondrocytes appear to be stratified into discrete resting, proliferative and hypertrophic zones corresponding to different stages of their maturation and characterised by unique cell morphology and gene expression signatures (Lui et al. 2010; Belluoccio et al. 2008). The resting zone houses small, round, irregularly arranged chondrocytes that will give rise to proliferative chondrocytes and is involved in directing overall growth plate orientation (Abad et al. 2002). In the proliferative zone, chondrocytes orientate under the control of β1 integrin in distinct columns of flattened cells as they promote longitudinal bone growth through rapid mitotic proliferation

(Aszodi et al. 2003), the extent of which is regulated by a feedback loop involving PTHrP expressed by cells of the resting zone and perichondrium and IHH produced by postmitotic, hypertrophic chondrocytes (Kronenberg 2006). FGF signalling is also critical in regulating the rate of growth plate chondrocyte proliferation and is dependent on the ligands FGF9 and FGF18 supplied by the perichondrium and periosteum and FGFR3 expressed in pre-hypertrophic growth plate cartilage (Ornitz and Marie 2015). As with hypertrophic chondrocytes generated during the formation of the presumptive cartilage anlage, growth plate hypertrophic chondrocytes orchestrate the ongoing remodelling and mineralisation of the growth plate through the secretion of matrix metalloproteinases (Stickens et al. 2004) and its vascular invasion through the secretion of proangiogenic factors (Gerber et al. 1999). Subsequent to the formation of the primary ossification centres, a secondary centre of ossification also forms within each cartilaginous epiphysis of growing long bones through essentially the same process of chondrocyte hypertrophy, cartilage remodelling and mineralisation that takes place at the primary centres of ossification.

6.3 Classificaton of Skeletal Dysplasias

The skeletal dysplasias (inherited disorders of bone and cartilage) are a heterogeneous collection of disorders caused by abnormalities in the development, growth and maintenance of the human skeleton. The current nosologic classification of these disorders (Bonafe et al. 2015) reflects their diverse manifestations, with over 450 conditions listed in 42 separate groups, some based on common molecular aetiology (i.e. type II collagen group), others on their predominant radiographic features (i.e. slender bone group) and others on their most striking clinical features (i.e. dysplasias with multiple joint dislocations) (see Table 6.1 for an abbreviated nosology). Skeletal dysplasias can present in myriad forms, ranging from lethal conditions in utero to late early adult-onset manifestations involving the spine and joints. Recently, with advances in genomic technologies, many of the genetic causes underlying the skeletal dysplasias have been elucidated, revealing the multiple pathogenetic mechanisms that can lead to skeletal defects.

6.3.1 Defects in Local Regulation of Cartilage Growth

Defects in local regulation of cartilage growth include disorders caused by abnormalities in growth factors and their receptors. Fibroblast growth factor receptor 3 plays a key role in the negative regulation of bone growth by inhibiting cell growth in cartilaginous growth plates. Almost all cases of achondroplasia (the most common form of human dwarfism) are due to one of two specific gain-of-function mutations in the *FGFR3* gene, which result in upregulation of the *FGFR3* pathway. The lethal dysplasia, thanatophoric dysplasia, is due to different mutations in the *FGFR3* gene.

Table 6.1 Abbreviated skeletal dysplasia nosology (for details see Bonafe et al. (2015))

Group/disorder	Gene/s
FGFR3 chondrodysplasia group	*FGFR3*
Type 2 collagen group	*COL2A1*
Type 11 collagen group	*COL11A1, COL11A2*
Sulphation disorders group	*DTDST, PAPSS2, IMPAD1, CHST3, CHST14*
Perlecan group	*HSPG2*
Aggrecan group	*AGC1*
Filamin group and related disorders	*FLNA, FLNB, SH3PXD2B*
TRPV4 group	*TRPV4*
Ciliopathies with major skeletal involvement	*EVC1, EVC2, DYNC2H1, WDR34, DYN2H1, IFT80, TTC21B, WDR19, IFT172, IFT140, NEK1, WDR35, TCTN3, IFT122, IFT43*
Multiple epiphyseal dysplasia and pseudoachondroplasia group	*COMP, COL9A1, COL9A2, COL9A3, MATN3*
Metaphyseal dysplasias	*COL10A1, RMRP, PDP1, PTHR1, SBDS, MMP13, MMP9, RUNX2*
Spondylometaphyseal dysplasias	*ACP5, PCY1A*
Spondylo-epi-(meta)physeal dysplasias	*DYM, RAB33B, SMARCAL1, EIF2AK3, MATN3, DDR2, SEDL, SLC39A13, LONP1*
Severe spondylodysplastic dysplasias	*TRIP11, SLC35D1, GPX4, SBDS, INPPL1, MAGMAS*
Acromelic dysplasias	*TRPS1, EXT1, IHH, ADAMTSL2, FBN1, ADAMTS10, ADAMTS17, LTBP2, SMAD4, PDE4D, PRKAR1A, GNAS*
Acromesomelic dysplasias	*NPR2, GDF5, BMPR1B, SHOX, GPC6, FZD2, ROR2, WNT5A, DVL1, SULF1, SLC05A1*
Campomelic dysplasia and related disorders	*SOX9, LIFR*
Slender bone dysplasia group	*CUL7, OBSL1, CCDC8, TBCE, FAM111A, RNU4ATAC, PCNT2, CDKN1C*
Dysplasias with multiple joint dislocations	*CANT1, XYLT1, KIF22, B3GALT6*
Chondrodysplasia punctata group	*EBP, ARSE, NSDHL, MGP, LBR, PEX7, DHPAT, AGPS*
Neonatal osteosclerotic dysplasias	*PTHR1, DHCR24, COL1A1, FAM20C*
Osteopetrosis and related disorders	*TCIRG1, CLCN7, SNX10, DSTM1, RANKL, PLEKHM1, CA2, LRP5, IKBKG, FERMT3, RASGRP2, CTSK, LEMD3, WTX, SLC29A3*
Other sclerosing bone disorders	*ANKH, TGFB1, TBXAS1, HPGD, GJA1, OPG, SOST, LRP4, DLX3, PTDSS1*
Osteogenesis imperfecta and decreased bone density group	*COL1A1, COL1A2, CRTAP, LEPRE1, PPIB, SERPINH1, BMP1, FKBP10, PLOD2, SP7, WNT1, TMEM38B, SEC24D, IFTIM5, PLS3, LRP5, P4HB, XYLT2, B4GALT7, GORAB, PYCR1, ATP6VOA2*

(continued)

Table 6.1 (continued)

Group/disorder	Gene/s
Abnormal mineralisation group	*ALPL, PHEX, FGF23, DMP1, ENPP!, CICN5, SLC34A3, CASR, ANKH*
Lysosomal storage diseases with skeletal involvement	*IDA, IDS, HSS, NAGLU, HSGNAT, GNS, GALNS, GLBI, ARSB, GUSB, FUCA, MANA, MANB, AGA, NEU1, SLC17A5, PPGB, SUMF1, GNPTAB, GNPTG*
Osteolysis group	*RANK, LMNA, ZMPSTE24, MMP2, NOTCH2, MAFB*
Disorganised development of skeletal components group	*EXT1, EXT2, SH3BP2, GNAS, TMEM16E, PTPN11, FGFR1, ACVR1, NF1, TREM2, TYROBP, IDH1, IDH2*
Overgrowth syndromes with skeletal involvement	*EZH2, NSD1, SETD2, NFIX, AKT1, PIK3CA, FBN1, FBN2, TGFBR1, TGFBR2, SMAD3, TGFB2, NPPC, NPR2*
Genetic inflammatory/rheumatoid-like osteoarthropathies	*WISP3, CIAS1, IL1RN, LPIN2, GALNT3, ANTXR2*
Cleidocranial dysplasia and related disorders	*RUNX2*, Fig. 6.4, *ALX4, MSX2*
Craniosynostosis syndromes	*FGFR1, FGFR2, FGFR3, POR, MSX2, TWIST1, SKI, RECOL4, RAB23, MEGF8, TCF12, ERF*
Dysostoses with predominant craniofacial involvement	*TCOF1, POLR1D, POLR1C, CXORF5, EVC1, EVC2, ICK, EFNB1, ALX3, ALX4, ALX1, DHODH, SF3B4, EFTUD2*
Dysostoses with predominant vertebral with and without costal involvement	*HLXB9, DLL3, MESP2, LFNG, HES7, TBX6, GDF6, GDF3, MEOX1, SNRBP, COG1, BMPER, NKX3-2*
Patellar dysostoses	*TBX4, LMX1B, KAT6B, ORC1, ORC4, ORC6, CDT1, CDC6*
Brachydactylies (without extraskeletal manifestations)	*IHH, BMPR1B, BMP2, GDF5, ROR2, NOG, HOXD13, PTHLH, SOX9*
Brachydactylies (with extraskeletal manifestations)	*HDAC4, PIGV, PED3A, MYCN, HOXA13, CREBBP, EP300, CHSY1, ARHGAP31, DOCK6, RBPJ, EOGT, TGDS*
Limb hypoplasia-reduction defects group	*TBX3, NIPBL, SMCA1, SMC3, RAD21, HDAC8, RMB8A, THPO, TBX5, SALL4, TBX15, ESCO2, BHLHA9, SHH-ZRS, LMBR1, WNT3, WNT7A, RECOL4*
Ectrodactyly with and without other manifestations	*P63, CDH3, DLX5, DLX6, WNT10B, FGFR1*
Polydactyly-syndactyly-triphalangism group	*SHH-ZRS, GLI3, FBLN1, HOXD13, SALL1, FGFR2, FGFR3, FGF10, KIF7, WNT6, LRP4, GREM1, FMN1, BHLHA9, FAM58A, IHH, GJA1, FGF16, CKAP2L, MKS1, TMEM216, TMEM67, CEP290, RPGRIP1L, CC2D2A*
Defects in joint formation and synostoses	*FGF9, NOG, GDF5, HOXA11, PITX1*

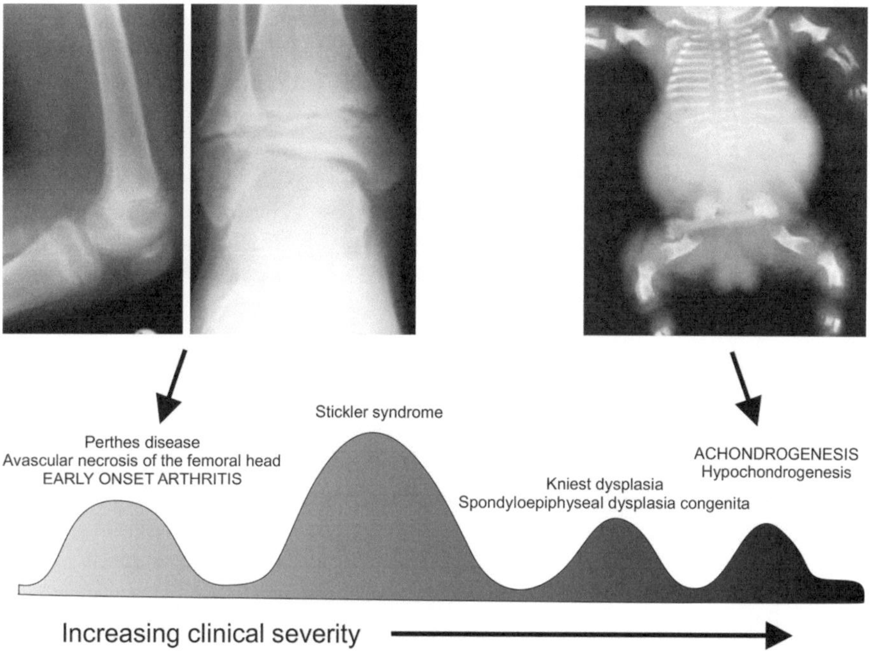

Fig. 6.1 Spectrum of type II collagen disorders. The diagram conceptualises phenotypic "islands" that vary in severity. At the peak of each island is the classic phenotype, with areas between the islands representing "bridging" phenotypes. The X-rays show examples of early-onset arthritis (*left panels*) and achondrogenesis (*right panel*) (Modified from (Kannu et al. 2010))

6.3.2 Defects in Structural Proteins of Cartilage

Defects in structural cartilage proteins such as collagen types I, II, IX, X and XI and extracellular matrix proteins such as COMP (cartilage oligomeric matrix protein) result in various forms of skeletal dysplasias that can be present at varying ages and with varying severity. Great examples of this are mutations in the gene encoding collagen type II that cause the group of disorders known collectively as the "type II collagenopathies", which comprises achondrogenesis type II, hypochondrogenesis, spondyloepiphyseal dysplasia congenita (SEDC), Kniest dysplasia, Stickler syndrome and early-onset arthritis (Kannu et al. 2010) (Fig. 6.1).

6.3.3 Defects in Cartilage Metabolic Pathways

Defects of enzymes, ion channels and transporters that are essential for cartilage metabolism and homeostasis have been identified as the cause of other skeletal dysplasias. An example of this is the *TRPV4* gene (transient receptor potential cation channel subfamily V, member 4), which encodes a calcium-permeable ion channel.

Dominant mutations in this gene result in activation of the channel leading to increased concentration of calcium within chondrocytes (Hurd et al. 2015). *TRPV4* mutations are the cause of a number of skeletal dysplasias including lethal and non-lethal metatropic dysplasia, which presents prenatally or in the neonatal period, and spondylometaphyseal dysplasia Kozlowski type, which presents in childhood (Andreucci et al. 2011; Kannu et al. 2007). The variable severity of clinical features seen in the TRPV4 groups of skeletal dysplasias can be correlated with the degree of activation of the channel (Loukin et al. 2011). Interestingly other mutations (likely diminishing function) in the *TRPV4* gene can present with an arthritic phenotype predominately involving the hands and feet (Lamande et al. 2011).

6.3.4 Diagnosis

Making a diagnosis of a specific skeletal dysplasia is often difficult given the vast number of recognised types. The involvement of physicians with specific skills in skeletal dysplasias, such as clinical geneticists, orthopaedic specialists and paediatric radiologists, is essential when trying to make an accurate diagnosis in the neonatal period and for subsequent management and monitoring.

Based on clinical history and examination, family history and expert evaluation of radiology (the single most powerful diagnostic tool), it is usually possible to make a diagnosis in the majority of those who present in childhood with a skeletal dysplasia. Molecular testing can provide confirmation of a clinical diagnosis. In some cases a skeletal dysplasia will be unclassifiable in the neonatal period but with time the development of additional clinical features and patterns may allow diagnosis. The advent of new diagnostic tools, such as exome sequencing and extensive skeletal gene panels, has greatly increased the power to detect the genetic basis of skeletal dysplasias.

6.3.5 Management

The management of skeletal dysplasias requires good understanding of the natural history and variability of the condition, a multidisciplinary approach and consideration of the psychosocial impacts of conditions that can alter appearance and height. Many skeletal dysplasias have nonskeletal manifestations (i.e. ocular features in type II collagen disorders), which require relevant subspecialty input. Ideally, patients and families should be managed in a centre with such experience and resources to maximise benefits.

6.4 Cartilage Components and ECMopathies

Cartilage extracellular matrix (ECM) production, architecture and homeostasis is orchestrated by chondrocytes, and it plays a fundamental role in cartilage morphogenesis, development and endochondral bone formation by regulating cell differentiation, proliferation, adhesion and migration. In the adult, the ECM of the articular cartilage endows to the articulating surfaces the key biomechanical properties required for diarthrodial joint function. While the complete landscape of critical ECM functional structures is not fully defined, it is clear that the correct temporal and spatial expression of ECM proteins, and the formation of architecturally precise interacting networks with unique functional and biological characteristics, is fundamental. The predominant cartilage ECM components are several members of the collagen family, proteoglycans and many noncollagenous proteins (Heinegard 2009). More detailed descriptions of cartilage ECM are provided in chapters 1, 2, and 3 in volume 1 (Aspberg 2016; Grässel 2016; Zaucke 2016).

The main collagenous component of cartilage is collagen II which is extensively covalently cross-linked in the characteristic fibrillar networks. These structural networks are critical for cartilage form and biomechanical function. Collagen II copolymerises with two minor cartilage collagens, types XI and IX, to form the mature heteropolymeric cartilage collagen fibrils (Eyre 2004; Eyre et al. 2002). Collagen XI molecules are localised in the interior of the collagen II fibril and act as a template for fibrillogenesis. Collagen XI molecules are cross-linked head to tail to other collagen XI molecules and laterally to the surrounding collagen II. Collagen IX is found on the surface of the fibril and is also cross-linked to collagen II. Unlike fibrillar collagens, II and XI, the collagen IX triple helix contains noncollagenous interruptions that allow its NC4 domain to project away from the lineal fibril and thus interact with the adjacent ECM. Collagen IX also has a post-translationally added glycosaminoglycan side chain, providing additional interaction sites for other ECM components. Collagen VI is also present in cartilage and its microfibrillar network is particularly rich in pericellular regions. Chondrodysplasias caused by mutations in the different cartilage collagen molecules are discussed in detail below.

In addition to the heteropolymeric collagen fibrils, a crucial component of the cartilage matrix is the large proteoglycan aggrecan. The aggrecan protein core is richly decorated with sulfated glycosaminoglycan side chains and large numbers of aggrecan molecules are linked via link protein interactions to hyaluronan to produce huge multimolecular aggregates. These aggrecan-hyaluron aggregates are trapped within the fibrillar collagen network, and because they are highly negatively charged, they have the potential to bind large amounts of water, causing the ECM to swell and thus sustain the high compressive loads that are imposed on cartilage during joint function (Fosang and Beier 2011). The integrity of the collagen fibril-aggrecan network is the key to cartilage function, and many noncollagenous proteins interact with or serve as adaptors in ECM protein interactions that are important in establishing network architecture or in maintaining its structure. Not unexpectedly, because of its key role in cartilage function, mutations in aggrecan cause several chondrodysplasias, spondyloepiphyseal dysplasia Kimberley type (Gleghorn et al. 2005), spondyloepimetaphyseal dysplasia aggrecan type (Tompson et al. 2009), osteochondritis dissecans (Stattin et al. 2010) and idiopathic short stature (Nilsson et al. 2014) (Table 6.2). Mutations in another large proteoglycan, perlecan, which is found in many basement membranes throughout the body but is also richly localised in the chondrocyte pericellular matrix, cause

Table 6.2 Chondrodysplasias caused by ECM gene mutations

Gene	Protein	Group: representative disorders
Collagens		
COL2A1	Collagen II	Achondrogenesis type II, hypochondrogenesis, spondyloepiphyseal dysplasia congenital, Kniest syndrome, Stickler syndrome type 1
COL9A1	Collagen IX	Multiple epiphyseal dysplasia type 6
COL9A2	Collagen IX	Multiple epiphyseal dysplasia type 2
COL9A3	Collagen IX	Multiple epiphyseal dysplasia type 3
COL10A1	Collagen X	Metaphyseal dysplasia Schmid type
COL11A1	Collagen XI	Stickler syndrome type 2, Marshall syndrome, fibrochondrogenesis type 1
COL11A2	Collagen XI	Stickler syndrome type 3, oto-spondylo-mega-epiphyseal dysplasia, fibrochondrogenesis type 2
ECM structural proteins		
HGPS2	Perlecan	Dyssegmental dysplasias, Schwartz-Jampel syndrome
AGC1	Aggrecan	Spondyloepiphyseal dysplasia Kimberley type, spondyloepimetaphyseal dysplasia aggrecan type
COMP	Cartilage oligomeric matrix protein	Pseudoachondroplasia, multiple epiphyseal dysplasia type 2
MATN3	Matrilin 3	Multiple epiphyseal dysplasia type 5
MGP	Matrix gamma-carboxyglutamic acid	Chondrodysplasia punctata group, Keutel syndrome
ECM post-translational modification		
DTDST	SLC6A2 sulphate transporter	Achondrogenesis type 1B, atelogenesis type 2, diastrophic dysplasia
PAPSS2	PAPS synthetase 2	Spondyloepimetaphyseal dysplasia PAPSS2 type, Brachyolmia recessive type
CHST3	Carbohydrate sulphotransferase 3	Chondrodysplasia with congenital joint dislocations CHST3 type
CHST14	Carbohydrate sulphotransferase 14	Ehlers-Danlos syndrome CHST14 type
ARSE	Arylsulfatase E	Chondrodysplasia punctata group: chondrodysplasia punctata, X-linked recessive, brachytelephalangic type

dyssegmental dysplasias (Arikawa-Hirasawa et al. 2001) and the Schwartz-Jampel syndrome (Nicole et al. 2000). While these mutations in the genes encoding proteoglycan protein backbones highlight the important structural role of proteoglycans, mutations in several of the enzymes involved in the post-translational modification pathways that generate the highly sulfated glycosaminoglycan side chains also cause chondrodysplasias (Table 6.2). Mutations in SLC26A2 sulphate transporter (*DTDST*) result in achondrogenesis type 1B, atelogenesis type 2 and diastrophic dysplasia (Rossi and Superti-Furga 2001); PAPS synthetase 2 (*PAPSS2*) mutations cause spondyloepimetaphyseal dysplasia PAPSS2 type (Faiyaz ul Haque et al. 1998) and brachyolmia recessive type (Miyake et al. 2012). Carbohydrate sulfotransferase 3 (*CHST3*) mutations cause chondrodysplasia with congenital joint dislocations CHST3 type (Thiele et al. 2004), and arylsulfatase E (*ARSE*) mutations underlie the chondrodysplasia punctata group: chondrodysplasia punctata, X-linked recessive and brachytelephalangic type (Brunetti-Pierri et al. 2003).

Members of the small leucine-rich protein (SLRP) family, including biglycan, chondroadherin, decorin, epiphycan, fibromodulin, lumican, asporin and PRELP, are found in cartilage. These proteins have a "horseshoe" shape (McEwan et al. 2006) that is important for protein interactions, allowing them to bind to growth factors and cytokines (Iozzo and Schaefer 2010; Iozzo and Karamanos 2010), cell surface receptors (Iozzo and Karamanos 2010) and ECM proteins (Wiberg et al. 2003) and affect signalling as well as ECM structure. Importantly SLRPs interact with fibril-forming collagens and play a critical role in collagen fibrillogenesis (Hedbom and Heinegard 1993; Douglas et al. 2006; Danielson et al. 1997). While human mutations in SLRPs causing chondrodysplasias have not been found to date, several SLRP knockout mice present cartilage phenotypes (Nuka et al. 2010; Hessle et al. 2014; Nikitovic et al. 2012) although because of their widespread expression other phenotypes predominate. Asporin has been linked to osteoarthritis (Xu et al. 2015).

In addition to SLRPs, matrilin-1 and matrilin-3 and COMP (cartilage oligomeric matrix protein) are important structural components of cartilage. Matrilin-1 can interact with the SLRPs biglycan and decorin and act as an adaptor protein connecting the collagen II/IX/XI fibrillar and collagen VI microfibrillar networks in cartilage (Wiberg et al. 2003). Matrilin-3 (*MATN3*) mutations have been described in multiple epiphyseal dysplasia type 5 and spondylo-epimetaphysial dysplasia MATN3 type implicating matrilin-3 oligomers (and matrilin-1/matrilin-3 hetero-oligomers) as key interacting structural and signalling components of the cartilage ECM (Jayasuriya et al. 2014; Cotterill et al. 2005). COMP mutations cause pseudoachondroplasia and multiple epiphyseal dysplasia type 2. The molecular pathology of COMP and matrilin-3 mutations is discussed in detail in Chap. 7.

6.5 Collagenopathies

The collagen fibrillar network forms the backbone of the cartilage ECM, directly providing the structure critical for resistance to biomechanical loading and also providing a scaffold for the myriad of interacting ECM proteins that add further functionality. It is therefore unsurprising that mutations impacting collagen quantity and quality are a significant cause of chondrodysplasias. In addition to underlying chondrodysplasias, more subtle collagen mutations can predispose cartilage to osteoarthritic damage. Mutations in the predominant cartilage fibrillar collagens II (*COL2A1*), XI (*COL11A1*, *COL11A2*), fibril-associated collagen IX (*COL9A1*, *COL9A2*, *COL9A3*) and hypertrophic cartilage network-forming collagen X (*COL10A1*) have been reported in a range of chondrodysplasias (Table 6.2). We will discuss the two mutation groups, null mutations and structural mutations, separately since recent data indicates that the pathophysiology is quite different.

Another collagen found in the cartilage pericellular matrix, collagen VI, has not been associated with human chondrodysplasias, although there is strong evidence that collagen VI is important in regulating the properties of the chondrocyte matrix and mechanotransduction (Zelenski et al. 2015). Collagen VI in the pericellular matrix is lost early in osteoarthritis (Felka et al. 2016) and collagen VI null mice exhibit skeletal abnormalities including delayed development and ossification, as well as accelerated osteoarthritic changes in the hip (Alexopoulos et al. 2009). Thus, while collagen VI is a good chondrodysplasia candidate, its widespread expression means that cartilage is not the key pathological target, and mutations in each of the three canonical collagen VI genes *COL6A1*, *COL6A2* and *COL6A3* result in the muscle disorders Ullrich congenital muscular dystrophy and Bethlem myopathy (Bushby et al. 2014). In these disorders the reduced biomechanical challenges to cartilage due to patient immobility may preclude the onset of a cartilage clinical phenotype.

6.6 Mutations Leading to Reduced Protein Synthesis

Mutations that introduce premature stop codons either via point mutations or frameshifting indels are a common cause of gene loss-of-function. In *COL2A1*, heterozygous premature termination mutations are often found in Stickler syndrome, type I, which is characterised by ocular, auditory, orofacial and skeletal abnormalities including mild spondyloepiphyseal dysplasia and early-onset osteoarthritis (Table 6.2). Premature termination mutations most commonly cause nonsense-mediated mRNA decay (NMD) resulting in functional haploinsufficiency of collagen II in the affected cartilage. This mechanism was confirmed in Stickler syndrome type I patients using the protein truncation test of in vitro transcribed and translated mutant and control alleles. Mutant allele protein was not produced unless NMD was inhibited with cycloheximide (Freddi et al. 2000). Heterozygous premature stop codon mutations in *COL11A2* cause Stickler syndrome, type III, which is a non-ocular form of Stickler syndrome. *COL11A2* is not expressed in the eye, accounting for this phenotype (Mayne et al. 1993). Homozygous *COL11A2* premature termination mutations have been described in otospondylomegaepiphyseal dysplasia, fibrochondrogenesis type 2 (Melkoniemi et al. 2000) and Weissenbacher-Zweymuller syndrome (Harel et al. 2005). While not tested directly, the assumption is that, in these patients with homozygous mutations, there would be little or no expression of the collagen α2(XI) chain. In *COL11A1*, heterozygous nonsense mutations in combination with a structural mutation on the other allele cause the severe recessive fibrochondrogenesis type 2 (Tompson et al. 2010). Early-onset hearing loss appears to be the major phenotype in heterozygous carriers of *COL11A1* loss-of-function alleles (Tompson et al. 2010). In the *cho/cho* mouse, homozygous for a *Col11a1* null allele, a lethal chondrodysplasia results (Li et al. 1995). In the absence of the α1(XI) chain, collagen II fibrillogenesis is severely disrupted and the resulting fibrils have larger diameters (Fernandes et al. 2007), indicating that collagen XI containing the α1(XI) chain, buried in the interior of collagen II fibrils, is required to maintain the thin diameter fibrils seen in vivo. The heterozygous *cho*/+ mouse, with α1(XI) haploinsufficiency, develops early-onset osteoarthritis (Xu et al. 2003).

Heterozygous null mutations in *COL10A1* cause metaphyseal chondrodysplasia-type Schmid which is characterised by irregularities of the metaphyseal ends of bones resulting in bow legs, coxa vara and short stature (Lachman et al. 1988). Collagen X forms a homotrimeric network restricted to the epiphyseal growth plate cartilages of long bones. Both frameshift mutations leading to premature translation termination and nonsense mutations have been characterised (Bateman et al. 2005). Interestingly, these are all localised within a relatively small area of exon 3, in the region coding for the C-terminal trimerisation domain. Studies on patient cartilage have shown that these mutations lead to complete NMD of mRNA from the mutant allele, so that *COL10A1* is truly haploinsufficient in the patient growth plate cartilage (Chan et al. 1998; Bateman et al. 2003). This is surprising for several reasons. Firstly, exon 3 is the most 3′ exon of the *COL10A1* gene, and the current NMD dogma is that mRNAs containing premature termination codons in the region encoded by the most 3′ exon are immune to NMD because these mRNAs do not have downstream exon-exon boundaries defined by protein deposition during splicing that define the stop codon as premature. In *COL10A1* NMD competency is determined by the position of the premature termination mutation relative to the 3′ UTR (Tan et al. 2008). It was also expected that more 5′ mutations in exon 3 and exon 2 would lead to NMD. However, studies using targeted mutagenesis demonstrated that this was not the case, and only mutations in the small region of exon 3 where all patient mutations have been localised were susceptible to NMD (Tan et al. 2008). These data on collagen X NMD suggest that the mutant mRNA surveillance mechanism might be fundamentally different to the current mechanistic models. This may be due, in part, to the structure of collagen X mRNA, where almost all of the protein is encoded by a large 3′ terminal exon. Understanding the molecular mechanisms of collagen X NMD will offer insights into how cells deal with premature termination mutations in other genes with a similar large 3' terminal exons (Fang et al. 2013).

6.7 Structural Mutations Affect Collagen Assembly at Multiple Levels

While the reduced collagen expression resulting from null mutations has the obvious effect of compromising the cartilage matrix arrangement, structural mutations in the collagen chains have more complex qualitative and quantitative impacts. To begin to understand these impacts, we need to consider the assembly framework of collagens. Collagens assemble via recognition sequences at the C-terminus (Khoshnoodi et al. 2006). In the case of collagen II, recognition directed by the noncollagenous C-propeptide sequence determines the selection of three proα1(II) chains to form collagen II homotrimers and in collagen XI selection of proα1(XI), proα2(XI) and proα1(II) [called proα3(XI) before it was shown to be the protein product of the *COL2A1* gene] to form collagen XI heterotrimers. In collagen IX assembly, proα1(IX), proα2(IX) and proα3(IX) chain selection information is contained within the C-terminal region of the COL1 domain and the C-terminal NC1 domains (Mechling et al. 1996). Correct registration of the proα-chains is critical to initiate collagen triple helix formation which occurs from the C- to N-terminus (Ishikawa and Bachinger 2013). While collagen sequences contain the information to fold into a triple helix in vitro, albeit slowly, the actual procollagen assembly and folding machinery in the cell are complex involving interplay between the collagen α-chains and a temporally and spatially regulated panel of molecular chaperones and post-translational processing enzymes. Collagen assembly and processing has been recently reviewed (Ishikawa and Bachinger 2013). Because of the strict requirements for correct collagen folding, it is unsurprising that mutations in the C-terminal domains that compromise initial chain assembly, or mutations in the triple helix that interrupt the essential Gly-X-Y triplet repeat sequence and impact triple helix formation, or mutations in the folding machinery itself result in many collagenopathies (Fig. 6.2). While autosomal recessive null mutations in components of the collagen assembly complex such as CRTAP, P3H (*LEPRE1*), cyclophilin B (*PPIB*), FKBP65 (*FKBP10*) and HSP47 (*SERPINH1*) have been described for osteogenesis imperfecta, predominantly impacting collagen I, mutations of these or comparable processing partners have not yet been implicated in cartilage disorders. As we learn more about tissue-specific and collagen type-specific foldases and chaperones, it seems certain that these components will be involved in chondrodysplasias, and mutations will be found that alter cartilage collagen assembly (Ishikawa and Bachinger 2013; Ishikawa et al. 2015). In this review we will focus on collagen structural mutations that predominantly affect cartilage biology.

The most common mutations found in collagens are missense and exon-skipping mutations that affect the amino acid sequence. The clinical severity depends on how the mutation impacts collagen assembly, stability and/or interactions and how important the collagen is for developing and maintaining matrix integrity and function.

With collagen IX the predominant mutations in *COL9A1*, *COL9A2* and *COL9A3* that cause forms of multiple epiphyseal dysplasia (Table 6.2) are exon-skipping changes that are concentrated in the gene region encoding the N-terminal COL3 domain (Jackson et al. 2012). These mutations can affect mRNA stability but also disrupt collagen IX heterotrimer assembly and secretion (Briggs and Chapman 2002). Since collagen IX on the collagen II/XI fibril surface interacts with other ECM networks, these quantitative and qualitative changes could alter cartilage structure. Importantly, collagen IX associates with matrilin-3 and COMP within the cell, so that secretion defects in collagen IX caused by misfolding and ER retention have the potential to also impact the secretion efficiency of these interacting partners (for further details, see Chap. 7).

Heterozygous mutations in *COL11A1* or *COL11A2*, which introduce structural changes such as in-frame exon deletions or glycine substitutions, disturb the obligatory Gly-X-Y repeat sequence of the helix and cause conditions such as Stickler

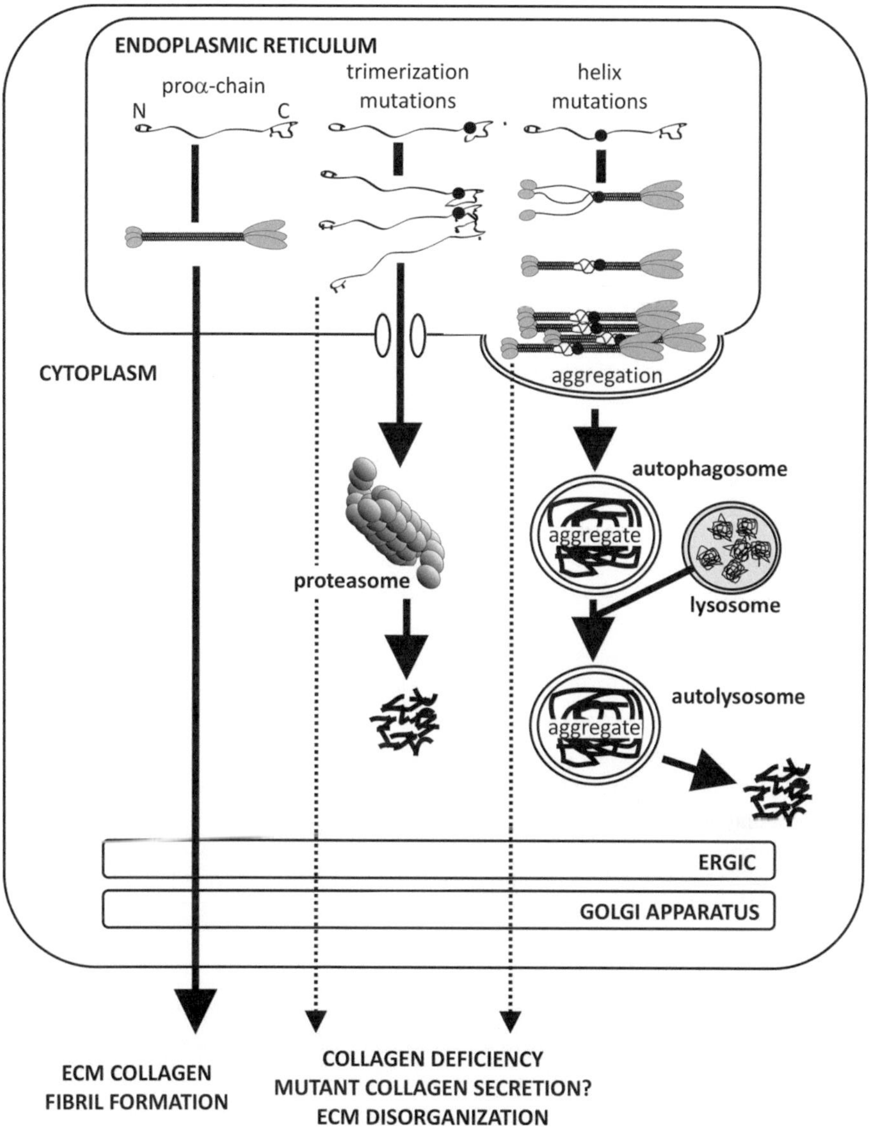

Fig. 6.2 Protein consequences of pathogenic structural collagen mutations. Structural mutations that lead to protein misfolding have a dominant negative effect. Mutations in the C-terminal propeptide compromise collagen chain trimerisation. Misfolded unassembled chains are retrotranslocated to the cytoplasm where they are degraded by proteasomes. This results in protein deficiency and, if some of the chains are incorporated into collagen trimers and are secreted, further deleterious effects on collagen extracellular assembly and matrix function. Mutations in the triple helical domain, commonly substitutions for the obligatory glycine residues in the Gly-X-Y collagen sequence or in-frame deletions as a result of exon-skipping mutations, allow initial chain selection and assembly, but the abnormal helical structure leads to retention in the ER and protein aggregation. The protein aggregates are degraded by the autophagy/lysosomal system and the result is a severe protein deficiency and ECM disorganisation

syndrome type 2 and Marshall syndrome (*COL11A1*) or Stickler syndrome type 3 and Weissenbacher-Zweymuller syndrome (*COL11A2*) (Table 6.2). Little direct data is available on how these structural mutations affect collagen XI assembly and stability, but dysfunction can be inferred from the other more extensively studied fibrillar collagen mutations, such as collagen II and I, discussed below. While in general the collagen XI structural mutations are less damaging than those in the more abundant collagen II, compound heterozygous *COL11A1* or *COL11A2* mutations can result fibrochondrosis type I and II (Tompson et al. 2010). In this condition the presence of a null allele in combination with in-frame exon deletions or helix glycine substitutions results in this very severe, often lethal, chondrodysplasia demonstrating the importance of the collagen XI template for collagen II fibrillogenesis. The third component of the collagen XI trimer is the α1(II) chain produced by the *COL2A1* gene [and formally called α3(XI)]; thus mutations in *COL2A1* have potential ramifications for collagen XI trimer stoichiometry and integrity.

Collagen II structural mutations cause cartilage disorders of varying severity from mild to severe and lethal conditions. Collagen II is structurally similar to the major non-cartilage fibrillar collagen, collagen I, and many of the genotype-phenotype correlations for the more intensely studied collagen I mutations in diseases such as osteogenesis imperfecta shed light on the molecular pathology of collagen II. Our discussion will focus on collagen II but will look to examples from studies on collagen I to provide mechanistic insights. As discussed above *COL2A1* mutations that result in haploinsufficiency cause mild phenotypes such as Stickler syndrome (Fig. 6.1). Likewise *COL1A1* haploinsufficiency causes the relatively mild dominant form of osteogenesis imperfecta. *COL2A1* missense mutations impact cartilage pathology according to their effect on collagen structure, folding, secretion and ability to form functional fibrils in the ECM. As an overarching concept, mutations that affect the C-propeptide and compromise the initial assembly of the three proα-chains and mutations that interfere with triple helix formation such as glycine substitutions in the obligatory Gly-X-Y sequence have serious clinical consequences, whereas other sequence changes are less severe. A full understanding of the relationship between the mutation type, sequence context and clinical outcome has not been achieved, but some general concepts are emerging including important differences in the cellular consequences of helix mutations and those that affect initial assembly. We will discuss what we know about the different mutation groups and how they may affect cartilage structure and function.

Numerous non-glycine polymorphisms are found in *COL2A1* in apparently normal individuals (Exome Aggregation Consortium (ExAC)) emphasising that the helix can accommodate X and Y position changes within the strict helical structure. These benign missense changes tend to conserve amino acid charge and size and thus would not be expected to affect helix structure or folding. Helix X or Y position missense mutations causing chondrodysplasias are rare; however, several substitutions of cysteine for arginine, p.R275C (Williams et al. 1993), p.R565C (Richards et al. 2000), p.R719C (Hoornaert et al. 2006), p.R904C (Hoornaert et al. 2006) and p.R989C (Chan et al. 1993) cause disorders that range from severe but non-lethal forms of spondyloepiphyseal dysplasia congenita through to Stickler syndrome and precocious osteoarthritis. The position of the mutation in the helix does not predict severity suggesting that local sequence contexts must be important in determining the phenotype. This is supported by normal sequence data (60,000 unrelated individuals; Exome Aggregation Consortium) which includes individuals with p.R1058C and p.R584C helix sequence variants (Exome Aggregation Consortium (ExAC)). The p.R1058C variant occurs in three individuals, highly suggestive that the arginine to cysteine change is tolerated in this region of the helix despite the fact that the cysteine, which does not normally occur in the helix, would be predicted to interfere with helix structure and introduce internal disulphide bonding into the collagen II heterotrimer.

Collagen II glycine to cysteine changes have been studied in patient samples to a limited extent, but more systematically in transfected cell model systems which reveal a complex pattern of how these mutations affect collagen II. Several mutations (p.R275C, p.R719C) occur in regions of the helix that are thermolabile and are accommodated in the helix without reducing helix thermal stability (Steplewski et al. 2004b). These mutations do not induce apoptosis in vitro (Hintze et al. 2008) and, consistent with only a small impact on collagen structure, they produce mild clinical phenotypes (Arnold and Fertala 2013). On the other hand, p.R989C is in a thermostable helix region and this substitution reduces helix thermal stability. The p.R989C mutation, and a comparable p.R1147C mutation found in a spondyloepiphyseal mouse model, cause chondrocyte apoptosis in vitro and in vivo and result in a more severe phenotype (Hintze et al. 2008). It has been suggested (Arnold and Fertala 2013) that the reduced thermostability in vitro may indicate that these mutations produce more misfolding at physiological temperatures than p.R275C and p.R719C. Importantly, p.R989C and p.R1147C also induce a cellular endoplasmic reticulum stress response, discussed in the next section. In contrast the p.R275C and p.R719C mutants do not appear to induce such cellular stress (Arnold and Fertala 2013). In addition to the cellular stress caused by p.R989C and p.R1147C mutations, culminating in apoptosis, p.R989C interacts with fibronectin inside the cell and extracellularly, and this binding likely inhibits collagen II fibril formation (Ito et al. 2005). Since collagen and fibronectin fibrillogenesis in vivo requires both proteins (Velling et al. 2002), strong fibronectin binding to p.R989C mutant collagen II could also be responsible for the observed disorganised fibronectin matrix and could potentially change the organisation of other extracellular matrix components. By contrast, the p.R275C mutant binds only very weakly to fibronectin and rapidly dissociates (Ito et al. 2005), and fibronectin deposited in the extracellular matrix of cells expressing this mutant has a normal fibrillar structure. The arginine to cysteine mutations do not obviously affect collagen II secretion (Ito et al. 2005), and the abnormal collagen likely impacts fibril assembly and structure (Steplewski et al. 2004a). These data emphasise the potential of both intracellular and extracellular impacts to contribute to the final functional phenotype of the cartilage.

The most common and most studied mutations in the collagen helix are heterozygous missense mutations that substitute other amino acids for glycine in the helical Gly-X-Y sequence. Approximately 100 of these substitution mutations in *COL2A1* have been reported in a range of severe chondrodysplasias including several forms of spondyloepiphyseal dysplasia, Kneist dysplasia, achondrogenesis type II and hypochondrogenesis (Kannu et al. 2012). Only one glycine substitution (p.G940R) has been reported in normal individuals (Exome Aggregation Consortium (ExAC)). With this single exception, the case for glycine helix-disrupting mutations as the major cause of cartilage collagenopathies is overwhelming, mirroring the predominance of *COL1A1* glycine mutations causing osteogenesis imperfecta. Aspects of the molecular pathology of helix glycine mutations are very clear and can be generalised. Studies on mutant collagen II parallel the wealth of data on mutant collagen I and demonstrate that helical glycine substitutions can delay helix formation and destabilise the helix, and as a result, the mutant collagen trimers have increased post-translational modifications (Bonaventure et al. 1995). As in collagen I, not all collagen II glycine mutations reduce thermostability and this likely reflects the mutation position within thermostable or thermolabile regions (Steplewski et al. 2004b; Ito et al. 2005). The disturbance to helix folding leads to intracellular retention of the mutant-containing molecules (Chan et al. 1995). Because collagen II is a homotrimer of three chains, 7/8 of the trimers will contain at least one mutant chain, so that the effect of heterozygous mutant expression has much more impact than it would if collagen chains did not form multimers. Intracellular retention of the mutant-containing trimers targets them for degradation by cellular quality control processes (Chan et al. 1995). While the extent of degradation may be influenced by the specific mutation, the extreme pathology that can result from this degradation was dramatically demonstrated in two lethal chondrodysplasia patients, one with hypochondrogenesis (p.G1113C [legacy mutation G913C] (Mundlos et al. 1996)) and one with achondrogenesis type II (p.G996S [legacy mutation G796S] (Chan et al. 1995)) where the mutations lead to complete degradation, and collagen II in the poorly formed cartilages is replaced with collagens I and III.

Mutations in the collagen II C-propeptide cause severe chondrodyplasias including spondyloepiphyseal dyspalsia, achondrogenesis II-hypochondrogenesis and spondyloperipheral dysplasia, Torrance type (Zankl et al. 2004, 2005; Nishimura et al. 2004). While there are no biochemical studies to directly determine the effect of the mutations on procollagen assembly, it is assumed based on mutations in this domain in collagen I (Lamande et al. 1995; Bateman et al. 1989) and collagen X (Bateman et al. 2005) that these compromise or prevent initial procollagen assembly (Fig. 6.2). Heterozygous mutations in *COL10A1* provide the most complete data set on how mutations in the C-terminal trimerisation domains result in failed collagen folding. To date 22 *COL10A1* missense mutations have been reported in metaphyseal chondrodysplasia Schmid-type patients (Table 6.2) (Bateman et al. 2005). These mutations cluster in a restricted region of exon 3 coding for the collagen X NC1 domain which is critical for the initial association of the chains into the trimeric collagen X structure. While crystal structures of this trimerisation domain (Bogin et al. 2002) predicted that mutations disrupting the hydrophobic core would exclude chains from assembly, in vitro and mouse studies have shown that disease-causing mutations are not restricted to the hydrophobic core sequence, and other mutations in the NC1 sequence from codon 555 to 677 impair collagen X trimerisation (Bateman et al. 2005). The unassembled collagen X chains are retained intracellularly and degraded (Wilson et al. 2002) suggesting that the reduced extracellular collagen X, like that caused by nonsense-mediated decay of mRNAs containing premature termination mutations, was the cause of the metaphyseal chondrodysplasia Schmid-type phenotype. However, this is not the case and a much more interesting scenario has emerged – the unassembled collagen X chains retained in the endoplasmic reticulum trigger a cellular stress response, the unfolded protein response (UPR) – and this is a major downstream effect that causes the pathology. The UPR, and how it applies to collagen X and other collagens, is discussed in the next section. The importance of the trimerisation domain and the UPR in *COL10A1* pathology is highlighted by the almost complete absence of disease-causing *COL10A1* missense mutations outside (i.e. more N-terminal) the NCI domain. Unlike other collagens, where glycine mutations are the major pathological change, in *COL10A1* only one glycine substitution in the triple helix has been reported in a patient, p.G288R (Park et al. 2015). The ExAc database (Exome Aggregation Consortium (ExAC)) reveals 70 glycine substitutions in *COL10A1* in normal individuals, including several close to p.G288, casting some doubt on whether this is the disease-causing mutation. Two mutations in the N-terminal signal peptide cleavage site (Ikegawa et al. 1997; Bateman et al. 2004) are predicted to prevent release of the mutant chains into the endoplasmic reticulum and result in haploinsufficiency (Chan et al. 2001).

6.8 Endoplasmic Reticulum Stress in Cartilage Collagenopathies

The endoplasmic reticulum (ER) imposes stringent quality control on protein folding, such that only folded functional proteins leave the ER (van Anken and Braakman 2005). The cellular response to misfolded proteins is called the unfolded protein response (UPR) [recently reviewed (Wang and Kaufman 2016)]. Activation of the UPR via canonical ER stress sensors, IRE1α, PERK and ATF6, results in a cascade of adaptive responses in a coordinated attempt to reduce the load of misfolded proteins (Hetz et al. 2011) (Fig. 6.3). These adaptive responses include attenuation of translation, degradation of misfolded proteins by proteasomal ER-associated degradation (ERAD) and autophagy. Initial UPR activation is cytoprotective, offering the cells an opportunity to return to protein-folding homeostasis. With increased duration of unfolded protein load, the balance in activities of the three pathways changes, and prolonged ER stress has deleterious effects for the cells such as apoptosis.

It is now clear that collagen misfolding mutations (as with other ECM gene mutations, e.g. matrilin-3, COMP) can induce a cellular UPR with consequences on tissue pathology (reviewed (Bateman et al. 2009; Boot-Handford and Briggs 2010; Patterson and Dealy 2014)). The most compelling data so far comes from studies on collagen X misfolding mutations in metaphyseal chondrodysplasia, Schmid type. In vitro studies show that the mutant collagen X misfolding results in retention of the mutant chains in the ER, aberrant disulphide-bonded dimer formation and degradation by ERAD, a key proteasomal degradation pathway activated by the UPR

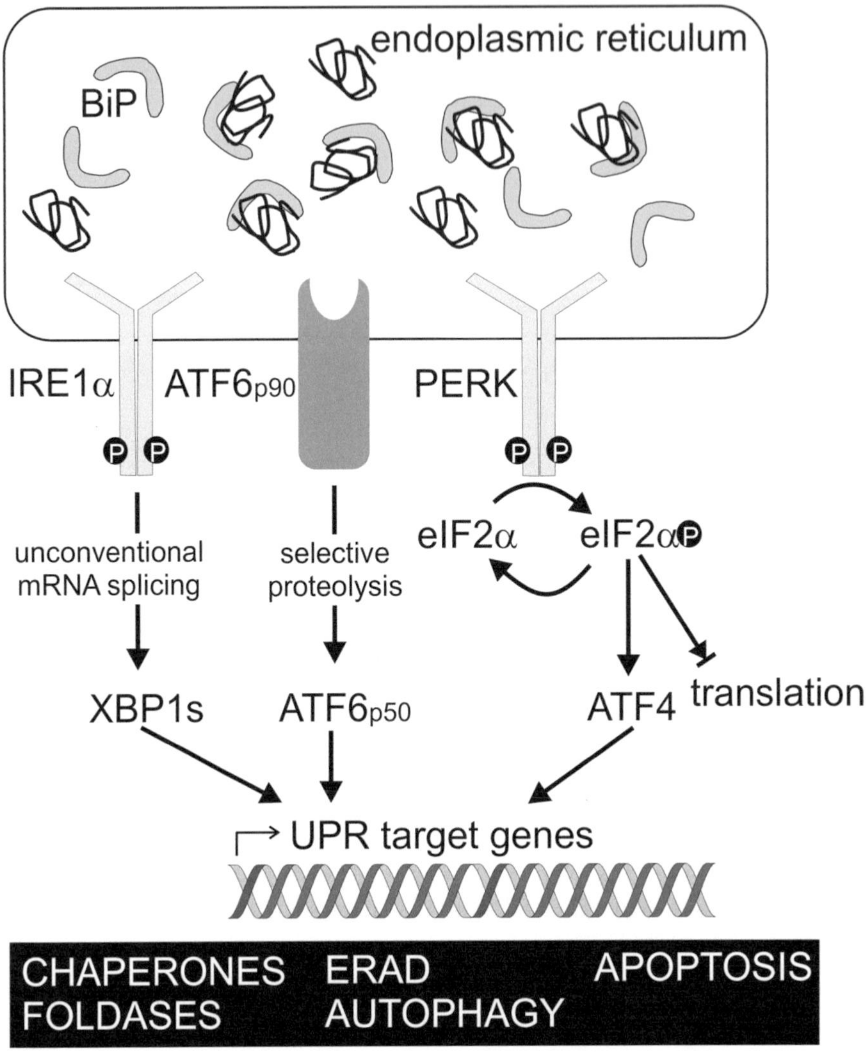

Fig. 6.3 The UPR signalling pathways. The UPR has three parallel branches activated by canonical UPR sensors embedded in the ER membrane, IRE1α, ATF6p90 and PERK. The molecular chaperone BiP is thought to bind constitutively to the sensors and render them inactive. Misfolded proteins in the ER bind to BiP, sequestering it away from the sensors and triggering their signalling pathways. IRE1α and PERK homodimerize and autophosphorylate and this activates XBP1 splicing and eIF2α phosphorylation. XBP1s produces a transcription factor that stimulates the synthesis of chaperones and foldases to increase the protein-folding capacity. Phosphorylated eIF2α inhibits general protein translation to reduce protein load and at the same time promotes translation of the transcription factor ATF4 which then increases production of apoptosis genes. Release of BiP from ATF6p90 allows the sensor to move to the Golgi where it is cleaved. The active cytosolic fragment ATF6p50 enters the nucleus and stimulates transcription of genes with an ER stress-responsive element

(Wilson et al. 2002, 2005). In vitro expression studies (Wilson et al. 2005) and mouse models (Tsang et al. 2007; Ho et al. 2007) where *Col10a1* trimerisation domain mutants were expressed in growth plate cartilage showed that the misfolded collagen X elicits an UPR where all three canonical arms IRE1α, ATF6 and PERK are activated, indicated by Xbp1 splicing, ATF6 cleavage and eIF2α phosphorylation (Rajpar et al. 2009; Cameron et al. 2011). UPR activation resulted in upregulation of downstream UPR targets, including molecular chaperones, foldases, ER-associated degradation machinery and CHOP (Tsang et al. 2007; Rajpar et al. 2009; Cameron et al. 2011). While chondrocyte apoptosis was increased from around 2 weeks of age (Cameron et al. 2015), increased apoptosis was not detected earlier in development (Tsang et al. 2007; Rajpar et al. 2009; Cameron et al. 2011). A key and unexpected consequence of the UPR were downstream effects on terminal differentiation, with the persistence of a proliferative chondrocyte-like expression profile in ER-stressed Schmid chondrocytes. It was postulated that CEBPβ dysregulation was a key component of this effect on the progression of chondrocytes from the proliferative to hypertrophic state (Cameron et al. 2011). A similar block in terminal differentiation was reported in another mouse model of collagen X misfolding, with a similar UPR upregulation (Tsang et al. 2007), although in this case it was proposed that the re-expression of proliferative chondrocyte genes was an adaptive response to the UPR; by downregulating collagen X expression and reverting to a less mature phenotype, the cells survive, although at a significant cost to their normal function. Whatever the precise mechanism of altered differentiation caused by the misfolded collagen X and the UPR is, the pathological impact of the collagen X-dependent UPR is disruption to cartilage maturation, remodelling, vascularisation and mineralisation, leading to cartilage hypertrophic zone expansion and dwarfism. These studies raise important questions about the relative impact of the two consequences of collagen X misfolding, the UPR and the quantitative and possible qualitative impact of structurally abnormal collagen X in the cartilage ECM (Fig. 6.4). For collagen X the consequences of the UPR alone are sufficient to cause the growth plate pathology. This was demonstrated by expressing a UPR-inducing protein, a mutant thyroglobulin, specifically in growth plate cartilage where it is normally not expressed and has therefore no role in cartilage structure and function (Rajpar et al. 2009). The mutant misfolded thyroglobulin induced an UPR in the growth plate cartilage of the transgenic mouse, resulting in the same metaphyseal chondrodysplasia phenotype as was caused by *Col10a1* missense mutations. These data unequivocally show that for collagen X misfolding mutations, the UPR is the key component of the molecular pathology.

The nature of the UPR and its role in the pathology caused by other cartilage collagen types is not yet resolved, but overall it seems that aspects of the UPR may be involved although the mutation specificity of the response remains to be defined. Collagen X is like virtually all other collagens in having a C-terminal domain that is involved in initial proα-chain selection and trimerisation; thus, it is expected that mutations in other collagens in these domains will also impact folding and cause a UPR which may contribute to the pathology. *COL2A1* missense and other structural mutations in the C-propeptide cause a range of chondrodysplasias including

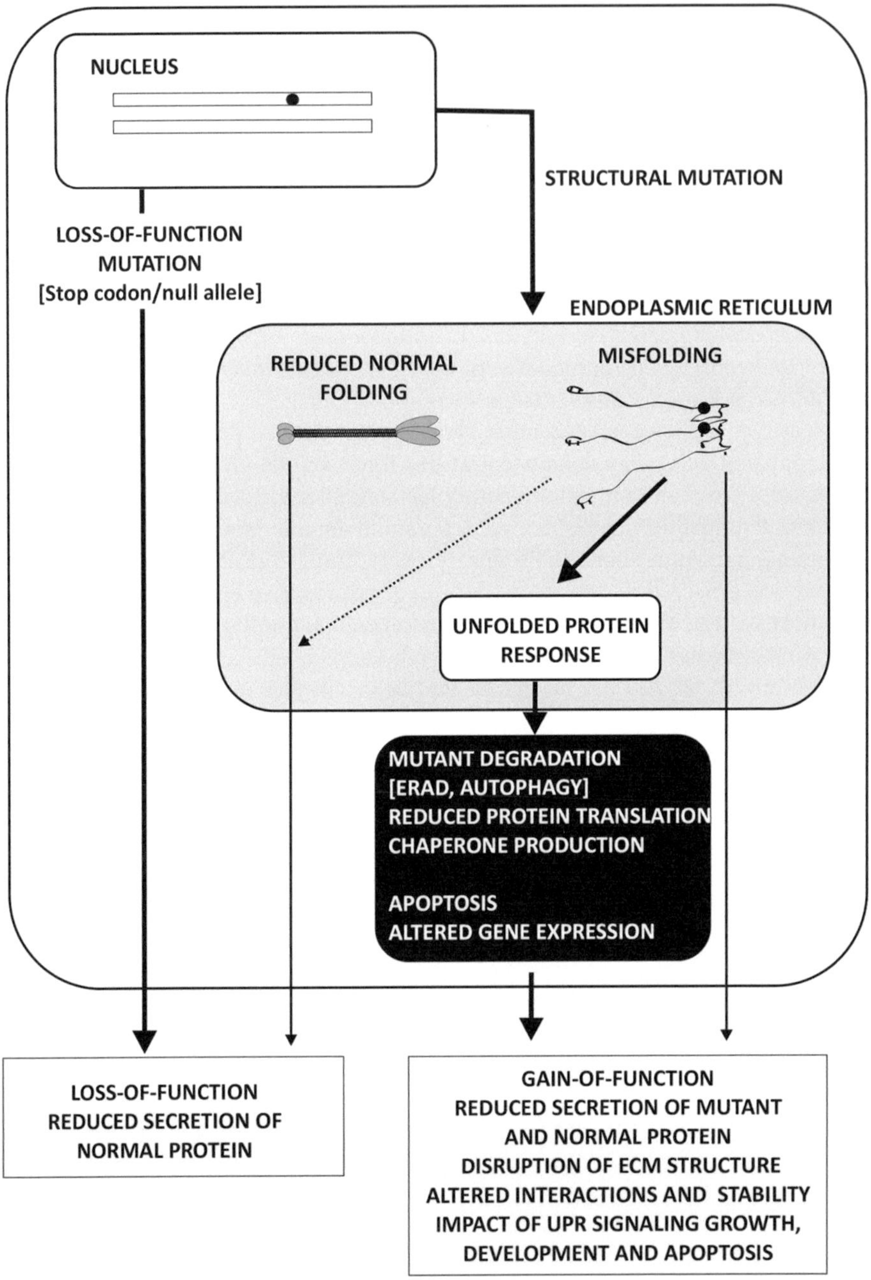
NUCLEUS
STRUCTURAL MUTATION
LOSS-OF-FUNCTION MUTATION [Stop codon/null allele]
ENDOPLASMIC RETICULUM
REDUCED NORMAL FOLDING
MISFOLDING
UNFOLDED PROTEIN RESPONSE
MUTANT DEGRADATION [ERAD, AUTOPHAGY]
REDUCED PROTEIN TRANSLATION
CHAPERONE PRODUCTION
APOPTOSIS
ALTERED GENE EXPRESSION
LOSS-OF-FUNCTION
REDUCED SECRETION OF NORMAL PROTEIN
GAIN-OF-FUNCTION
REDUCED SECRETION OF MUTANT AND NORMAL PROTEIN
DISRUPTION OF ECM STRUCTURE
ALTERED INTERACTIONS AND STABILITY
IMPACT OF UPR SIGNALING GROWTH, DEVELOPMENT AND APOPTOSIS

spondyloepiphyseal dysplasia (Mortier et al. 2000), spondyloperipheral dysplasia (Zankl et al. 2004), platyspondylic skeletal dysplasia Torrance type (Zankl et al. 2005; Nishimura et al. 2004), avascular necrosis of the femoral head (Kannu et al. 2011) and vitreoretinopathy with phalangeal epiphyseal dysplasia (Richards et al. 2002). The functional effects of these different mutations on assembly and how this can explain the clinical phenotypic differences have not been determined. However, mouse models with *Col2a1* p.S1386P, p.Y1391S and p.D1469A C-propeptide mutations produce chondrodysplasias with the hallmarks of the canonical unfolded protein response including protein retention in the ER (Esapa et al. 2012; Furuichi et al. 2011), an expanded ER (Esapa et al. 2012; Furuichi et al. 2011; Kimura et al. 2015), increased mRNA expression of ER stress markers *Grp94* and *Chop* (Kimura et al. 2015; Furuichi et al. 2011) and chondrocyte apoptosis (Kimura et al. 2015). While this strongly suggests that the canonical UPR is activated by collagen II C-propeptide mutations, there is also supporting evidence from *COL1A1* C-propeptide mutations in osteogenesis imperfecta (OI). In a mouse model of OI caused by a frameshift mutation in the C-propeptide of collagen I, there was ER retention of the mutant collagen in heterozygous mice, increased BiP and the collagen-specific chaperone Hsp47, increased Chop with consequent caspase-12 and caspase-3 activation and apoptosis of osteoblasts both in vitro and in vivo (Lisse et al. 2008). The engagement of a canonical UPR with *COL1A1* trimerisation mutations in osteogenesis imperfecta is supported by early biochemical studies demonstrating BiP binding and proteasomal degradation in patient fibroblasts and transfected cells (Lamande et al. 1995; Fitzgerald et al. 1999; Chessler and Byers 1993).

While there appears to be involvement of a canonical UPR downstream of collagen II mutations that affect initial assembly, it is not yet known how much this contributes to the clinical phenotype, in addition to the reduced collagen secretion due to retention and degradation. Given that the phenotypes for heterozygous C-propeptide mutations are generally more severe than null allele mutations, it is reasonable to assume the UPR may contribute to the pathology in many instances. However, the consequences may well depend on whether the mutation completely prevents assembly or allows partial assembly and has further dominant negative effects on the collagen II population.

Fig. 6.4 Cellular and extracellular consequences of collagen mutations. Reduced protein synthesis caused by heterozygous premature stop codon mutations and mRNA decay or null alleles results in a protein deficiency in the ECM and relatively mild clinical phenotypes. Structural mutations have dominant negative effects. Misfolded proteins are partially or completely retained in the ER and, depending on the folding defect, are degraded by either ERAD/proteasomes or autophagy. If abnormal proteins are secreted, they have a further deleterious effect on ECM structure and function. Misfolded proteins can also induce the cellular unfolded protein response, which if unresolved can lead to changes in gene expression and differentiation and eventually apoptosis. The relative impact of extracellular and intracellular consequences for different mutations in different collagen types has not been fully resolved

The most common collagen II (and other collagen) mutations are helix-disrupting mutations such as glycine substitutions. Given that glycine mutations result in intracellular retention and degradation, it would seem logical that some form of ER stress would result which may contribute to the pathology. In collagen II there is good evidence of an UPR resulting from disruptions in the helix caused by p.G1182A (achondrogenesis type II), p.G450R (hypochondrogenesis) and an exon 41 skipping mutation (achondrogenesis type II) (Okada et al. 2015). Patient fibroblasts were directly converted into chondrocyte-like cells in vitro by transducing cells with c-MYC, KLF4 and SOX9. In these cells there was increased BiP, GRP94 and CHOP and increased XBP1 splicing, ATF6 cleavage, eIF2α phosphorylation and apoptosis, indicating activation of the UPR. Treating the mutant chondrocytes with TMAO (trimethylamine N-oxide), which is a chemical chaperone that can rescue protein-folding defects, reduced the elevated BiP and CHOP and reduced apoptosis (Okada et al. 2015). Similarly, BiP upregulation, ER stress, UPR activation and apoptosis are evident in the *Col2a1* p.G1170S-mutated mouse chondrodysplasia model (Liang et al. 2014). While these data provide strong evidence for a canonical UPR for these mutations, other data discussed below suggests that the cellular response to collagens containing helix mutations may be more complex and dependent on additional factors such as the mutation context. One key to this may be if the helix disruption is in a region that is thermostable or thermolabile (Steplewski et al. 2004a) as discussed above for arginine to cysteine mutations where only those mutations that reduce collagen helix thermal stability result in an UPR (Chung et al. 2009). In support of this hypothesis, the *Col2a1* p.G1053E mutation does not reduce helix thermal stability or elicit apoptosis (Hintze et al. 2008); however, the thermal stability of only a relatively few collagen II helix mutations has been determined, so correlating the destabilising effect of the mutation with an UPR is not yet possible and so far has only been clearly demonstrated with the arginine to cysteine mutations (Arnold and Fertala 2013). It is of interest that studies on the helix destabilising p.R989C and p.R1147C mutations demonstrated that the chemical chaperone TMAO stabilised the mutant collagen and increased mutant secretion (Gawron et al. 2010), presumably as a consequence of reducing the UPR.

For collagen I, the only studies to look at BiP binding to helix-mutant collagen I failed to show any interaction (Lamande et al. 1995) that would be indicative of helix-mutant collagen I triggering a conventional ER stress response. Recent studies on a mouse model of OI with a *Col1a2* p.G610C mutation also failed to show BiP upregulation despite intracellular retention of the mutant collagen (Mirigian et al. 2016). However, consistent with other studies, autophagy was upregulated, and there was clearly a "stress" response that resulted in osteoblast malfunction. Further studies are required to determine how collagen I and collagen II helix mutations engage with the osteoblast/chondrocyte quality control systems and how this may depend on the mutation context and effect on helix structure.

Whatever the precise nature of the UPR is, and what downstream pathological pathways may be affected, one key question in understanding the molecular pathology and envisaging new therapeutic approaches is how do the chondrocytes degrade the mutant collagens. With mutations that prevent assembly, such as C-propeptide collagen mutations discussed above, it seems clear from inhibitor studies in several cell model systems that the unassembled procollagen chains are retro-translocated to the cytoplasm and degraded by the proteasome system (Lamande et al. 1995; Fitzgerald et al. 1999; Ishida et al. 2009) (Fig. 6.3). This ERAD is one of the well-characterised cellular responses directed by the UPR, a targeted response to reduce the mutant protein load. In the case of collagen I with mutations in the helix that allow initial assembly of the procollagen trimer but affect downstream helix formation and stability, the partially assembled molecules are not retro-translocated for proteasomal breakdown and form aggregates in the ER. These aggregates are degraded by the autophagosome-lysosomal system (Ishida et al. 2009; Mirigian et al. 2016), and although a stress response results in mouse osteoblasts expressing *Col1a1* p.G610C mutant chains, it is an unconventional stress response that does not involve BiP upregulation or *Xbp1* splicing (Mirigian et al. 2016). Importantly, stimulating autophagy with rapamycin or nutrient deprivation prevented accumulation of the misfolded collagen I, eliminated the ER distention and restored collagen I extracellular matrix deposition (Mirigian et al. 2016). While these studies suggest potential therapeutic strategies for collagen I mutations, further studies on collagen II mutants are required to understand the cellular responses and degradation pathways downstream of different mutations.

6.9 Therapeutically Targeting the UPR

In a number of protein-folding diseases, therapeutic approaches that target protein misfolding at multiple levels are being deployed (Cao and Kaufman 2013). In general, these approaches attempt to restore proteostasis by stimulating degradation of the misfolded protein, or correcting the mutant protein folding, or act downstream of protein folding to manipulate different UPR signalling components. While many questions remain to be answered about how collagen misfolding mutations interact with the cellular unfolded protein response machinery, these approaches are a promising avenue to explore with collagen mutations.

Studies with the chemical chaperone, TMAO, have shown in vitro that the mutant collagen II helix is stabilised (Gawron et al. 2010), and secretion is increased in cell disease models (Okada et al. 2015). Studies with another chemical chaperone sodium phenylbutyrate on cells expressing mutant collagen IV have shown comparable rescue of misfolding and secretion (Murray et al. 2014). However, in cartilage in vivo, the possible effect of rescuing mutant collagen secretion on extracellular matrix formation and stability must be considered. In this context the altered interaction of collagen II p.R989C with fibronectin and the ability of the mutant-"kinked" collagen to impact both collagen and fibronectin fibrillogenesis (Steplewski et al. 2004a) suggest caution in this approach until details of both the intracellular and extracellular consequences of the mutant collagen are clarified experimentally.

Increasing breakdown of the mutant misfolded intracellular collagens by stimulating the proteasomal and/or autophagosome-lysosome system is another promising approach. While this has not been tested with cartilage collagen misfolding mutations, rapamycin treatment of osteoblasts isolated from a mouse model of OI with a *Col1a2* p.G610C mutation rescued the altered osteoblast differentiation and apoptosis that contributed to the pathology (Mirigian et al. 2016). While stimulating mutant degradation will reduce ER stress, it will also reduce mutant collagen secretion, thus overcoming any possible dominant negative effects that secreted mutant could have on the ECM. However, clearly with this approach, there will be an overall reduction in the total collagen deposited in the matrix. While 50 % of the normal level of collagen II caused by heterozygous null alleles leads to a relatively mild Stickler phenotype, a much greater reduction might result from stimulating breakdown depending on how many of the mutant and normal collagen II heterotrimers are degraded. This is likely to be mutation specific and it will be important to experimentally determine the ECM impacts of this approach.

A third therapeutic approach that is being explored in several diseases including cancer, neurodegenerative disorders, metabolic diseases, heart ischaemia, inflammation and liver disease is targeting the UPR sensing and downstream signalling pathways (Maly and Papa 2014; Rivas et al. 2015; Jiang et al. 2015; Hetz et al. 2013). Both IRE1α and PERK have druggable enzyme activities and are the main focus for drug development. Small molecules that modulate PERK signalling and the IRE1α-XBP1 arm are generating a lot of interest, and some are already FDA approved or in preclinical trials for cancer (Maly and Papa 2014; Jiang et al. 2015; Hetz et al. 2013). Natural compounds that upregulate or downregulate ATF6 expression have also been identified and promise therapeutic potential but their mode of action is not currently known (Rivas et al. 2015).

The downstream cellular and extracellular consequences of different collagen mutations are clearly complex and much work remains to be done before we have a sophisticated and detailed understanding. A major challenge to advancing effective therapies for the collagen chondrodysplasias that appropriately target the different signalling branches of the UPR will be developing accurate human in vitro disease models that can be used to fully define the molecular pathology and used for high-throughput drug screening. A powerful emerging approach is using patient-derived induced pluripotent stem cells (iPSCs) and differentiating them into chondrocytes (Okada et al. 2015; Tsumaki et al. 2015) to produce in vitro cartilage. The utility of this approach was recently demonstrated in elegant experiments that showed statins could rescue the poor chondrogenic differentiation of *FGFR3* mutant iPSCs and then went on to show that lovastatin treatment of *Fgfr3* mutant mice significantly improved their bone growth (Yamashita et al. 2014). While these approaches offer hope for treating genetic chondrodysplasias once thought to be intractable, we need to go forward with a systematic approach that assesses the consequences of manipulating the UPR on global cellular pathways as well as the organisation and structure of the extracellular matrix.

References

Abad V, Meyers JL, Weise M, Gafni RI, Barnes KM, Nilsson O, Bacher JD, Baron J (2002) The role of the resting zone in growth plate chondrogenesis. Endocrinology 143(5):1851–1857. doi:10.1210/endo.143.5.8776

Akiyama H, Chaboissier MC, Martin JF, Schedl A, de Crombrugghe B (2002) The transcription factor Sox9 has essential roles in successive steps of the chondrocyte differentiation pathway and is required for expression of Sox5 and Sox6. Genes Dev 16(21):2813–2828. doi:10.1101/gad.1017802

Alexopoulos LG, Youn I, Bonaldo P, Guilak F (2009) Developmental and osteoarthritic changes in Col6a1-knockout mice: biomechanics of type VI collagen in the cartilage pericellular matrix. Arthritis Rheum 60(3):771–779. doi:10.1002/art.24293

Andreucci E, Aftimos S, Alcausin M, Haan E, Hunter W, Kannu P, Kerr B, McGillivray G, McKinlay Gardner RJ, Patricelli MG, Sillence D, Thompson E, Zacharin M, Zankl A, Lamande SR, Savarirayan R (2011) TRPV4 related skeletal dysplasias: a phenotypic spectrum highlighted byclinical, radiographic, and molecular studies in 21 new families. Orphanet J Rare Dis 6:37. doi:10.1186/1750-1172-6-37

Arikawa-Hirasawa E, Wilcox WR, Le AH, Silverman N, Govindraj P, Hassell JR, Yamada Y (2001) Dyssegmental dysplasia, Silverman-Handmaker type, is caused by functional null mutations of the perlecan gene. Nat Genet 27(4):431–434. doi:10.1038/86941

Arnold WV, Fertala A (2013) Skeletal diseases caused by mutations that affect collagen structure and function. Int J Biochem Cell Biol 45(8):1556–1567. doi:10.1016/j.biocel.2013.05.017

Aspberg A (2016) Cartilage proteoglycans. In: Grässel S, Aszódi A (eds) Cartilage: volume 1: physiology and development. Springer International Publishing, Cham, pp 1–22. doi:10.1007/978-3-319-29568-8_1

Aszodi A, Hunziker EB, Brakebusch C, Fassler R (2003) Beta1 integrins regulate chondrocyte rotation, G1 progression, and cytokinesis. Genes Dev 17(19):2465–2479. doi:10.1101/gad.277003

Bateman JF, Lamande SR, Dahl HH, Chan D, Mascara T, Cole WG (1989) A frameshift mutation results in a truncated nonfunctional carboxyl-terminal pro alpha 1(I) propeptide of type I collagen in osteogenesis imperfecta. J Biol Chem 264(19):10960–10964

Bateman JF, Freddi S, Nattrass G, Savarirayan R (2003) Tissue-specific RNA surveillance? Nonsense-mediated mRNA decay causes collagen X haploinsufficiency in Schmid metaphyseal chondrodysplasia cartilage. Hum Mol Genet 12(3):217–225

Bateman JF, Freddi S, McNeil R, Thompson E, Hermanns P, Savarirayan R, Lamande SR (2004) Identification of four novel COL10A1 missense mutations in schmid metaphyseal chondrodysplasia: further evidence that collagen X NC1 mutations impair trimer assembly. Hum Mutat 23(4):396

Bateman JF, Wilson R, Freddi S, Lamande SR, Savarirayan R (2005) Mutations of COL10A1 in Schmid metaphyseal chondrodysplasia. Hum Mutat 25(6):525–534

Bateman JF, Boot-Handford RP, Lamande SR (2009) Genetic diseases of connective tissues: cellular and extracellular effects of ECM mutations. Nat Rev 10(3):173–183

Baur ST, Mai JJ, Dymecki SM (2000) Combinatorial signaling through BMP receptor IB and GDF5: shaping of the distal mouse limb and the genetics of distal limb diversity. Development 127(3):605–619

Belluoccio D, Bernardo BC, Rowley L, Bateman JF (2008) A microarray approach for comparative expression profiling of the discrete maturation zones of mouse growth plate cartilage. Biochim Biophys Acta 1779(5):330–340

Bi W, Deng JM, Zhang Z, Behringer RR, de Crombrugghe B (1999) Sox9 is required for cartilage formation. Nat Genet 22(1):85–89. doi:10.1038/8792

Bogin O, Kvansakul M, Rom E, Singer J, Yayon A, Hohenester E (2002) Insight into Schmid metaphyseal chondrodysplasia from the crystal structure of the collagen X NC1 domain trimer. Structure 10(2):165–173

Bonafe L, Cormier-Daire V, Hall C, Lachman R, Mortier G, Mundlos S, Nishimura G, Sangiorgi L, Savarirayan R, Sillence D, Spranger J, Superti-Furga A, Warman M, Unger S (2015) Nosology and classification of genetic skeletal disorders: 2015 revision. Am J Med Genet A 167(12):2869–2892. doi:10.1002/ajmg.a.37365

Bonaventure J, Cohen-Solal L, Ritvaniemi P, Van Maldergem L, Kadhom N, Delezoide AL, Maroteaux P, Prockop DJ, Ala-Kokko L (1995) Substitution of aspartic acid for glycine at position 310 in type II collagen produces achondrogenesis II, and substitution of serine at position

805 produces hypochondrogenesis: analysis of genotype-phenotype relationships. Biochem J 307(Pt 3):823–830

Boot-Handford RP, Briggs MD (2010) The unfolded protein response and its relevance to connective tissue diseases. Cell Tissue Res 339(1):197–211. doi:10.1007/s00441-009-0877-8

Briggs MD, Chapman KL (2002) Pseudoachondroplasia and multiple epiphyseal dysplasia: mutation review, molecular interactions, and genotype to phenotype correlations. Hum Mutat 19(5): 465–478

Brunetti-Pierri N, Andreucci MV, Tuzzi R, Vega GR, Gray G, McKeown C, Ballabio A, Andria G, Meroni G, Parenti G (2003) X-linked recessive chondrodysplasia punctata: spectrum of arylsulfatase E gene mutations and expanded clinical variability. Am J Med Genet A 117A(2): 164–168

Bushby KM, Collins J, Hicks D (2014) Collagen type VI myopathies. Adv Exp Med Biol 802:185–199. doi:10.1007/978-94-007-7893-1_12

Buxton P, Edwards C, Archer CW, Francis-West P (2001) Growth/differentiation factor-5 (GDF-5) and skeletal development. J Bone Joint Surg Am Vol 83-A(Suppl 1 (Pt 1)):S23–S30

Cameron TL, Belluoccio D, Farlie PG, Brachvogel B, Bateman JF (2009) Global comparative transcriptome analysis of cartilage formation in vivo. BMC Dev Biol 9:20

Cameron TL, Bell KM, Tatarczuch L, Mackie EJ, Rajpar MH, McDermott BT, Boot-Handford RP, Bateman JF (2011) Transcriptional profiling of chondrodysplasia growth plate cartilage reveals adaptive ER-stress networks that allow survival but disrupt hypertrophy. PLoS ONE 6(9):e24600. doi:10.1371/journal.pone.0024600, PONE-D-11-07777 [pii]

Cameron TL, Bell KM, Gresshoff IL, Sampurno L, Mullan L, Ermann J, Glimcher LH, Boot-Handford RP, Bateman JF (2015) XBP1-independent UPR pathways suppress C/EBP-beta mediated chondrocyte differentiation in ER-stress related skeletal disease. PLoS Genet 11(9):e1005505. doi:10.1371/journal.pgen.1005505

Cao SS, Kaufman RJ (2013) Targeting endoplasmic reticulum stress in metabolic disease. Expert Opin Ther Targets 17(4):437–448. doi:10.1517/14728222.2013.756471

Carlson HL, Quinn JJ, Yang YW, Thornburg CK, Chang HY, Stadler HS (2015) LncRNA-HIT functions as an epigenetic regulator of chondrogenesis through its recruitment of p100/CBP complexes. PLoS Genet 11(12):e1005680. doi:10.1371/journal.pgen.1005680

Chan D, Taylor TK, Cole WG (1993) Characterization of an arginine 789 to cysteine substitution in alpha 1 (II) collagen chains of a patient with spondyloepiphyseal dysplasia. J Biol Chem 268(20):15238–15245

Chan D, Cole WG, Chow CW, Mundlos S, Bateman JF (1995) A COL2A1 mutation in achondrogenesis type II results in the replacement of type II collagen by type I and III collagens in cartilage. J Biol Chem 270(4):1747–1753

Chan D, Weng YM, Graham HK, Sillence DO, Bateman JF (1998) A nonsense mutation in the carboxyl-terminal domain of type X collagen causes haploinsufficiency in schmid metaphyseal chondrodysplasia. J Clin Invest 101(7):1490–1499

Chan D, Ho MS, Cheah KS (2001) Aberrant signal peptide cleavage of collagen X in Schmid metaphyseal chondrodysplasia. Implications for the molecular basis of the disease. J Biol Chem 276(11):7992–7997. doi:10.1074/jbc.M003361200

Chessler SD, Byers PH (1993) BiP binds type I procollagen pro alpha chains with mutations in the carboxyl-terminal propeptide synthesized by cells from patients with osteogenesis imperfecta. J Biol Chem 268(24):18226–18233

Chung HJ, Jensen DA, Gawron K, Steplewski A, Fertala A (2009) R992C (p.R1192C) substitution in collagen II alters the structure of mutant molecules and induces the unfolded protein response. J Mol Biol 390(2):306–318. doi:10.1016/j.jmb.2009.05.004

Colnot C, Lu C, Hu D, Helms JA (2004) Distinguishing the contributions of the perichondrium, cartilage, and vascular endothelium to skeletal development. Dev Biol 269(1):55–69. doi:10.1016/j.ydbio.2004.01.011

Cotterill SL, Jackson GC, Leighton MP, Wagener R, Makitie O, Cole WG, Briggs MD (2005) Multiple epiphyseal dysplasia mutations in MATN3 cause misfolding of the A-domain and prevent secretion of mutant matrilin-3. Hum Mutat 26(6):557–565. doi:10.1002/humu.20263

Danielson KG, Baribault H, Holmes DF, Graham H, Kadler KE, Iozzo RV (1997) Targeted disruption of decorin leads to abnormal collagen fibril morphology and skin fragility. J Cell Biol 136(3):729–743

Day TF, Guo X, Garrett-Beal L, Yang Y (2005) Wnt/beta-catenin signaling in mesenchymal progenitors controls osteoblast and chondrocyte differentiation during vertebrate skeletogenesis. Dev Cell 8(5):739–750, doi:S1534-5807(05)00106-1 [pii] 10.1016/j.devcel.2005.03.016

Douglas T, Heinemann S, Bierbaum S, Scharnweber D, Worch H (2006) Fibrillogenesis of collagen types I, II, and III with small leucine-rich proteoglycans decorin and biglycan. Biomacromolecules 7(8):2388–2393. doi:10.1021/bm0603746

Eames BF, de la Fuente L, Helms JA (2003) Molecular ontogeny of the skeleton. Birth Defects Res C Embryo Today Rev 69(2):93–101. doi:10.1002/bdrc.10016

Esapa CT, Hough TA, Testori S, Head RA, Crane EA, Chan CP, Evans H, Bassett JH, Tylzanowski P, McNally EG, Carr AJ, Boyde A, Howell PG, Clark A, Williams GR, Brown MA, Croucher PI, Nesbit MA, Brown SD, Cox RD, Cheeseman MT, Thakker RV (2012) A mouse model for spondyloepiphyseal dysplasia congenita with secondary osteoarthritis due to a Col2a1 mutation. J Bone Miner Res 27(2):413–428. doi:10.1002/jbmr.547

Exome Aggregation Consortium (ExAC) C, MA (URL: http://exac.broadinstitute.org) [date (month, year) accessed]

Eyre DR (2004) Collagens and cartilage matrix homeostasis. Clin Orthop Relat Res (427 Suppl):S118–S122

Eyre DR, Wu JJ, Fernandes RJ, Pietka TA, Weis MA (2002) Recent developments in cartilage research: matrix biology of the collagen II/IX/XI heterofibril network. Biochem Soc Trans 30(Pt 6):893–899. doi:10.1042/

Faiyaz ul Haque M, King LM, Krakow D, Cantor RM, Rusiniak ME, Swank RT, Superti-Furga A, Haque S, Abbas H, Ahmad W, Ahmad M, Cohn DH (1998) Mutations in orthologous genes in human spondyloepimetaphyseal dysplasia and the brachymorphic mouse. Nat Genet 20(2):157–162. doi:10.1038/2458

Fang Y, Bateman JF, Mercer JF, Lamande SR (2013) Nonsense-mediated mRNA decay of collagen -emerging complexity in RNA surveillance mechanisms. J Cell Sci 126(Pt 12):2551–2560. doi:10.1242/jcs.120220

Felka T, Rothdiener M, Bast S, Uynuk-Ool T, Zouhair S, Ochs BG, De Zwart P, Stoeckle U, Aicher WK, Hart ML, Shiozawa T, Grodzinsky AJ, Schenke-Layland K, Venkatesan JK, Cucchiarini M, Madry H, Kurz B, Rolauffs B (2016) Loss of spatial organization and destruction of the pericellular matrix in early osteoarthritis in vivo and in a novel in vitro methodology. Osteoarthr Cartil/OARS Osteoarthr Res Soc 24:1200. doi:10.1016/j.joca.2016.02.001

Fernandes RJ, Weis M, Scott MA, Seegmiller RE, Eyre DR (2007) Collagen XI chain misassembly in cartilage of the chondrodysplasia (cho) mouse. Matrix Biol 26(8):597–603

Fitzgerald J, Lamande SR, Bateman JF (1999) Proteasomal degradation of unassembled mutant type I collagen pro-alpha1(I) chains. J Biol Chem 274(39):27392–27398

Fosang AJ, Beier F (2011) Emerging frontiers in cartilage and chondrocyte biology. Best Pract Res Clin Rheumatol 25(6):751–766. doi:10.1016/j.berh.2011.11.010

Freddi S, Savarirayan R, Bateman JF (2000) Molecular diagnosis of Stickler syndrome: a COL2A1 stop codon mutation screening strategy that is not compromised by mutant mRNA instability. Am J Med Genet 90(5):398–406

Furuichi T, Masuya H, Murakami T, Nishida K, Nishimura G, Suzuki T, Imaizumi K, Kudo T, Ohkawa K, Wakana S, Ikegawa S (2011) ENU-induced missense mutation in the C-propeptide coding region of Col2a1 creates a mouse model of platyspondylic lethal skeletal dysplasia, Torrance type. Mamm Genome 22(5–6):318–328. doi:10.1007/s00335-011-9329-3

Gawron K, Jensen DA, Steplewski A, Fertala A (2010) Reducing the effects of intracellular accumulation of thermolabile collagen II mutants by increasing their thermostability in cell culture conditions. Biochem Biophys Res Commun 396(2):213–218. doi:10.1016/j.bbrc.2010.04.056

Gerber HP, Vu TH, Ryan AM, Kowalski J, Werb Z, Ferrara N (1999) VEGF couples hypertrophic cartilage remodeling, ossification and angiogenesis during endochondral bone formation. Nat Med 5(6):623–628. doi:10.1038/9467

Gleghorn L, Ramesar R, Beighton P, Wallis G (2005) A mutation in the variable repeat region of the aggrecan gene (AGC1) causes a form of spondyloepiphyseal dysplasia associated with severe, premature osteoarthritis. Am J Hum Genet 77(3):484–490

Grässel S (2016) Collagens in hyaline cartilage. In: Grässel S, Aszódi A (eds) Cartilage: volume 1: physiology and development. Springer International Publishing, Cham, pp 23–53. doi:10.1007/978-3-319-29568-8_2

Hall BK, Miyake T (1995) Divide, accumulate, differentiate: cell condensation in skeletal development revisited. Int J Dev Biol 39(6):881–893

Harel T, Rabinowitz R, Hendler N, Galil A, Flusser H, Chemke J, Gradstein L, Lifshitz T, Ofir R, Elbedour K, Birk OS (2005) COL11A2 mutation associated with autosomal recessive Weissenbacher-Zweymuller syndrome: molecular and clinical overlap with otospondylomegaepiphyseal dysplasia (OSMED). Am J Med Genet A 132A(1):33–35. doi:10.1002/ajmg.a.30371

Hedbom E, Heinegard D (1993) Binding of fibromodulin and decorin to separate sites on fibrillar collagens. J Biol Chem 268(36):27307–27312

Heinegard D (2009) Proteoglycans and more – from molecules to biology. Int J Exp Pathol 90(6):575–586. doi:10.1111/j.1365-2613.2009.00695.x

Hessle L, Stordalen GA, Wenglen C, Petzold C, Tanner E, Brorson SH, Baekkevold ES, Onnerfjord P, Reinholt FP, Heinegard D (2014) The skeletal phenotype of chondroadherin deficient mice. PLoS ONE 8(6):e63080. doi:10.1371/journal.pone.0063080

Hetz C, Martinon F, Rodriguez D, Glimcher LH (2011) The unfolded protein response: integrating stress signals through the stress sensor IRE1alpha. Physiol Rev 91(4):1219–1243, doi:91/4/1219 [pii] 10.1152/physrev.00001.2011

Hetz C, Chevet E, Harding HP (2013) Targeting the unfolded protein response in disease. Nat Rev Drug Discov 12(9):703–719. doi:10.1038/nrd3976

Hintze V, Steplewski A, Ito H, Jensen DA, Rodeck U, Fertala A (2008) Cells expressing partially unfolded R789C/p.R989C type II procollagen mutant associated with spondyloepiphyseal dysplasia undergo apoptosis. Hum Mutat 29(6):841–851

Ho MS, Tsang KY, Lo RL, Susic M, Makitie O, Chan TW, Ng VC, Sillence DO, Boot-Handford RP, Gibson G, Cheung KM, Cole WG, Cheah KS, Chan D (2007) COL10A1 nonsense and frame-shift mutations have a gain-of-function effect on the growth plate in human and mouse metaphyseal chondrodysplasia type Schmid. Hum Mol Genet 16(10):1201–1215

Hoornaert KP, Dewinter C, Vereecke I, Beemer FA, Courtens W, Fryer A, Fryssira H, Lees M, Mullner-Eidenbock A, Rimoin DL, Siderius L, Superti-Furga A, Temple K, Willems PJ, Zankl A, Zweier C, De Paepe A, Coucke P, Mortier GR (2006) The phenotypic spectrum in patients with arginine to cysteine mutations in the COL2A1 gene. J Med Genet 43(5):406–413. doi:10.1136/jmg.2005.035717

Hurd L, Kirwin SM, Boggs M, Mackenzie WG, Bober MB, Funanage VL, Duncan RL (2015) A mutation in TRPV4 results in altered chondrocyte calcium signaling in severe metatropic dysplasia. Am J Med Genet A 167A(10):2286–2293. doi:10.1002/ajmg.a.37182

Ikegawa S, Nakamura K, Nagano A, Haga N, Nakamura Y (1997) Mutations in the N-terminal globular domain of the type X collagen gene (COL10A1) in patients with Schmid metaphyseal chondrodysplasia. Hum Mutat 9(2):131–135. doi:10.1002/(SICI)1098-1004(1997)9:2<131::AID-HUMU5>3.0.CO;2-C

Iozzo RV, Karamanos N (2010) Proteoglycans in health and disease: emerging concepts and future directions. FEBS J 277(19):3863. doi:10.1111/j.1742-4658.2010.07796.x

Iozzo RV, Schaefer L (2010) Proteoglycans in health and disease: novel regulatory signaling mechanisms evoked by the small leucine-rich proteoglycans. FEBS J 277(19):3864–3875. doi:10.1111/j.1742-4658.2010.07797.x

Ishida Y, Yamamoto A, Kitamura A, Lamande SR, Yoshimori T, Bateman JF, Kubota H, Nagata K (2009) Autophagic elimination of misfolded procollagen aggregates in the endoplasmic reticulum as a means of cell protection. Mol Biol Cell 20(11):2744–2754, doi:E08-11-1092 [pii] 10.1091/mbc.E08-11-1092

Ishikawa Y, Bachinger HP (2013) A molecular ensemble in the rER for procollagen maturation. Biochim Biophys Acta 1833(11):2479–2491. doi:10.1016/j.bbamcr.2013.04.008

Ishikawa Y, Boudko S, Bachinger HP (2015) Ziploc-ing the structure: triple helix formation is coordinated by rough endoplasmic reticulum resident PPIases. Biochim Biophys Acta 1850(10):1983–1993. doi:10.1016/j.bbagen.2014.12.024

Ito H, Rucker E, Steplewski A, McAdams E, Brittingham RJ, Alabyeva T, Fertala A (2005) Guilty by association: some collagen II mutants alter the formation of ECM as a result of atypical interaction with fibronectin. J Mol Biol 352(2):382–395. doi:10.1016/j.jmb.2005.07.019

Jayasuriya CT, Zhou FH, Pei M, Wang Z, Lemme NJ, Haines P, Chen Q (2014) Matrilin-3 chondrodysplasia mutations cause attenuated chondrogenesis, premature hypertrophy and aberrant response to TGF-beta in chondroprogenitor cells. Int J Mol Sci 15(8):14555–14573. doi:10.3390/ijms150814555

Jackson GC, Mittaz-Crettol L, Taylor JA, Mortier GR, Spranger J, Zabel B, Le Merrer M, Cormier-Daire V, Hall CM, Offiah A, Wright MJ, Savarirayan R, Nishimura G, Ramsden SC, Elles R, Bonafe L, Superti-Furga A, Unger S, Zankl A, Briggs MD (2012) Pseudoachondroplasia and multiple epiphyseal dysplasia: a 7-year comprehensive analysis of the known disease genes identify novel and recurrent mutations and provides an accurate assessment of their relative contribution. Hum Mutat 33(1):144–157. doi: 10.1002/humu.21611.

Jiang D, Niwa M, Koong AC (2015) Targeting the IRE1alpha-XBP1 branch of the unfolded protein response in human diseases. Semin Cancer Biol 33:48–56. doi:10.1016/j.semcancer.2015.04.010

Kamiya N, Watanabe H, Habuchi H, Takagi H, Shinomura T, Shimizu K, Kimata K (2006) Versican/PG-M regulates chondrogenesis as an extracellular matrix molecule crucial for mesenchymal condensation. J Biol Chem 281(4):2390–2400. doi:10.1074/jbc.M509341200

Kannu P, Aftimos S, Mayne V, Donnan L, Savarirayan R (2007) Metatropic dysplasia: clinical and radiographic findings in 11 patients demonstrating long-term natural history. Am J Med Genet A 143A(21):2512–2522. doi:10.1002/ajmg.a.31941

Kannu P, Bateman JF, Randle S, Cowie S, du Sart D, McGrath S, Edwards M, Savarirayan R (2010) Premature arthritis is a distinct type II collagen phenotype. Arthritis Rheum 62(5):1421–1430. doi:10.1002/art.27354

Kannu P, O'Rielly DD, Hyland JC, Kokko LA (2011) Avascular necrosis of the femoral head due to a novel C propeptide mutation in COL2A1. Am J Med Genet A 155A(7):1759–1762. doi:10.1002/ajmg.a.34056

Kannu P, Bateman J, Savarirayan R (2012) Clinical phenotypes associated with type II collagen mutations. J Paediatr Child Health 48(2):E38–E43. doi:10.1111/j.1440-1754.2010.01979.x

Khoshnoodi J, Cartailler JP, Alvares K, Veis A, Hudson BG (2006) Molecular recognition in the assembly of collagens: terminal noncollagenous domains are key recognition modules in the formation of triple helical protomers. J Biol Chem 281(50):38117–38121

Kielty CM, Kwan AP, Holmes DF, Schor SL, Grant ME (1985) Type X collagen, a product of hypertrophic chondrocytes. Biochem J 227(2):545–554

Kimura M, Ichimura S, Sasaki K, Masuya H, Suzuki T, Wakana S, Ikegawa S, Furuichi T (2015) Endoplasmic reticulum stress-mediated apoptosis contributes to a skeletal dysplasia resembling platyspondylic lethal skeletal dysplasia, Torrance type, in a novel Col2a1 mutant mouse line. Biochem Biophys Res Commun 468(1–2):86–91. doi:10.1016/j.bbrc.2015.10.160

Kronenberg HM (2006) PTHrP and skeletal development. Ann N Y Acad Sci 1068:1–13. doi:10.1196/annals.1346.002

Lachman RS, Rimoin DL, Spranger J (1988) Metaphyseal chondrodysplasia, Schmid type. Clinical and radiographic delineation with a review of the literature. Pediatr Radiol 18(2):93–102

Lamande SR, Chessler SD, Golub SB, Byers PH, Chan D, Cole WG, Sillence DO, Bateman JF (1995) Endoplasmic reticulum-mediated quality control of type I collagen production by cells from osteogenesis imperfecta patients with mutations in the pro alpha 1 (I) chain carboxyl-terminal propeptide which impair subunit assembly. J Biol Chem 270(15):8642–8649

Lamande SR, Yuan Y, Gresshoff IL, Rowley L, Belluoccio D, Kaluarachchi K, Little CB, Botzenhart E, Zerres K, Amor DJ, Cole WG, Savarirayan R, McIntyre P, Bateman JF (2011)

Mutations in TRPV4 cause an inherited arthropathy of hands and feet. Nat Genet 43(11):1142–1146, doi:ng.945 [pii] 10.1038/ng.945

Lehmann K, Seemann P, Stricker S, Sammar M, Meyer B, Suring K, Majewski F, Tinschert S, Grzeschik KH, Muller D, Knaus P, Nurnberg P, Mundlos S (2003) Mutations in bone morphogenetic protein receptor 1B cause brachydactyly type A2. Proc Natl Acad Sci U S A 100(21):12277–12282. doi:10.1073/pnas.2133476100

Li Y, Lacerda DA, Warman ML, Beier DR, Yoshioka H, Ninomiya Y, Oxford JT, Morris NP, Andrikopoulos K, Ramirez F et al (1995) A fibrillar collagen gene, Col11a1, is essential for skeletal morphogenesis. Cell 80(3):423–430

Liang G, Lian C, Huang D, Gao W, Liang A, Peng Y, Ye W, Wu Z, Su P, Huang D (2014) Endoplasmic reticulum stress-unfolding protein response-apoptosis cascade causes chondrodysplasia in a col2a1 p.Gly1170Ser mutated mouse model. PLoS ONE 9(1):e86894. doi:10.1371/journal.pone.0086894

Lim J, Tu X, Choi K, Akiyama H, Mishina Y, Long F (2015) BMP-Smad4 signaling is required for precartilaginous mesenchymal condensation independent of Sox9 in the mouse. Dev Biol 400(1):132–138. doi:10.1016/j.ydbio.2015.01.022

Lisse TS, Thiele F, Fuchs H, Hans W, Przemeck GK, Abe K, Rathkolb B, Quintanilla-Martinez L, Hoelzlwimmer G, Helfrich M, Wolf E, Ralston SH, de Angelis MH (2008) ER stress-mediated apoptosis in a new mouse model of osteogenesis imperfecta. PLoS Genet 4(2):e7

Loukin S, Su Z, Kung C (2011) Increased basal activity is a key determinant in the severity of human skeletal dysplasia caused by TRPV4 mutations. PLoS ONE 6(5):e19533. doi:10.1371/journal.pone.0019533

Lui JC, Andrade AC, Forcinito P, Hegde A, Chen W, Baron J, Nilsson O (2010) Spatial and temporal regulation of gene expression in the mammalian growth plate. Bone 46(5):1380–1390. doi:10.1016/j.bone.2010.01.373

Maly DJ, Papa FR (2014) Druggable sensors of the unfolded protein response. Nat Chem Biol 10(11):892–901. doi:10.1038/nchembio.1664

Mayne R, Brewton RG, Mayne PM, Baker JR (1993) Isolation and characterization of the chains of type V/type XI collagen present in bovine vitreous. J Biol Chem 268(13):9381–9386

McEwan PA, Scott PG, Bishop PN, Bella J (2006) Structural correlations in the family of small leucine-rich repeat proteins and proteoglycans. J Struct Biol 155(2):294–305. doi:10.1016/j.jsb.2006.01.016

Mechling DE, Gambee JE, Morris NP, Sakai LY, Keene DR, Mayne R, Bachinger HP (1996) Type IX collagen NC1 domain peptides can trimerize in vitro without forming a triple helix. J Biol Chem 271(23):13781–13785

Melkoniemi M, Brunner HG, Manouvrier S, Hennekam R, Superti-Furga A, Kaariainen H, Pauli RM, van Essen T, Warman ML, Bonaventure J, Miny P, Ala-Kokko L (2000) Autosomal recessive disorder otospondylomegaepiphyseal dysplasia is associated with loss-of-function mutations in the COL11A2 gene. Am J Hum Genet 66(2):368–377. doi:10.1086/302750

Mirigian LS, Makareeva E, Mertz EL, Omari S, Roberts-Pilgrim AM, Oestreich AK, Phillips CL, Leikin S (2016) Osteoblast malfunction caused by cell stress response to procollagen misfolding in alpha2(I)-G610C mouse model of osteogenesis imperfecta. J Bone Miner Res. doi:10.1002/jbmr.2824

Miyake N, Elcioglu NH, Iida A, Isguven P, Dai J, Murakami N, Takamura K, Cho TJ, Kim OH, Hasegawa T, Nagai T, Ohashi H, Nishimura G, Matsumoto N, Ikegawa S (2012) PAPSS2 mutations cause autosomal recessive brachyolmia. J Med Genet 49(8):533–538. doi:10.1136/jmedgenet-2012-101039

Mortier GR, Weis M, Nuytinck L, King LM, Wilkin DJ, De Paepe A, Lachman RS, Rimoin DL, Eyre DR, Cohn DH (2000) Report of five novel and one recurrent COL2A1 mutations with analysis of genotype-phenotype correlation in patients with a lethal type II collagen disorder. J Med Genet 37(4):263–271

Mundlos S, Chan D, McGill J, Bateman JF (1996) An alpha 1(II) Gly913 to Cys substitution prevents the matrix incorporation of type II collagen which is replaced with type I and III

collagens in cartilage from a patient with hypochondrogenesis. Am J Med Genet 63(1):129–136
Murray LS, Lu Y, Taggart A, Van Regemorter N, Vilain C, Abramowicz M, Kadler KE, Van Agtmael T (2014) Chemical chaperone treatment reduces intracellular accumulation of mutant collagen IV and ameliorates the cellular phenotype of a COL4A2 mutation that causes haemorrhagic stroke. Hum Mol Genet 23(2):283–292. doi:10.1093/hmg/ddt418
Nicole S, Davoine CS, Topaloglu H, Cattolico L, Barral D, Beighton P, Hamida CB, Hammouda H, Cruaud C, White PS, Samson D, Urtizberea JA, Lehmann-Horn F, Weissenbach J, Hentati F, Fontaine B (2000) Perlecan, the major proteoglycan of basement membranes, is altered in patients with Schwartz-Jampel syndrome (chondrodystrophic myotonia). Nat Genet 26(4):480–483. doi:10.1038/82638
Nikitovic D, Aggelidakis J, Young MF, Iozzo RV, Karamanos NK, Tzanakakis GN (2012) The biology of small leucine-rich proteoglycans in bone pathophysiology. J Biol Chem 287(41):33926–33933. doi:10.1074/jbc.R112.379602
Nilsson O, Guo MH, Dunbar N, Popovic J, Flynn D, Jacobsen C, Lui JC, Hirschhorn JN, Baron J, Dauber A (2014) Short stature, accelerated bone maturation, and early growth cessation due to heterozygous aggrecan mutations. J Clin Endocrinol Metab 99(8):E1510–E1518. doi:10.1210/jc.2014-1332
Nishimura G, Nakashima E, Mabuchi A, Shimamoto K, Shimamoto T, Shimao Y, Nagai T, Yamaguchi T, Kosaki R, Ohashi H, Makita Y, Ikegawa S (2004) Identification of COL2A1 mutations in platyspondylic skeletal dysplasia, Torrance type. J Med Genet 41(1):75–79
Nuka S, Zhou W, Henry SP, Gendron CM, Schultz JB, Shinomura T, Johnson J, Wang Y, Keene DR, Ramirez-Solis R, Behringer RR, Young MF, Hook M (2010) Phenotypic characterization of epiphycan-deficient and epiphycan/biglycan double-deficient mice. Osteoarthr Cartil/OARS Osteoarthr Res Soc 18(1):88–96. doi:10.1016/j.joca.2009.11.006
Oberlender SA, Tuan RS (1994) Expression and functional involvement of N-cadherin in embryonic limb chondrogenesis. Development 120(1):177–187
Okada M, Ikegawa S, Morioka M, Yamashita A, Saito A, Sawai H, Murotsuki J, Ohashi H, Okamoto T, Nishimura G, Imaizumi K, Tsumaki N (2015) Modeling type II collagenopathy skeletal dysplasia by directed conversion and induced pluripotent stem cells. Hum Mol Genet 24(2):299–313. doi:10.1093/hmg/ddu444
Ornitz DM, Marie PJ (2015) Fibroblast growth factor signaling in skeletal development and disease. Genes Dev 29(14):1463–1486. doi:10.1101/gad.266551.115
Park H, Hong S, Cho SI, Cho TJ, Choi IH, Jin DK, Sohn YB, Park SW, Cho HH, Cheon JE, Kim SY, Kim JY, Park SS, Seong MW (2015) Case of mild Schmid-type metaphyseal chondrodysplasia with novel sequence variation involving an unusual mutational site of the COL10A1 gene. Eur J Med Genet 58(3):175–179. doi:10.1016/j.ejmg.2014.12.011
Patterson SE, Dealy CN (2014) Mechanisms and models of endoplasmic reticulum stress in chondrodysplasia. Dev Dyn: Off Publ Am Assoc Anatomists 243(7):875–893. doi:10.1002/dvdy.24131
Rajpar MH, McDermott B, Kung L, Eardley R, Knowles L, Heeran M, Thornton DJ, Wilson R, Bateman JF, Poulsom R, Arvan P, Kadler KE, Briggs MD, Boot-Handford RP (2009) Targeted induction of endoplasmic reticulum stress induces cartilage pathology. PLoS Genet 5(10):e1000691
Richards AJ, Baguley DM, Yates JR, Lane C, Nicol M, Harper PS, Scott JD, Snead MP (2000) Variation in the vitreous phenotype of Stickler syndrome can be caused by different amino acid substitutions in the X position of the type II collagen Gly-X-Y triple helix. Am J Hum Genet 67(5):1083–1094. doi:10.1016/S0002-9297(07)62938-3
Richards AJ, Morgan J, Bearcroft PW, Pickering E, Owen MJ, Holmans P, Williams N, Tysoe C, Pope FM, Snead MP, Hughes H (2002) Vitreoretinopathy with phalangeal epiphyseal dysplasia, a type II collagenopathy resulting from a novel mutation in the C-propeptide region of the molecule. J Med Genet 39(9):661–665
Rivas A, Vidal RL, Hetz C (2015) Targeting the unfolded protein response for disease intervention. Expert Opin Ther Targets 19(9):1203–1218. doi:10.1517/14728222.2015.1053869

Rossi A, Superti-Furga A (2001) Mutations in the diastrophic dysplasia sulfate transporter (DTDST) gene (SLC26A2): 22 novel mutations, mutation review, associated skeletal phenotypes, and diagnostic relevance. Hum Mutat 17(3):159–171. doi:10.1002/humu.1

Stattin EL, Wiklund F, Lindblom K, Onnerfjord P, Jonsson BA, Tegner Y, Sasaki T, Struglics A, Lohmander S, Dahl N, Heinegard D, Aspberg A (2010) A missense mutation in the aggrecan C-type lectin domain disrupts extracellular matrix interactions and causes dominant familial osteochondritis dissecans. Am J Hum Genet 86(2):126–137. doi:10.1016/j.ajhg.2009.12.018

Steplewski A, Ito H, Rucker E, Brittingham RJ, Alabyeva T, Gandhi M, Ko FK, Birk DE, Jimenez SA, Fertala A (2004a) Position of single amino acid substitutions in the collagen triple helix determines their effect on structure of collagen fibrils. J Struct Biol 148(3):326–337. doi:10.1016/j.jsb.2004.07.006

Steplewski A, Majsterek I, McAdams E, Rucker E, Brittingham RJ, Ito H, Hirai K, Adachi E, Jimenez SA, Fertala A (2004b) Thermostability gradient in the collagen triple helix reveals its multi-domain structure. J Mol Biol 338(5):989–998. doi:10.1016/j.jmb.2004.03.037

Stickens D, Behonick DJ, Ortega N, Heyer B, Hartenstein B, Yu Y, Fosang AJ, Schorpp-Kistner M, Angel P, Werb Z (2004) Altered endochondral bone development in matrix metalloproteinase 13-deficient mice. Development 131(23):5883–5895. doi:10.1242/dev.01461

Storm EE, Huynh TV, Copeland NG, Jenkins NA, Kingsley DM, Lee SJ (1994) Limb alterations in brachypodism mice due to mutations in a new member of the TGF beta-superfamily. Nature 368(6472):639–643. doi:10.1038/368639a0

Tan JT, Kremer F, Freddi S, Bell KM, Baker NL, Lamande SR, Bateman JF (2008) Competency for nonsense-mediated reduction in collagen X mRNA is specified by the 3′ UTR and corresponds to the position of mutations in Schmid metaphyseal chondrodysplasia. Am J Hum Genet 82(3):786–793

Thiele H, Sakano M, Kitagawa H, Sugahara K, Rajab A, Hohne W, Ritter H, Leschik G, Nürnberg P, Mundlos S (2004) Loss of chondroitin 6-O-sulfotransferase-1 function results in severe human chondrodysplasia with progressive spinal involvement. Proc Natl Acad Sci U S A 101(27):10155–10160. doi:10.1073/pnas.0400334101

Tompson SW, Merriman B, Funari VA, Fresquet M, Lachman RS, Rimoin DL, Nelson SF, Briggs MD, Cohn DH, Krakow D (2009) A recessive skeletal dysplasia, SEMD aggrecan type, results from a missense mutation affecting the C-type lectin domain of aggrecan. Am J Hum Genet 84(1):72–79. doi:10.1016/j.ajhg.2008.12.001

Tompson SW, Bacino CA, Safina NP, Bober MB, Proud VK, Funari T, Wangler MF, Nevarez L, Ala-Kokko L, Wilcox WR, Eyre DR, Krakow D, Cohn DH (2010) Fibrochondrogenesis results from mutations in the COL11A1 type XI collagen gene. Am J Hum Genet 87(5):708–712. doi:10.1016/j.ajhg.2010.10.009

Tsang KY, Chan D, Cheslett D, Chan WC, So CL, Melhado IG, Chan TW, Kwan KM, Hunziker EB, Yamada Y, Bateman JF, Cheung KM, Cheah KS (2007) Surviving endoplasmic reticulum stress is coupled to altered chondrocyte differentiation and function. PLoS Biol 5(3):e44

Tsang KY, Chan D, Cheah KS (2015) Fate of growth plate hypertrophic chondrocytes: death or lineage extension? Dev Growth Differ 57(2):179–192. doi:10.1111/dgd.12203

Tsumaki N, Okada M, Yamashita A (2015) iPS cell technologies and cartilage regeneration. Bone 70:48–54. doi:10.1016/j.bone.2014.07.011

van Anken E, Braakman I (2005) Versatility of the endoplasmic reticulum protein folding factory. Crit Rev Biochem Mol Biol 40(4):191–228

Velling T, Risteli J, Wennerberg K, Mosher DF, Johansson S (2002) Polymerization of type I and III collagens is dependent on fibronectin and enhanced by integrins alpha 11beta 1 and alpha 2beta 1. J Biol Chem 277(40):37377–37381. doi:10.1074/jbc.M206286200

Wang M, Kaufman RJ (2016) Protein misfolding in the endoplasmic reticulum as a conduit to human disease. Nature 529(7586):326–335. doi:10.1038/nature17041

Wang JR, Wang CJ, Xu CY, Wu XK, Hong D, Shi W, Gong Y, Chen HX, Long F, Wu XM (2016) Signaling cascades governing Cdc42-mediated chondrogenic differentiation and mensenchymal condensation. Genetics 202(3):1055–1069. doi:10.1534/genetics.115.180109

Wiberg C, Klatt AR, Wagener R, Paulsson M, Bateman JF, Heinegard D, Morgelin M (2003) Complexes of matrilin-1 and biglycan or decorin connect collagen VI microfibrils to both collagen II and aggrecan. J Biol Chem 278(39):37698–37704

Widelitz RB, Jiang TX, Murray BA, Chuong CM (1993) Adhesion molecules in skeletogenesis: II. Neural cell adhesion molecules mediate precartilaginous mesenchymal condensations and enhance chondrogenesis. J Cell Physiol 156(2):399–411. doi:10.1002/jcp.1041560224

Williams CJ, Considine EL, Knowlton RG, Reginato A, Neumann G, Harrison D, Buxton P, Jimenez S, Prockop DJ (1993) Spondyloepiphyseal dysplasia and precocious osteoarthritis in a family with an Arg75-->Cys mutation in the procollagen type II gene (COL2A1). Hum Genet 92(5):499–505

Wilson R, Freddi S, Bateman JF (2002) Collagen X chains harboring Schmid metaphyseal chondrodysplasia NC1 domain mutations are selectively retained and degraded in stably transfected cells. J Biol Chem 277(15):12516–12524

Wilson R, Freddi S, Chan D, Cheah KS, Bateman JF (2005) Misfolding of collagen X chains harboring Schmid metaphyseal chondrodysplasia mutations results in aberrant disulfide bond formation, intracellular retention, and activation of the unfolded protein response. J Biol Chem 280(16):15544–15552

Xu L, Flahiff CM, Waldman BA, Wu D, Olsen BR, Setton LA, Li Y (2003) Osteoarthritis-like changes and decreased mechanical function of articular cartilage in the joints of mice with the chondrodysplasia gene (cho). Arthritis Rheum 48(9):2509–2518. doi:10.1002/art.11233

Xu L, Li Z, Liu SY, Xu SY, Ni GX (2015) Asporin and osteoarthritis. Osteoarthr Cartil/OARS Osteoarthr Res Soc 23(6):933–939. doi:10.1016/j.joca.2015.02.011

Yamashita A, Morioka M, Kishi H, Kimura T, Yahara Y, Okada M, Fujita K, Sawai H, Ikegawa S, Tsumaki N (2014) Statin treatment rescues FGFR3 skeletal dysplasia phenotypes. Nature 513(7519):507–511. doi:10.1038/nature13775

Zankl A, Zabel B, Hilbert K, Wildhardt G, Cuenot S, Xavier B, Ha-Vinh R, Bonafe L, Spranger J, Superti-Furga A (2004) Spondyloperipheral dysplasia is caused by truncating mutations in the C-propeptide of COL2A1. Am J Med Genet A 129A(2):144–148. doi:10.1002/ajmg.a.30222

Zankl A, Neumann L, Ignatius J, Nikkels P, Schrander-Stumpel C, Mortier G, Omran H, Wright M, Hilbert K, Bonafe L, Spranger J, Zabel B, Superti-Furga A (2005) Dominant negative mutations in the C-propeptide of COL2A1 cause platyspondylic lethal skeletal dysplasia, torrance type, and define a novel subfamily within the type 2 collagenopathies. Am J Med Genet A 133A(1):61–67. doi:10.1002/ajmg.a.30531

Zaucke F (2016) Cartilage glycoproteins. In: Grässel S, Aszódi A (eds) Cartilage: volume 1: physiology and development. Springer International Publishing, Cham, pp 55–81. doi:10.1007/978-3-319-29568-8_3

Zelenski NA, Leddy HA, Sanchez-Adams J, Zhang J, Bonaldo P, Liedtke W, Guilak F (2015) Type VI collagen regulates pericellular matrix properties, chondrocyte swelling, and mechanotransduction in mouse articular cartilage. Arthritis Rheum (Hoboken, NJ) 67(5):1286–1294. doi:10.1002/art.39034

Zheng Q, Zhou G, Morello R, Chen Y, Garcia-Rojas X, Lee B (2003) Type X collagen gene regulation by Runx2 contributes directly to its hypertrophic chondrocyte-specific expression in vivo. J Cell Biol 162(5):833–842. doi:10.1083/jcb.200211089

7 Pseudoachondroplasia and Multiple Epiphyseal Dysplasia: Molecular Genetics, Disease Mechanisms and Therapeutic Targets

Michael D. Briggs, Peter Bell, and Katarzyna A. Piróg

Abstract

Genetic skeletal diseases (GSDs) are a diverse and complex group of over 450 rare diseases that affect the development and homoeostasis of the skeleton. Although individually rare, as a group of related genetic skeletal diseases, they have an overall prevalence of at least 1 per 4000 children, which extrapolates to a minimum of 225,000 people in the European Union, and this extensive burden in pain and disability leads to poor quality of life and high healthcare costs.

Dominant-negative (qualitative) defects in numerous cartilage structural proteins result in a broad range of GSDs, and this chapter will focus on a disease spectrum resulting from mutations in the glycoproteins, cartilage oligomeric matrix protein (COMP), type IX collagen and matrilin-3, which together cause a continuum of phenotypes that are amongst the most common of the autosomal dominant GSDs.

Pseudoachondroplasia (PSACH) and autosomal dominant multiple epiphyseal dysplasia (MED) define a disease spectrum typified by varying degrees of short-limbed dwarfism, joint pain with stiffness and early-onset osteoarthritis (OA). The generation and deep phenotyping of a range of genetic cell and mouse models of the PSACH and MED disease spectrum has allowed the disease mechanisms to be characterised in detail. Furthermore, the generation of novel phenocopies to model specific disease mechanisms has confirmed the importance of endoplasmic reticulum stress, reduced chondrocyte proliferation and increased dysregulated apoptosis as key indicators of growth plate dysplasia and eventually reduced bone growth. Lastly, new insight into disease-related musculoskeletal complications such as myopathy, ligamentous laxity and tendinopathy has been gained through the analysis of mouse models of the PSACH and MED disease spectrum.

M.D. Briggs (✉) • P. Bell • K.A. Piróg
Institute of Genetic Medicine, Newcastle University, International Centre for Life, Central Parkway, Newcastle upon Tyne NE1 3BZ, UK
e-mail: michael.briggs@newcastle.ac.uk; Peter.Bell3@newcastle.ac.uk; Katarzyna.Pirog@newcastle.ac.uk

S. Grässel, A. Aszódi (eds.), *Cartilage*, DOI 10.1007/978-3-319-45803-8_7

7.1 Endochondral Ossification and the Skeletal Dysplasias

All long bones in the skeleton grow by endochondral ossification (EO), a complex process in which chondrocytes differentiate forming a growth plate (GP), which subsequently results in a remodelling and mineralisation of the cartilage anlagen (Karsenty 2003; Provot and Schipani 2005). Chondrocytes within the GP undergo a coordinated and highly regulated process of cell proliferation, maturation and hypertrophy, at which stage osteoblasts begin to replace the cartilage matrix with the trabecular bone (Fig. 7.1) (Ortega et al. 2004; Buckwalter et al. 1986). At each stage of this process, the chondrocytes synthesise and secrete specific proteins that are incorporated into the ECM extracellular matrix biology. Chondrocyte proliferation and hypertrophy in the growth plate are vital for correct long bone growth, whilst apoptosis of terminal hypertrophic chondrocytes plays a critical role in the transition from chondrogenesis to osteogenesis (Cowell et al. 1987). Reduced chondrocyte proliferation, decreased hypertrophy and increased/dysregulated apoptosis are therefore likely to have a profound effect on EO and the rate of long bone growth.

Disruption to EO leads to delayed and irregular bone formation and results in a heterogeneous group of diseases known as the skeletal dysplasias (Superti-Furga et al. 2001). There are over 300 unique skeletal dysplasia phenotypes, which range in severity from mild to lethal conditions and have an overall prevalence of at least 1/4000 (International nomenclature and classification of the osteochondrodysplasias (1997). International Working Group on Constitutional Diseases of Bone 1998). Many of these individual phenotypes have been grouped

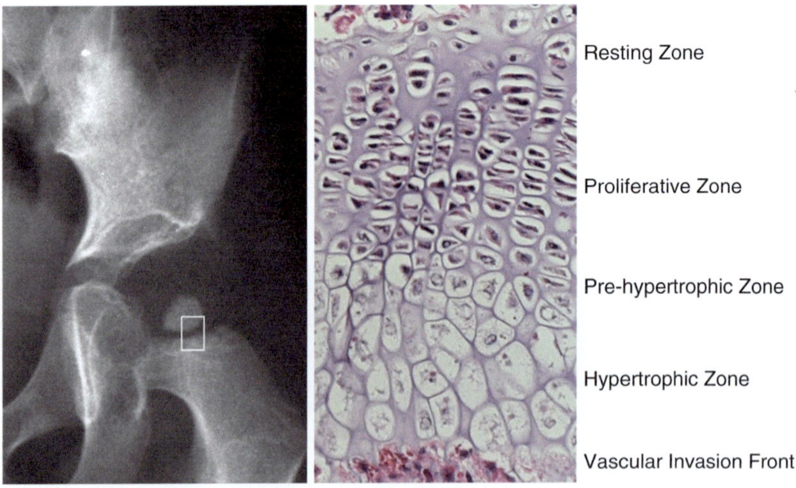

Fig. 7.1 Hip radiograph of a child with pseudoachondroplasia (*left*) demonstrating a small and irregular secondary centre of ossification (epiphyses) and a 3-week-old mouse growth plate highlighting the individual zones

into 'bone dysplasia families' on the basis of similar clinical and radiographic features (Spranger 1988) and are postulated to share common disease mechanisms. One such bone dysplasia family is the pseudoachondroplasia (PSACH) and multiple epiphyseal dysplasia (MED) group of diseases, which are a collection of various phenotypes that are characterised by varying degrees of disproportionate short stature resulting from epimetaphyseal dysplasia (Briggs and Chapman 2002; Bateman et al. 2005).

7.2 The Pseudoachondroplasia and Multiple Epiphyseal Dysplasia Disease Group

PSACH and MED are members of the same bone dysplasia family and were postulated to share a common disease pathophysiology based on clinical and radiographic similarities (Spranger 1988; Briggs and Chapman 2002). This family comprises a broad spectrum of clinically related diseases that manifest with varying degrees of short stature, joint laxity and early-onset osteoarthritis (Table 7.1) (Fairbank 1947; Barrie et al. 1958; Rimoin et al. 1994). In addition, there are allelic conditions and/or phenotypic outliers of the well-established PSACH and MED disease spectrum that share some common clinical, radiographic and genetic characteristics. These diseases include recessive spondyloepimetaphyseal dysplasia, matrilin-3 related (rSEMD, matrilin-3), pseudo-pseudoachondroplasia (pPSACH), Meyers dysplasia, Beukes hip dysplasia (BHD), bilateral Perthes disease (BPD) and dominant osteochondritis dissecans (OCD) (Table 7.1). In addition, mutations in aggrecan can cause phenotypes such as dominant osteochondritis dissecans (with short stature and early-onset osteoarthritis) and idiopathic short stature (with advanced bone age early-onset osteoarthritis) which may be reminiscent of mild MED (Gibson and Briggs in press). Indeed, the carrier of a recessive *ACAN* mutation in a single family with recessive spondyloepimetaphyseal dysplasia aggrecan related (rSEMD, aggrecan) shows some clinical evidence of a mild MED-like phenotype (Tompson et al. 2009).

7.3 Pseudoachondroplasia and Multiple Epiphyseal Dysplasia (PSACH-MED): Known Disease Genes and Their Relative Frequency

PSACH results exclusively from mutations in the gene-encoding cartilage oligomeric matrix protein (COMP) (Briggs et al. 1995; Hecht et al. 1995). Although several studies have failed to identify *COMP* mutations in a small proportion of clinically confirmed cases of PSACH, no alternative gene has yet been identified (Jackson et al. 2012). In addition, a form of pseudo-pseudoachondroplasia has been described that does not result from a *COMP* mutation, but the causative molecular defect also remains unresolved in this single family (Spranger et al. 2005).

Table 7.1 Phenotypes comprising the pseudoachondroplasia and multiple epiphyseal dysplasia disease spectrum, allelic disorders and phenotypic outliers

Gene	Protein	Phenotype	Genetic symbol	OMIN	ORPHA
COMP	Cartilage oligomeric matrix protein	PSACH	PSACH	177170	ORPHA750
		MED	EDM1	132400	ORPHA93308
Unknown	Not known	pPSACH			
MATN3	Matrilin-3	Epiphyseal dysplasia, multiple, 5 (microepiphyseal dysplasia, bilateral hereditary)	EDM5 (BHMD)	607078	ORPHA93311
		rSEMD matrilin-3 related		608728	ORPHA156728
		OA susceptibility 2	OS2	140600	
COL9A1	Type IX collagen	Epiphyseal dysplasia, multiple, 6	EDM6	614135	ORPHA166002
COL9A2		Epiphyseal dysplasia, multiple, 2	EDM2	600204	ORPHA166002
COL9A3		Epiphyseal dysplasia, multiple, 3 (with myopathy included)	EDM3	600969	ORPHA166002
SLC26A2	Solute carrier family 26 (anion exchanger), Member 2	Epiphyseal dysplasia, multiple, 4	EDM4	226900	ORPHA93307
COL2A1	Type II collagen	MED variants that have clinical and radiographic overlap with SED congenital, including EDMMD			ORPHA166011
					ORPHA2380
					ORPHA93279
					ORPHA86820
ACAN	Aggrecan	dOCD	OD	165800	ORPHA251262
		rSEMD, aggrecan type		612813	ORPHA171866
		Short stature-advanced bone age-early-onset osteoarthritis syndrome		n/a	ORPHA435804
		SED Kimberley type	SEDK	608361	ORPHA93283

Table 7.1 (continued)

Gene	Protein	Phenotype	Genetic symbol	OMIN	ORPHA
UFSP2	Ubiquitin-fold modifier-1	Beukes familial hip dysplasia	BHD	142669	ORPHA2114

Each of the confirmed members of the PSACH–MED disease spectrum (PSACH; EMD1-6) is shown along with other phenotypes that are either allelic or share some phenotypic similarities. When available the genetic symbol, On-line Mendelian Inheritance in Man (OMIM: http://www.omim.org) and Orphanet (ORPHA: http://www.orpha.net) reference numbers are shown

Key: *PSACH* pseudoachondroplasia, *MED* multiple epiphyseal dysplasia, *pPSACH* pseudo-pseudoachondroplasia, *rSEMD* recessive spondyloepimetaphyseal dysplasia, *OA* osteoarthritis, *SED* spondyloepiphyseal dysplasia, *EDMMD* epiphyseal dysplasia, multiple with myopia and conductive deafness, *dOCD* dominant osteochondritis dissecans, *BHMD* bilateral hereditary microepiphyseal dysplasia, *BHD* Beukes familial hip dysplasia

Autosomal dominant MED is a genetically heterogeneous disease, and autosomal dominant forms can result from mutations in the gene-encoding COMP, matrilin-3 (*MATN3*) (Chapman et al. 2001) and type IX collagen (*COL9A1*, *COL9A2* and *COL9A3*) (Muragaki et al. 1996; Czarny-Ratajczak et al. 2001; Bonnemann et al. 2000). More recently, mutations in exon 50 of *COL2A1* have been identified in two families with a MED-like phenotype (Jackson et al. 2012). Finally, a recessive form of MED can result from mutations in the 26A2 sulphate transporter (*SLC26A2*) (Rossi and Superti-Furga 2001). Regardless of the individual genes, the majority of MED mutations cluster in distinct regions and usually affect conserved residues that are important for the structure and/or function of the relevant proteins (Briggs and Chapman 2002).

Taken together mutations in the six known genes can account for between 50 and 70% of MED depending on ascertainment criteria (Table 7.1) (Itoh et al. 2006; Zankl et al. 2007; Newton et al. 1994; Jackson et al. 2012). These data suggest that additional MED-disease loci exist, but their identity has not been determined to date, and mutations in these unknown genes do not account for the majority of MED cases as previously suggested (Jakkula et al. 2005). It is interesting to speculate which other genes might be associated with the development of MED; for example, one study has excluded matrilin-1 mutations as a cause of MED (Jackson et al. 2012). However, it remains possible that mutations in other structural proteins expressed in the cartilage growth plate can cause MED, but their identity remains elusive. The use of next-generation sequencing on small families and cohorts of patients, who do not have mutations in the known genes, will provide a panel of novel candidate variants for functional validation.

The clinical presentation of MED caused by type IX collagen gene mutations can also be associated with osteochondritis dissecans (OCD) and mild myopathy (Jackson et al. 2010). However, like PSACH and MED caused by *COMP* mutations, the myopathy does not appear to be a consistent feature and is poorly documented. Furthermore, the lack of suitable tissue obtained through surgery has limited our insight into these secondary consequences of type IX collagen and COMP gene mutations. The study of relevant animal models of PSACH and MED has provided new insight into these additional musculoskeletal complications of PSACH and MED and is discussed elsewhere in this chapter.

7.4 PSACH and MED–COMP Related (EDM1: Genetic and Cell-Based Studies

COMP is the fifth member of the thrombospondin (TSP) protein family and is a modular protein that comprises a coiled-coil oligomerisation domain, four TSP type II repeats (TSP2), eight TSP type III repeats (TSP3) and a large C-terminal globular domain (CTD) (Newton et al. 1994). The precise role of COMP in the cartilage ECM has not been fully determined, but it can interact in vitro with matrilin-3, type II and IX collagen molecules via the CTD (Rosenberg et al. 1998; Holden et al. 2001; Thur et al. 2001; Mann et al. 2004). COMP has been shown to act as a catalyst in collagen fibrillogenesis (Halasz et al. 2007) and binds to chondrocytes via interactions with integrins, suggesting that it also plays a role in cell-matrix signalling (Chen et al. 2005).

Of the many different *COMP* mutations identified to date, the majority (85 %) are found within exons 8–14, which encode the Ca^{2+}-binding TSP3 of COMP; however, key mutations (~15 %) have also been identified within the CTD (Briggs et al. 2014) that appear to cluster in distinct regions. Approximately 50 % of the TSP3 mutations are small in-frame deletions or duplications, whilst the common p.Asp473del mutation accounts for ~30 % of all PSACH (Briggs et al. 2014).

Over the last 20 years, the effect of COMP mutations on chondrocyte function and ECM assembly has been studied using three different cell model systems. These have been (a) a limited microscopic analysis of patient cartilage obtained from iliac crest biopsy (Hecht et al. 2001; Hecht et al. 2004; Cotterill et al. 2005), (b) the redifferentiation and culture of patient chondrocytes in vitro (Hecht et al. 1998, 2001, 2005) and (c) the use of cell culture systems employing a variety of chondrocyte and non-chondrocyte cells (Chen et al. 2004; Hashimoto et al. 2003; Cotterill et al. 2005; Otten et al. 2005). These various studies have consistently demonstrated that mutant COMP harbouring type III mutations is retained within the rough endoplasmic reticulum (rER) of chondrocytes along with other ECM proteins such as type IX collagen and matrilin-3 (Hecht et al. 2005). The retained mutant COMP is also associated with a number of chaperone proteins that are involved in either protein folding or the unfolded protein response (UPR) (Hecht et al. 2001). Furthermore, several studies have demonstrated that the retention of mutant COMP eventually results in increased cell death (Hecht et al. 2004; Hashimoto et al. 2003).

Interestingly, in contrast to the type III mutations, the expression of COMP harbouring a CTD mutation (Thr587Arg) has shown that this mutant protein is efficiently secreted in vitro and therefore likely to exert a dominant-negative effect within the ECM (Schmitz et al. 2006; Spitznagel et al. 2004). However, very little is known to date about the precise effect of these mutant COMP molecules on collagen fibrillogenesis, ECM assembly and cell-matrix signalling.

7.5 MED–MATN3 Related (EDM5): Genetic and Cell-Based Studies

Matrilin-3 is the third member of a family of four ECM proteins; matrilin-1 and matrilin-3 are primarily expressed in the cartilage, whilst matrilin-2 and matrilin-4 have a wider pattern of expression in a variety of ECM including nonskeletal tissues (Wagener et al. 2005). Matrilin-3 consists of a single von Willebrand factor A-like domain (A-domain), four EGF-like motifs and a coiled-coil oligomerisation domain (Wagener et al. 1997). Matrilin-3 can form hetero-oligomers with matrilin-1 (Zhang and Chen 2000; Wu and Eyre 1998) and has been shown to bind to COMP and type II and type IX collagen in vitro (Budde et al. 2005; Mann et al. 2004) (for further details see Chap. 3 in volume 1 (Zaucke 2016)).

All of the MED mutations identified in *MATN3* are missense mutations, which primarily affect conserved residues that comprise the single A-domain of matrilin-3. Of the numerous *MATN3* mutations reported to date, most affect residues in the β-strands that comprise the single β-sheet (Jackson et al. 2004; Cotterill et al. 2005; Mabuchi et al. 2004; Mostert et al. 2003; Maeda et al. 2005). The exceptions to this rule are several missense mutations located in the α-helical regions of the A-domain (Maeda et al. 2005; Mabuchi et al. 2004; Fresquet 2007a). Overall, these data confirm that mutations, which affect important and/or conserved residues in both the α-helices and β-sheets of the matrilin-3 A-domain, can cause MED (Jackson et al. 2012).

The effect that A-domain mutations have on the folding and trafficking of matrilin-3 has been studied in vitro using surrogate cell models. Mutant matrilin-3 is retained within the rER of cells (Otten et al. 2005; Cotterill et al. 2005), where it exists as an unfolded intermediate and is associated with ERp72, a chaperone protein known to be involved in mediating disulphide bond formation (Cotterill et al. 2005). In contrast, the effect of α-helix mutations on the folding and trafficking of matrilin-3 and ultimately disease mechanism(s) has not been fully determined. An in vitro expression system, previously used to characterise β-sheet mutations (Cotterill et al. 2005), was used to study the effect of α-helical mutations on the structure and function of mutant A-domains (Fresquet 2007a). Interestingly, two of the three α-helical mutations (p.F105S and pA173D) prevented the secretion of mutant A-domain in vitro, whilst p.K231N did not prevent the secretion of the matrilin-3 A-domain, nor did it disrupt the structure of this domain or inhibit its binding to type II or type IX collagen (Fresquet 2007a). Therefore, despite extensive biochemical analysis, the disease mechanism of several *MATN3* variants remains unresolved.

In addition to the MED mutations in *MATN3*, two mutations have been identified in the first EGF-like repeat of matrilin-3 that result in either a recessive form of spondyloepiphyseal dysplasia, matrilin-3 related (rSEMD: Cys304Ser) (Borochowitz et al. 2004), or confer susceptibility to osteoarthritis (OA: Thr303Met) (Pullig et al. 2007; Min et al. 2006; Stefansson et al. 2003). The biochemical effect of the EGF-like mutations has been studied in vitro, and whilst the Cys304Ser mutation results in the retention of some misfolded mutant protein, the Thr303Met mutation allows the mutant protein to be efficiently secreted and participate in the formation of filamentous networks surrounding the cells (Otten et al. 2005). However, the structural and functional consequence of this mutated form of matrilin-3 in the ECM has not been determined, and it is critical therefore to understand these effects in order to provide insight into its role in OA.

7.6 MED: Type IX Collagen Related (EDM2, EDM3 and EDM4): Genetic and Cell-Based Studies

Type IX collagen gene mutations are perhaps the most remarkable of all the MED mutations reported to date because all but one cause splicing defects that have a very restricted distribution. These mutations are all postulated to cause in-frame deletions from an equivalent region of the COL3 domain in the α1(IX), α2(IX) or α3(IX) chains (Fig. 7.2) (Briggs and Chapman 2002). The one reported exception is c.104G>A in exon 2 of *COL9A3* that causes Gly35Asp in α3(IX) and was identified in an affected father and son with typical MED.

It is interesting to note that of all the different forms of MED, the least is known about the disease mechanisms of MED caused by type IX collagen gene mutations. For example, it has not been resolved as to whether mutant type IX collagen is retained within the rER (Bonnemann et al. 2000) or secreted into the ECM (Spayde et al. 2000). Furthermore, the precise grouping of type IX collagen gene mutations has yet to be fully explained; however it is possible that these deletions may disrupt important interactions with other components of the ECM. Indeed it has been shown that the A-domain of matrilin-3

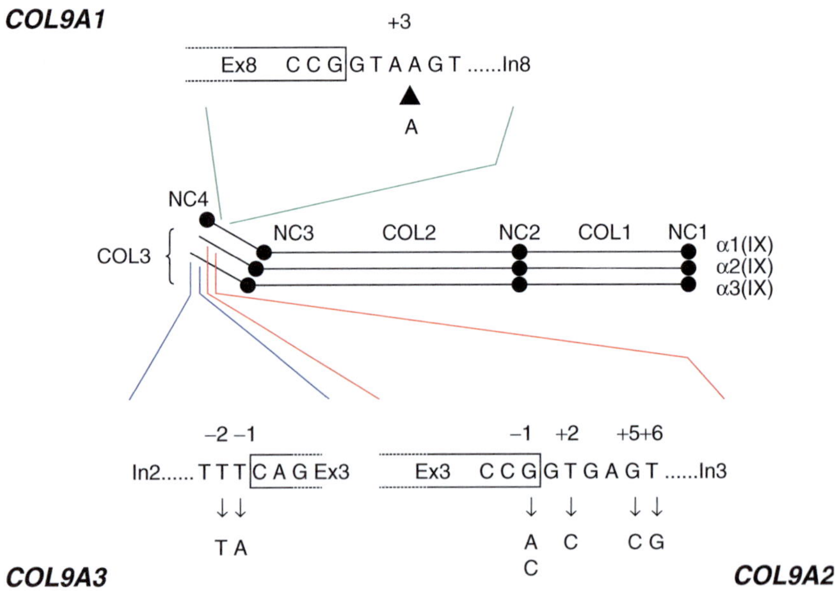

Fig. 7.2 Schematic of a type IX collagen molecule showing the exon skipping mutations found in the COL9A1, COL9A2 and COL9A3 genes that cause multiple epiphyseal dysplasia. All exon skipping results in the in-frame deletion of amino acid residues from the equivalent region of the COL3 domain. Key: *NC* non-collagenous, *COL* collagenous, *Ex* exon, *In* intron

interacts with the COL3 domain of type IX collagen and that this interaction is abolished when type IX collagen harbours an MED mutation (Fresquet 2007b). These studies have provided the first biochemical insight into why MED mutations might cluster in the COL3 domain of type IX collagen and provide some rationale for the non-allelic genetic heterogeneity of these forms of MED (EDM2, EDM3, EDM4). The Gly35Asp substitution affects the third Gly-Xxx-Yyy repeat of the α3(IX) chain, which is situated amongst a region of contiguous Gly-Pro-Pro triplets suggesting that this area may be important for the stability of the collagen IX triple helix.

7.7 Determining the Mechanistic Link Between a Mutant Gene Product and Disturbed Endochondral Ossification Through the Use of Targeted Mouse Models of PSACH and MED

It is understandable that in vitro studies have been unable to determine exactly how the expression of a mutant gene product results in disturbed endochondral ossification and reduced bone growth in vivo. Unfortunately, the use of knockout technology has also proven to be unsuitable since both the *Comp* and *Matn3* knockout mice have no overt skeletal phenotype (Ko et al. 2004; Svensson et al. 2002). Therefore, to determine the disease mechanisms that underpin the pathophysiology of PSACH–MED in vivo, homologous recombination and *Cre*-lox technology were used to generate a series of mouse lines that genetically modelled phenotypes within the full PSACH and MED disease spectrum. These mice provided a valuable resource for describing key pathological findings, identifying and studying disease mechanisms in detail and as preclinical models for the validation of therapeutic targets.

7.7.1 Severe PSACH Resulting from Asp469del in the Type III Repeat Region of COMP

A knock-in mouse model of typical severe PSACH was generated with the Asp469del mutation in the TBSP3 repeat region of COMP (Table 7.2). Mutant animals were normal at birth but grow slower than their wild-type littermates, and by 9 weeks of age, they had short-limbed dwarfism. In the growth plates of mutant mice, the chondrocyte columns were sparse and poorly organised, whilst mutant COMP was retained within the rER of cells and elicited a novel cell stress response. Ultimately chondrocyte proliferation was significantly reduced, whilst apoptosis was both increased and spatially dysregulated in the mutant growth plates.

7.7.2 Mild PSACH/MED Resulting from a Thr585Met Missense Mutation in the CTD of COMP

A knock-in mouse model of mild PSACH was generated with the Thr585Met in the C-terminal domain of COMP (Table 7.2) (Pirog-Garcia et al. 2007). Mutant animals were normal at birth but grow slower than their wild-type littermates and developed

Table 7.2 Mouse models of the PSACH–MED disease spectrum and novel phenocopies to model ER in the cartilage growth plate

Disease	Gene	Mutation	Experimental approach taken	Promoter
PSACH	*COMP*	Asp469del	Homologous recombination (knock in)	Native mouse
PSACH-MED	*COMP*	Thr585Met	Homologous recombination (knock in)	Native mouse
MED	*MATN3*	Val194Asp	Homologous recombination (knock in)	Native mouse
Chondrodysplasia	*TG*	*Rdw* (G2320R)	Transgenic overexpressing phenocopy	*Col2a1*
Chondrodysplasia	*TG*	*Cog* (L2293P)	Transgenic overexpressing phenocopy	*Col2a1*

Key: *COMP* cartilage oligomeric matrix protein, *MATN3* matrilin-3, *PSACH* pseudoachondroplasia, *MED* multiple epiphyseal dysplasia, *Tg* thyroglobulin, *Rdw* congenital dwarfism mutation, *Cog* congenital goitre mutation

mild short-limbed dwarfism (Pirog-Garcia et al. 2007). In the growth plates of mutant mice, the chondrocyte columns were sparse and poorly organised. Mutant COMP was secreted into the extracellular matrix, but its localisation was disrupted along with the distribution of several COMP-binding proteins (Pirog-Garcia et al. 2007). Although mutant COMP was not retained within the ER, there was a mild UPR/cell stress response, and chondrocyte proliferation was significantly reduced, whilst apoptosis was both increased and spatially dysregulated (Pirog-Garcia et al. 2007). Overall, these data suggest a mutation in the CTD of COMP which exerts a dominant-negative effect on both intra- and extracellular processes.

7.7.3 MED Resulting from a Val194Asp Missense Mutation in the A-domain of Matrilin-3

A knock-in mouse model with a Val194Asp mutation in the β-sheet of the A-domain of matrilin-3 was generated by homologous recombination (Table 7.2). Mice that are homozygous for the mutation developed a progressive growth plate dysplasia and short-limbed dwarfism that was consistent in severity with the relevant human phenotype (Leighton et al. 2007). Mutant matrilin-3 was retained within the ER of chondrocytes (Figs. 7.3 and 7.4) and associated with an UPR characterised by the upregulation of classical early markers of the UPR such as binding immunoglobulin protein (BiP) and calreticulin (Leighton et al. 2007). Eventually, there was reduced proliferation and spatially dysregulated apoptosis of chondrocytes in the cartilage growth plate (Leighton et al. 2007).

Fig. 7.4 Electron micrograph showing enlarged ER in mutant chondrocytes (**a**) compared to wild type (**b**). *Scale bars* represent 8 μm

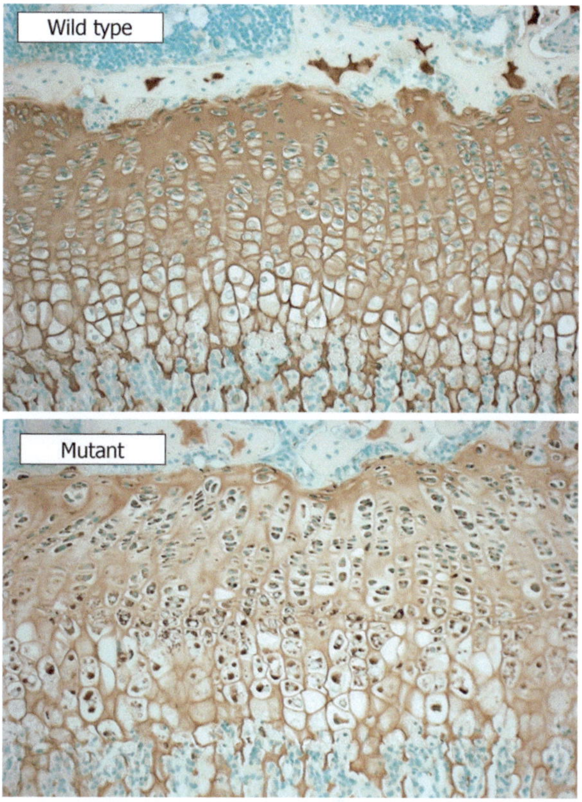

Fig. 7.3 The organisation of the growth plate in a V194D Matn3 mutant mouse (*bottom*) is disrupted by 3 weeks of age and shows mislocalisation of matrilin-3 compared to wild type (*top*). Mutant matrilin-3 (brown staining) is retained within the cell and not secreted into the extracellular matrix

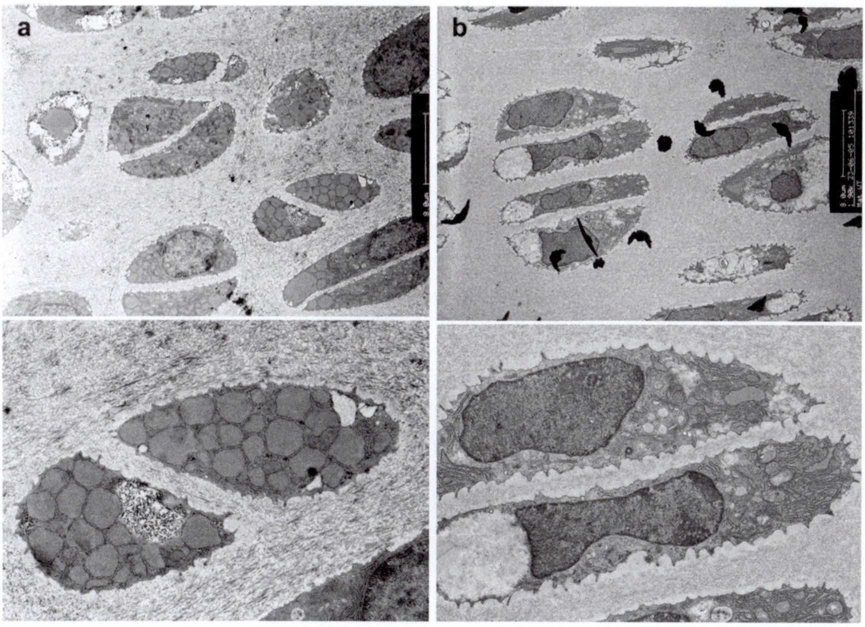

7.8 Common Disease Mechanisms in the PSACH and MED Disease Spectrum

Studies to date had demonstrated that whilst there are clear differences in the trafficking of mutant forms of COMP and matrilin-3, the expression of all three mutant proteins resulted in ER stress that had downstream consequences in terms of chondrocyte proliferation and apoptosis (Table 7.3) (Briggs et al. 2015a, b). It had been hypothesised that reduced proliferation and increased spatially dysregulated apoptosis of chondrocytes was the most likely cause of disrupted linear bone growth and could therefore be defined as 'core disease mechanisms' in the PSACH and MED bone dysplasia family.

Therefore, to further delineate the relative contributions of (1) mutant protein retention vs. disturbed cartilage ECM and (2) reduced chondrocyte proliferation vs. increased apoptosis, to growth plate dysplasia and reduced bone growth, a novel transgenic approach was adopted. In these experiments two different forms of mutant thyroglobulin (Tg) were expressed in the cartilage under the *Col2a1* promoter (Table 7.3). Tg is a 660-kDa homodimeric protein produced exclusively by the thyroid gland and is the major secretory glycoprotein produced by thyrocytes. *Cog* (congenital goitre mutation) and *Rdw* (congenital dwarfism) are naturally occurring thyroglobulin mutations in mouse and rat, respectively. The *Cog* mutation causes ER stress in thyrocytes, whilst *Rdw* leads to increased cell death. Expression of both of these mutant proteins in the cartilage produces a chondrodysplasia phenotype resulting from ER stress that was independent of an endogenous cartilage protein Gualeni 2013 and Kung 2015. Significantly, neither mutation caused increased chondrocyte apoptosis in the growth plate, but both

Table 7.3 Key pathological and quantitative measures of disease severity in various knock-in mouse models of PSACH–MED and novel ER stress phenocopies

Mouse model	Age of onset	Cell proliferation	Apoptosis	Mutant protein retention	ER stress	Stress pathway activated
Comp^{D469del}	~6 weeks	Decreased	Increased and spatially dysregulated	Yes	Yes	Novel (APR/EOR)
Comp^{T585M}	~9 weeks	Decreased	Increased and spatially dysregulated	Slight	Mild	Mild UPR
Matn3^{V194D}	~2 weeks	Decreased	Increased and spatially dysregulated	Yes	Yes	UPR
Col2-Tg^{Rdw}	At birth	Decreased	No differences	Yes	Yes	Novel (APR/EOR)
Col2-Tg^{cog}	~3 weeks	Decreased	No differences	Yes	Yes	UPR

EOR ER overload response, *APR* aggregated protein response, *UPR* unfolded protein response

mutations did result in a significant reduction in chondrocyte proliferation, thereby confirming this as the likely mechanism of reduced long bone growth (Table 7.3).

7.9 The Induction of Genotype-Specific Stress Pathways in Growth Plate Chondrocytes

The retention and/or delayed processing and trafficking of mutant structural proteins causes ER stress in PSACH and MED cell and mouse models that has been well documented. In addition to the accumulation of misfolded mutant proteins, ER stress can also be induced by a variety of causes including viral infection, chronic inflammation, hypoxia and ischemia.

Although mutant forms of COMP, matrilin-3 and thyroglobulin all induce ER stress, the specific downstream genetic pathways that are activated appear to be genotype specific (Table 7.3). For example, the expression of the Val194Asp mutant form of matrilin-3 induces classical UPR characterised by the upregulation of BiP (a sentinel marker of UPR) and a plethora of other chaperones, foldases and protein disulphide isomerases (Leighton et al. 2007). To a certain extent, the Thr585Met mutant form of COMP also induces mild UPR due to the delayed secretion of the mutant protein. In contrast, the Asp469del mutant form of COMP induces novel stress pathways that are characterised by changes in the expression of genes implicated in oxidative stress, cell cycle regulation and apoptosis (Suleman et al. 2012). This disease signature is similar to other UPR-independent pathways that are activated by the aggregation of mutant proteins as a result of toxic gain of function. These novel stress pathways have been termed either an aggregated protein response (APR) or an ER overload response (EOR). For example, the serpinopathies are due to the aggregation of insoluble mutant α1-antitrypsin that triggers EOR, which is independent of the UPR and involves activation of nuclear factor-kB signalling. Interestingly, the two different mutant forms of thyroglobulin (Cog and Rdw) induced different stress pathways when expressed in chondrocytes in vivo. The Cog mutation induced a canonical UPR characterised by increased BiP expression, phosphorylation of eIF2α and splicing of Xbp1 (Kung et al. 2015). In contrast, the Rdw mutation did not induce transcriptional activation of a classical UPR but a downregulation in the expression of genes involved in cell proliferation (Gualeni et al. 2013).

7.10 Genetic Modifiers of Disease Severity

Patients with PSACH and MED present with a high degree of inter- and/or intra-familial variability with regard to disease severity and phenotypic features. Various studies have suggested that this variability is most likely due to the influence of genetic modifiers that can influence the penetrance, expressivity and severity of the primary disease-causing mutation. These modifiers might directly influence the structure and/or function of the mutant gene product or act to modify the cell's response to its expression. For example, a non-synonymous polymorphism in *MATN3* found in a family with MED, although in itself not disease causing, may act as a modifier of disease severity. The affected individuals, described by Jackson and colleagues (Jackson et al. 2004), carried a disease mutation but with an additional change in the matrilin-3 (V245M) that was subsequently shown to influence the trafficking of matrilin-3 in vitro.

Genetic modifiers of PSACH and MED severity might also affect the cell's response to the expression of mutant protein. For example, increasing the relative proportion of a C57BL6/J background in the COMP Thr585Met mouse model resulted in a more severe chondrodysplasia phenotype (Pirog et al. 2014). Studies in mouse models of other genetic conditions such as Treacher Collins syndrome (TCS), Pelizaeus–Merzbacher disease (PMD) and spastic paraplegia type 2 (SPG2) have shown that the C57BL6/J background in particular has a heightened reaction to a misfolded protein stimulus, specifically showing increased levels of pro-apoptotic protein CHOP. Indeed, abnormal chondrocyte apoptosis in the growth plates was further increased in the Thr585Met COMP on an enriched C57BL6/J background (Pirog et al. 2014).

Understanding further the role of genetic modifiers in PSACH and MED pathobiology is of utmost importance for the discovery and tailoring of future therapies. Recent developments in genomic technologies as well as advancements in systems biology will enable the discovery of these modifiers in the future.

7.11 Soft Tissue Complications in PSACH/MED

Matrilin-3 is expressed almost exclusively in the cartilage; however, COMP and type IX collagen are also present at significant levels in the tendon, ligament and skeletal muscle. It is therefore not surprising that a subset of PSACH and MED patients presents with a variety of musculoskeletal complications such as mild myopathy and joint laxity. Ligamentous abnormalities associated with PSACH and MED disease spectrum include differences in collagen fibre diameters and mutant protein retention within the tenocytes of affected individuals.

However, the mild myopathy proved more difficult to elucidate, with variable phenotypic presentation and sporadic recording of the clinical case reports. The in-depth investigation of the knock-in mouse models of the PSACH and MED disease spectrum (T585M COMP, D469del COMP and V194D Matn3) finally revealed that mild myopathy associated with these disease spectra derives from structural abnormalities in fibrous tissues at the myotendinous and perimysial junctions, which led to changes in biomechanical properties of these tissues. This correlation was observed for both COMP genotypes (T585M and D469del) but not for the V194D *Matn3* model MED. This finding possibly explains the milder clinical manifestation of MED caused by *MATN3* mutations and provides important information for the clinical management.

7.12 Conclusions and Future Perspectives

Extensive studies on both the knock-in PSACH and MED mouse models and transgenic 'ER stress phenocopies' confirmed that ER stress in growth plate chondrocytes, induced by the retention and/or processing of mutant proteins, results in decreased chondrocyte proliferation and increased apoptosis in the growth plate that ultimately causes short-limbed dwarfism. Therefore, a therapeutic approach to target the ER stress is an attractive proposal that is independent of genome editing and manipulation of the cartilage ECM. Such chemical chaperones are already in use for some rER stress-related conditions such as type II diabetes and cystic fibrosis and may provide a suitable treatment for a broad range of human connective tissue diseases.

References

Barrie H, Carter C, Sutcliffe J (1958) Multiple epiphyseal dysplasia. Br Med J 2:133–137

Bateman JF, Wilson R, Freddi S, Lamande SR, Savarirayan R (2005) Mutations of COL10A1 in Schmid metaphyseal chondrodysplasia. Hum Mutat 25(6):525–534

Bonnemann CG, Cox GF, Shapiro F, Wu JJ, Feener CA, Thompson TG, Anthony DC, Eyre DR, Darras BT, Kunkel LM (2000) A mutation in the alpha 3 chain of type IX collagen causes autosomal dominant multiple epiphyseal dysplasia with mild myopathy. Proc Natl Acad Sci U S A 97(3):1212–1217

Borochowitz ZU, Scheffer D, Adir V, Dagoneau N, Munnich A, Cormier-Daire V (2004) Spondylo-epi-metaphyseal dysplasia (SEMD) matrilin 3 type: homozygote matrilin 3 mutation in a novel form of SEMD. J Med Genet 41(5):366–372

Briggs MD, Chapman KL (2002) Pseudoachondroplasia and multiple epiphyseal dysplasia: mutation review, molecular interactions, and genotype to phenotype correlations. Hum Mutat 19(5):465–478

Briggs MD, Hoffman SM, King LM, Olsen AS, Mohrenweiser H, Leroy JG, Mortier GR, Rimoin DL, Lachman RS, Gaines ES et al (1995) Pseudoachondroplasia and multiple epiphyseal dysplasia due to mutations in the cartilage oligomeric matrix protein gene. Nat Genet 10(3): 330–336

Briggs MD, Brock J, Ramsden SC, Bell PA (2014) Genotype to phenotype correlations in cartilage oligomeric matrix protein associated chondrodysplasias. Eur J Hum Genet 22(11):1278–1282. doi:10.1038/ejhg.2014.30

Briggs MD, Bell PA, Pirog KA (2015a) The utility of mouse models to provide information regarding the pathomolecular mechanisms in human genetic skeletal diseases: the emerging role of endoplasmic reticulum stress (review). Int J Mol Med. doi:10.3892/ijmm.2015.2158

Briggs MD, Bell PA, Wright MJ, Pirog KA (2015b) New therapeutic targets in rare genetic skeletal diseases. Expert Opin Orphan Drugs 3(10):1137–1154. doi:10.1517/21678707.2015.1083853

Buckwalter JA, Mower D, Ungar R, Schaeffer J, Ginsberg B (1986) Morphometric analysis of chondrocyte hypertrophy. J Bone Joint Surg Am 68(2):243–255

Budde B, Blumbach K, Ylostalo J, Zaucke F, Ehlen HW, Wagener R, Ala-Kokko L, Paulsson M, Bruckner P, Grassel S (2005) Altered integration of matrilin-3 into cartilage extracellular matrix in the absence of collagen IX. Mol Cell Biol 25(23):10465–10478

Chapman KL, Mortier GR, Chapman K, Loughlin J, Grant ME, Briggs MD (2001) Mutations in the region encoding the von Willebrand factor A domain of matrilin-3 are associated with multiple epiphyseal dysplasia. Nat Genet 28(4):393–396

Chen TL, Stevens JW, Cole WG, Hecht JT, Vertel BM (2004) Cell-type specific trafficking of expressed mutant COMP in a cell culture model for PSACH. Matrix Biol 23(7):433–444

Chen FH, Thomas AO, Hecht JT, Goldring MB, Lawler J (2005) Cartilage oligomeric matrix protein/thrombospondin 5 supports chondrocyte attachment through interaction with integrins. J Biol Chem 280(38):32655–32661

Cotterill SL, Jackson GC, Leighton MP, Wagener R, Makitie O, Cole WG, Briggs MD (2005) Multiple epiphyseal dysplasia mutations in MATN3 cause misfolding of the A-domain and prevent secretion of mutant matrilin-3. Hum Mutat 26(6):557–565

Cowell HR, Hunziker EB, Rosenberg L (1987) The role of hypertrophic chondrocytes in endochondral ossification and in the development of secondary centers of ossification. J Bone Joint Surg Am 69(2):159–161

Czarny-Ratajczak M, Lohiniva J, Rogala P, Kozlowski K, Perala M, Carter L, Spector TD, Kolodziej L, Seppanen U, Glazar R, Krolewski J, Latos-Bielenska A, Ala-Kokko L (2001) A mutation in COL9A1 causes multiple epiphyseal dysplasia: further evidence for locus heterogeneity. Am J Hum Genet 69(5):5

Fairbank T (1947) Dysplasia epiphysialis multiplex. Proc R Soc Med (Ortho Sec) 39:315–317

Fresquet (2007a) Novel mutations in exon 2 of MATN3 affect residues within the a-helices of the A-domain and may act through different disease mechanisms to those of the more common b-sheet mutations. Manuscript under review.

Fresquet (2007b) Structural and functional characterisation of recombinant matrilin-3 A-domain and implications for human genetic disease. J Biol Chem. Manuscript accepted.

Gibson BG, Briggs MD. The aggrecanopathies; an evolving phenotypic spectrum of human genetic skeletal diseases. Orphanet J Rare Dis. 2016 Jun 28;11(1):86. doi: 10.1186/s13023-016-0459-2. Review. PubMed PMID: 27353333; PubMed Central PMCID: PMC4924316.

Gualeni B, Rajpar MH, Kellogg A, Bell PA, Arvan P, Boot-Handford RP, Briggs MD (2013) A novel transgenic mouse model of growth plate dysplasia reveals that decreased chondrocyte proliferation due to chronic ER stress is a key factor in reduced bone growth. Dis Model Mech 6(6):1414–1425. doi:10.1242/dmm.013342

Halasz K, Kassner A, Morgelin M, Heinegard D (2007) Comp acts as a catalyst in collagen fibrillogenesis. J Biol Chem 282:31166–31173

Hashimoto Y, Tomiyama T, Yamano Y, Mori H (2003) Mutation (D472Y) in the type 3 repeat domain of cartilage oligomeric matrix protein affects its early vesicle trafficking in endoplasmic reticulum and induces apoptosis. Am J Pathol 163(1):101–110

Hecht JT, Nelson LD, Crowder E, Wang Y, Elder FF, Harrison WR, Francomano CA, Prange CK, Lennon GG, Deere M et al (1995) Mutations in exon 17B of cartilage oligomeric matrix protein (COMP) cause pseudoachondroplasia. Nat Genet 10(3):325–329

Hecht JT, Montufar-Solis D, Decker G, Lawler J, Daniels K, Duke PJ (1998) Retention of cartilage oligomeric matrix protein (COMP) and cell death in redifferentiated pseudoachondroplasia chondrocytes. Matrix Biol 17(8–9):625–633

Hecht JT, Hayes E, Snuggs M, Decker G, Montufar-Solis D, Doege K, Mwalle F, Poole R, Stevens J, Duke PJ (2001) Calreticulin, PDI, Grp94 and BiP chaperone proteins are associated with retained COMP in pseudoachondroplasia chondrocytes. Matrix Biol 20(4):251–262

Hecht JT, Makitie O, Hayes E, Haynes R, Susic M, Montufar-Solis D, Duke PJ, Cole WG (2004) Chondrocyte cell death and intracellular distribution of COMP and type IX collagen in the pseudoachondroplasia growth plate. J Orthop Res 22(4):759–767

Hecht JT, Hayes E, Haynes R, Cole WG (2005) COMP mutations, chondrocyte function and cartilage matrix. Matrix Biol 23(8):525–533. doi:10.1016/j.matbio.2004.09.006, doi:S0945-053X(04)00123-4 [pii]

Holden P, Meadows RS, Chapman KL, Grant ME, Kadler KE, Briggs MD (2001) Cartilage oligomeric matrix protein interacts with type IX collagen, and disruptions to these interactions identify a pathogenetic mechanism in a bone dysplasia family. J Biol Chem 276(8):6046–6055

International nomenclature and classification of the osteochondrodysplasias (1997). International Working Group on Constitutional Diseases of Bone (1998) Am J Med Genet 79(5):376–382

Itoh T, Shirahama S, Nakashima E, Maeda K, Haga N, Kitoh H, Kosaki R, Ohashi H, Nishimura G, Ikegawa S (2006) Comprehensive screening of multiple epiphyseal dysplasia mutations in Japanese population. Am J Med Genet A 140(12):1280–1284

Jackson GC, Barker FS, Jakkula E, Czarny-Ratajczak M, Makitie O, Cole WG, Wright MJ, Smithson SF, Suri M, Rogala P, Mortier GR, Baldock C, Wallace A, Elles R, Ala-Kokko L, Briggs MD (2004) Missense mutations in the beta strands of the single A-domain of matrilin-3 result in multiple epiphyseal dysplasia. J Med Genet 41(1):52–59

Jackson GC, Marcus-Soekarman D, Stolte-Dijkstra I, Verrips A, Taylor JA, Briggs MD (2010) Type IX collagen gene mutations can result in multiple epiphyseal dysplasia that is associated with osteochondritis dissecans and a mild myopathy. Am J Med Genet A 152A(4):863–869. doi:10.1002/ajmg.a.33240

Jackson GC, Mittaz-Crettol L, Taylor JA, Mortier GR, Spranger J, Zabel B, Le Merrer M, Cormier-Daire V, Hall CM, Offiah A, Wright MJ, Savarirayan R, Nishimura G, Ramsden SC, Elles R, Bonafe L, Superti-Furga A, Unger S, Zankl A, Briggs MD (2012) Pseudoachondroplasia and multiple epiphyseal dysplasia: a 7-year comprehensive analysis of the known disease genes identify novel and recurrent mutations and provides an accurate assessment of their relative contribution. Hum Mutat 33(1):144–157. doi:10.1002/humu.21611

Jakkula E, Makitie O, Czarny-Ratajczak M, Jackson GC, Damignani R, Susic M, Briggs MD, Cole WG, Ala-Kokko L (2005) Mutations in the known genes are not the major cause of MED; distinctive phenotypic entities among patients with no identified mutations. Eur J Hum Genet 13(3):292–301. doi:10.1038/sj.ejhg.5201314

Karsenty G (2003) The complexities of skeletal biology. Nature 423(6937):316–318

Ko Y, Kobbe B, Nicolae C, Miosge N, Paulsson M, Wagener R, Aszodi A (2004) Matrilin-3 is dispensable for mouse skeletal growth and development. Mol Cell Biol 24(4):1691–1699

Kung LH, Rajpar MH, Preziosi R, Briggs MD, Boot-Handford RP (2015) Increased classical endoplasmic reticulum stress is sufficient to reduce chondrocyte proliferation rate in the growth plate and decrease bone growth. PLoS One 10(2):e0117016. doi:10.1371/journal.pone.0117016

Leighton MP, Nundlall S, Starborg T, Meadows RS, Suleman F, Knowles L, Wagener R, Thornton DJ, Kadler KE, Boot-Handford RP, Briggs MD (2007) Decreased chondrocyte proliferation and dysregulated apoptosis in the cartilage growth plate are key features of a murine model of epiphyseal dysplasia caused by a matn3 mutation. Hum Mol Genet 16:1728–1741

Mabuchi A, Haga N, Maeda K, Nakashima E, Manabe N, Hiraoka H, Kitoh H, Kosaki R, Nishimura G, Ohashi H, Ikegawa S (2004) Novel and recurrent mutations clustered in the von Willebrand factor A domain of MATN3 in multiple epiphyseal dysplasia. Hum Mutat 24(5):439–440

Maeda K, Nakashima E, Horikoshi T, Mabuchi A, Ikegawa S (2005) Mutation in the von Willebrand factor-A domain is not a prerequisite for the MATN3 mutation in multiple epiphyseal dysplasia. Am J Med Genet A 136(3):285–286

Mann HH, Ozbek S, Engel J, Paulsson M, Wagener R (2004) Interactions between the cartilage oligomeric matrix protein and matrilins. Implications for matrix assembly and the pathogenesis of chondrodysplasias. J Biol Chem 279(24):25294–25298

Min JL, Meulenbelt I, Riyazi N, Kloppenburg M, Houwing-Duistermaat JJ, Seymour AB, van Duijn CM, Slagboom PE (2006) Association of matrilin-3 polymorphisms with spinal disc degeneration and osteoarthritis of the first carpometacarpal joint of the hand. Ann Rheum Dis 65(8):1060–1066

Mostert AK, Dijkstra PF, Jansen BR, van Horn JR, de Graaf B, Heutink P, Lindhout D (2003) Familial multiple epiphyseal dysplasia due to a matrilin-3 mutation: further delineation of the phenotype including 40 years follow-up. Am J Med Genet A 120(4):490–497

Muragaki Y, Mariman EC, van Beersum SE, Perala M, van Mourik JB, Warman ML, Olsen BR, Hamel BC (1996) A mutation in the gene encoding the alpha 2 chain of the fibril- associated collagen IX, COL9A2, causes multiple epiphyseal dysplasia (EDM2). Nat Genet 12(1):103–105

Newton G, Weremowicz S, Morton CC, Copeland NG, Gilbert DJ, Jenkins NA, Lawler J (1994) Characterization of human and mouse cartilage oligomeric matrix protein. Genomics 24(3):435–439

Ortega N, Behonick DJ, Werb Z (2004) Matrix remodeling during endochondral ossification. Trends Cell Biol 14(2):86–93
Otten C, Wagener R, Paulsson M, Zaucke F (2005) Matrilin-3 mutations that cause chondrodysplasias interfere with protein trafficking while a mutation associated with hand osteoarthritis does not. J Med Genet 42(10):774–779
Pirog KA, Irman A, Young S, Halai P, Bell PA, Boot-Handford RP, Briggs MD (2014) Abnormal chondrocyte apoptosis in the cartilage growth plate is influenced by genetic background and deletion of CHOP in a targeted mouse model of pseudoachondroplasia. PLoS One 9(2):e85145. doi:10.1371/journal.pone.0085145
Pirog-Garcia KA, Meadows RS, Knowles L, Heinegard D, Thornton DJ, Kadler KE, Boot-Handford RP, Briggs MD (2007) Reduced cell proliferation and increased apoptosis are significant pathological mechanisms in a murine model of mild pseudoachondroplasia resulting from a mutation in the C-terminal domain of COMP. Hum Mol Genet 16:2072–2088
Provot S, Schipani E (2005) Molecular mechanisms of endochondral bone development. Biochem Biophys Res Commun 328(3):658–665
Pullig O, Tagariello A, Schweizer A, Swoboda B, Schaller P, Winterpacht A (2007) MATN3 (matrilin-3) sequence variation (pT303M) is a risk factor for osteoarthritis of the CMC1 joint of the hand, but not for knee osteoarthritis. Ann Rheum Dis 66(2):279–280
Rimoin DL, Rasmussen IM, Briggs MD, Roughley PJ, Gruber HE, Warman ML, Olsen BR, Hsia YE, Yuen J, Reinker K et al (1994) A large family with features of pseudoachondroplasia and multiple epiphyseal dysplasia: exclusion of seven candidate gene loci that encode proteins of the cartilage extracellular matrix. Hum Genet 93(3):236–242
Rosenberg K, Olsson H, Morgelin M, Heinegard D (1998) Cartilage oligomeric matrix protein shows high affinity zinc-dependent interaction with triple helical collagen. J Biol Chem 273(32):20397–20403
Rossi A, Superti-Furga A (2001) Mutations in the diastrophic dysplasia sulfate transporter (DTDST) gene (SLC26A2): 22 novel mutations, mutation review, associated skeletal phenotypes, and diagnostic relevance. Hum Mutat 17(3):159–171
Schmitz M, Becker A, Schmitz A, Weirich C, Paulsson M, Zaucke F, Dinser R (2006) Disruption of extracellular matrix structure may cause pseudoachondroplasia phenotypes in the absence of impaired cartilage oligomeric matrix protein secretion. J Biol Chem 281(43):32587–32595
Spayde EC, Joshi AP, Wilcox WR, Briggs M, Cohn DH, Olsen BR (2000) Exon skipping mutation in the COL9A2 gene in a family with multiple epiphyseal dysplasia. Matrix Biol 19(2):121–128
Spitznagel L, Nitsche DP, Paulsson M, Maurer P, Zaucke F (2004) Characterization of a pseudoachondroplasia-associated mutation (His587→Arg) in the C-terminal, collagen-binding domain of cartilage oligomeric matrix protein (COMP). Biochem J 377(Pt 2):479–487
Spranger J (1988) Bone dysplasia 'families'. Pathol Immunopathol Res 7(1–2):76–80
Spranger JW, Zabel B, Kennedy J, Jackson G, Briggs M (2005) A disorder resembling pseudoachondroplasia but without COMP mutation. Am J Med Genet A 132A(1):20–24
Stefansson SE, Jonsson H, Ingvarsson T, Manolescu I, Jonsson HH, Olafsdottir G, Palsdottir E, Stefansdottir G, Sveinbjornsdottir G, Frigge ML, Kong A, Gulcher JR, Stefansson K (2003) Genomewide scan for hand osteoarthritis: a novel mutation in matrilin-3. Am J Hum Genet 72(6):1448–1459
Suleman F, Gualeni B, Gregson HJ, Leighton MP, Pirog KA, Edwards S, Holden P, Boot-Handford RP, Briggs MD (2012) A novel form of chondrocyte stress is triggered by a COMP mutation causing pseudoachondroplasia. Hum Mutat 33(1):218–231. doi:10.1002/humu.21631
Superti-Furga A, Bonafe L, Rimoin DL (2001) Molecular-pathogenetic classification of genetic disorders of the skeleton. Am J Med Genet 106(4):282–293
Svensson L, Aszodi A, Heinegard D, Hunziker EB, Reinholt FP, Fassler R, Oldberg A (2002) Cartilage oligomeric matrix protein-deficient mice have normal skeletal development. Mol Cell Biol 22(12):4366–4371

Thur J, Rosenberg K, Nitsche DP, Pihlajamaa T, Ala-Kokko L, Heinegard D, Paulsson M, Maurer P (2001) Mutations in cartilage oligomeric matrix protein causing pseudoachondroplasia and multiple epiphyseal dysplasia affect binding of calcium and collagen I, II, and IX. J Biol Chem 276(9):6083–6092

Tompson SW, Merriman B, Funari VA, Fresquet M, Lachman RS, Rimoin DL, Nelson SF, Briggs MD, Cohn DH, Krakow D (2009) A recessive skeletal dysplasia, SEMD aggrecan type, results from a missense mutation affecting the C-type lectin domain of aggrecan. Am J Hum Genet 84(1):72–79. doi:10.1016/j.ajhg.2008.12.001

Wagener R, Kobbe B, Paulsson M (1997) Primary structure of matrilin-3, a new member of a family of extracellular matrix proteins related to cartilage matrix protein (matrilin-1) and von Willebrand factor. FEBS Lett 413(1):129–134

Wagener R, Ehlen HW, Ko YP, Kobbe B, Mann HH, Sengle G, Paulsson M (2005) The matrilins--adaptor proteins in the extracellular matrix. FEBS Lett 579(15):3323–3329

Wu JJ, Eyre DR (1998) Matrilin-3 forms disulfide-linked oligomers with matrilin-1 in bovine epiphyseal cartilage. J Biol Chem 273(28):17433–17438

Zankl A, Jackson GC, Crettol LM, Taylor J, Elles R, Mortier GR, Spranger J, Zabel B, Unger S, Merrer ML, Cormier-Daire V, Hall CM, Wright MJ, Bonafe L, Superti-Furga A, Briggs MD (2007) Preselection of cases through expert clinical and radiological review significantly increases mutation detection rate in multiple epiphyseal dysplasia. Eur J Hum Genet 15(2):150–154

Zaucke F (2016) Cartilage glycoproteins. In: Grässel S, Aszódi A (eds) Cartilage: volume 1: physiology and development. Springer International Publishing, Cham, pp 55–81. doi:10.1007/978-3-319-29568-8_3

Zhang Y, Chen Q (2000) Changes of matrilin forms during endochondral ossification. Molecular basis of oligomeric assembly. J Biol Chem 275(42):32628–32634

Integrin-Mediated Interactions in Cartilage Physiology and Pathophysiology

8

Attila Aszódi

Abstract

The development and homeostasis of cartilaginous tissues are regulated by diverse microenvironmental cues including integrin-mediated interactions between chondrocytes and the extracellular matrix (ECM). Integrins are membrane receptors responsible for bi-directional communication between the cells and the surrounding by transmitting physicochemical signals through adhesion complexes. In addition, integrins are involved in sensing mechanical stress signals generated by the ECM and transduce them into the cell interior converting physical stimuli to biochemical signaling. Integrin-activated signaling cascades modulate various chondrocyte functions and play important roles in cartilage morphogenesis, homeostasis, and repair. This chapter will summarize and discuss the role of integrins in the physiology and pathophysiology of the growth plate and articular cartilage.

8.1 Introduction

In vertebrates, the skeletal system derives from mesodermal progenitors through a complex morphogenetic process laying down bone, cartilage, and ligaments/tendons, components of the skeleton which all together provide body support and enable movement. Endochondral bones of the axial and appendicular skeleton develop through cartilaginous intermediates which intensively grow until the end of puberty owing to the coordinated proliferation and differentiation activities of chondrocytes residing in a specialized structure called growth plate. During

A. Aszódi, PhD
Laboratory of Experimental Surgery and Regenerative Medicine, Clinic for General, Trauma and Reconstruction Surgery, Ludwig-Maximilians-University,
Nussbaumstrasse 20, D-80336 Munich, Germany
e-mail: attila.aszodi@med.uni-muenchen.de

S. Grässel, A. Aszódi (eds.), *Cartilage*, DOI 10.1007/978-3-319-45803-8_8

morphogenesis, the transient cartilaginous templates are gradually replaced by bone, except at the articulating surfaces where cartilage exists permanently as articular cartilage that enables load distribution and frictionless movements of synovial joints. Skeletal tissues are adopted for specific mechanical demands through distinct organization of their hierarchical composite material consisting of a network of collagen fibrils embedded into a heterogeneous mixture of extracellular matrix (ECM) molecules. In the hyaline cartilage, the fibrillar meshwork composed of mixed fibrils of collagen types II, IX, and XI provides tensile strength, whereas the osmotic swelling pressure generated by the hydrated proteoglycans resists compressive forces. The composition of cartilaginous matrices is reflected in the synthetic activities of chondrocytes and is maintained with the help of carefully balanced anabolic and catabolic processes. Changes in ECM structure, composition, and chondrocyte metabolic activities, either in the growth plate or in the articular cartilage, caused by aging, environmental, or genetic factors, could lead to cartilage-associated skeletal pathologies such as chondrodysplasias and osteoarthritis (for further details, see Chaps. 1, 6, and 7). Injuries are another challenge of the cartilaginous tissues (see also Chap. 2), especially in the articular cartilage which has a very limited intrinsic healing capacity. Differentiation, function, and repair of cartilaginous tissues crucially depend on the continuous interactions between cells and the immediate pericellular compartment of the ECM. Integrins are the major class of adhesion and signaling receptors on the surface of chondrocytes, and their role for numerous aspects of skeletal tissue properties is increasingly recognized. In this chapter, we will focus on integrin expression and function in the cartilaginous growth plate and the articular surface and emphasize the importance of integrin signaling in response to chemical and mechanical stress signals.

8.2 Integrins

Integrins are heterodimeric, type I transmembrane glycoproteins composed of non-covalently associated α and β subunits (Legate et al. 2009). Each subunits contains a large extracellular domain, a single transmebrane α helix, and a usually short cytoplasmic domain (Wegener and Campbell 2008). In mammals, 18 α and 8 β subunits have been identified, which can combine into 24 different heterodimers with specific or partially overlapping binding capacity to ECM components (Fig. 8.1a). Integrins can be classified into four main categories based on the molecular properties of integrin-ligand interactions (Humphries et al. 2006; Campbell and Humphries 2011). The subunits α5, α8, αv, and αIIb in combination with the β1 or β3 subunits interact with the arginine-glycine-aspartate (RGD) tripeptide motif found in a large number of ECM molecules including fibronectin, tenascin, and vitronectin. Eight heterodimers including the four members of the β2 subfamily bind to an RGD-related LDV or LDV-homologous site. Four α integrins (α1, α2, α10, and 11), which dimerize with the β1 subunit, are collagen-binding receptors carrying an "inserted" or "αI domain." This inserted domain, which is present in five

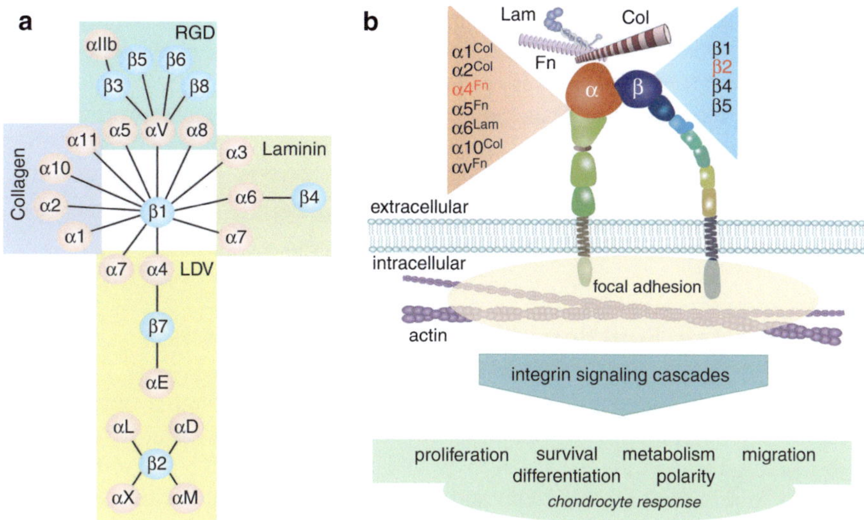

Fig. 8.1 Integrins (**a**) In mammals, 18 α and 8 β subunits can combine into 24 heterodimers. Based on the predominant binding specificity, integrin-ligand interactions are classified into four main classes. (**b**) Integrin subunits expressed on normal and osteoarthritic (in *red*) chondrocytes. Chondrocyte integrins interact with the major proteins of the cartilage extracellular matrix (*ECM*) including type II collagen (*Col*), fibronectin (*Fn*) and laminin (*Lam*). Integrins link the ECM to the actin cytoskeleton and control chondrocyte behavior through signaling cascades initiated at the focal adhesion sites

additional α subunits, carries metal ion-dependent adhesive sites essential for divalent cation-dependent ligand binding and shows homology to von Willebrand factor A domains. Finally, integrins α3β1, α6β1, α7β1, and α6β4 are selective laminin-binding heterodimers without αI-domain.

Following a series of intracellular events referred as inside-out signaling, integrins are activated and able to bind ECM molecules. Integrin-ligand interactions mediate cell adhesion to the ECM and are pivotal for development, tissue remodeling, tissue repair processes, host defense, and hemostasis. Integrins link the extracellular microenvironment to the intracellular actin cytoskeleton providing the mechanical basis for anchorage, cell shape determination, force transmission, and migration (Fig. 8.1b). Integrins transmit extracellularly generated chemical and mechanical signals into the cell interior in complex process called outside-in signaling via focal adhesion sites. These focal sites by which the cells attach to the ECM are multiprotein complexes associated with the integrin cytoplasmic tails and are composed of numerous signaling, adaptor, and actin-binding molecules (Geiger et al. 2009). The multiple signaling cascades induced by integrin engagement include phosphorylation by kinases (e.g., focal adhesion kinase (FAK), Src family tyrosine kinases, mitogen-activated protein (MAP) kinases), calcium influx, activation of Rho GTPases, and cytoskeletal rearrangements which all together regulate various aspects of cell behavior including proliferation, survival, and differentiation (Legate et al. 2009). In addition, integrins cooperate and share common signaling pathways with growth factor receptors to generate a fine-tuned cellular response for extracellular cues. At least 12 integrin receptors contain the β1 subunit (Fig. 8.1a) and form the largest integrin subfamily. Ablation of the β1 integrin gene in mice leads to peri-implantation lethality (Fassler and Meyer 1995; Bouvard et al. 2001), which demonstrates the essential role of β1 integrins in early embryonic development.

8.3 Integrin Function in Cartilage Morphogenesis

Most elements of the mammalian skeleton are generated through a cartilaginous template by the multistep developmental process of endochondral bone formation (Yeung Tsang et al. 2014; Lefebvre and Bhattaram 2010). The initial cartilaginous molds of the future axial and appendicular skeleton are formed by the condensation and chondrogenic differentiation of the committed mesenchymal cells. The molds, which are surrounded by the perichondrium, consist of immature, proliferative chondroblasts which synthesize an extracellular matrix (ECM) composed of type II collagen fibrils, the large proteoglycan aggrecan, and various small proteoglycans and non-collagenous glycoproteins. Next, centrally located cells exit from the cell cycle (maturation), increase in cell volume, and became hypertrophic chondrocytes followed by the differentiation of perichondrial osteoprogenitor cells into osteoblasts, which produce osteoid to form a bony collar around the shaft (diaphysis) of the bone. Chondrocyte hypertrophy triggers ECM mineralization, blood vessel invasion, and the recruitment of chondroclasts/osteoclasts, which gradually degrade the cartilage matrix, and osteoblasts, which replace the cartilage scaffold by bone. The program of proliferation, maturation, hypertrophy, and remodeling takes place within a specialized structure, called growth plate (for further details, see Chap. 4 in volume 1 (Aszódi 2016)). Growth plate chondrocytes arrange into horizontal zones of resting, proliferative, prehypertrophic, and hypertrophic chondrocytes according to the metabolic and/or differentiation stages. Vertically, proliferative chondrocytes form columnar stacks in which the flattened cells are oriented with right angle to the longitudinal axis of the growing bone. Proliferation, matrix production, and hypertrophy of growth plate chondrocytes are the major factors contributing for proper and timely elongation of endochondral bones. Chondrodysplasias, a diverse group of genetic disorders with primary effect on cartilage development and growth, commonly result in short stature (dwarfism) and, occasionally, lethality (Krakow and Rimoin 2010; Krakow 2015; Warman et al. 2011) (for details, see Chaps. 6 and 7). The coordinated process of endochondral bone formation is under the control of multiple molecular mechanisms involving transcription and growth factors, ECM proteins, and integrin-mediated cell-matrix interactions (Docheva et al. 2014; Aszodi et al. 2000).

8.3.1 Integrin Expression on Fetal and Growth Plate Chondrocytes

Several studies investigating cartilages in the developing mammalian limb have demonstrated that chondrocytes express various integrin receptors for cartilage ECM proteins including α1β1, α2β1, and α10β1 (for collagen II); α5β1, αvβ3, and αvβ5 (for fibronectin); and α6β1 (for laminin). Fluorescence-activated cell sorter (FACS) analysis of isolated human fetal knee chondrocytes has showed the surface expression of α2, α5, α6, αv, and β1 integrins, whereas only the α5, αv, and β1 subunits were detected on intact tissue sections by immunohistochemistry (Durr et al. 1993). In a more detailed immunohistochemical study, α5, α6, αv, and β1 integrins were found throughout the human fetal growth plate, whereas α1, α2, and α3 were absent in growth plate chondrocytes (Salter et al. 1995). In adolescent human growth plate, immunohistochemistry indicated the presence of α6, αv, and β1 subunits in all growth plate zones; the α1, α2, and α5 subunits in the proliferative and hypertrophic zones; and the β5 subunit in the hypertrophic zone (Hausler et al. 2002). The integrin subunit α10, which was isolated using its capacity to bind collagen type II (Camper et al. 1998; Lundgren-Akerlund and Aszodi 2014), is predominantly present in collagen II expressing human tissues at the mRNA level (Gullberg and Lundgren-Åkerlund 2002).

The spatially variable expression pattern of integrin subunits was also revealed in the mouse cartilage. Integrin β1 was found throughout the cartilage anlage at embryonic day 18.5 (just before birth), while α2, α3, and α5 subunits were predominantly expressed in the proliferative zone (Ma et al. 1998). The distribution of the three collagen-binding integrins α1β1, α2β1, and α10β1 was further analyzed by immunohistochemistry in the developing mouse limb at various embryonic and postnatal stages (Camper et al. 2001). The integrin subunit α10 co-localized with collagen II in the initial cartilaginous molds at embryonic day 11.5. The integrin subunit α10 was the predominantly expressing collagen-binding integrin in epiphyseal and growth plate cartilage until birth, while α1 was detected only in the articular surface and the hypertrophic zone. Integrin α2β1 was undetectable in any region of the developing limb cartilage. At 4 weeks after birth, α10 was strongly, and α1β1 was weakly detectable in growth plate chondrocytes.

The aforementioned experiments showed good agreement between the dynamic localization of integrins on chondrocytes and the deposition of their respective ligands into the cartilage ECM. Laminin and fibronectin are weakly and/or zone-specifically present in the growth plate cartilage, while collagen II is being the most abundant throughout the growth plate (Durr et al. 1993, 1996; Hausler et al. 2002; Lee et al. 1997; Salter et al. 1995; Camper et al. 2001). The importance of collagen II in the structural and functional properties of the developing cartilage and the prominent expression of integrins on the surface of chondrocytes thus clearly imply that cell-matrix interactions mediated by integrin heterodimers are important for cartilage differentiation and chondrocyte function in the growth plate. Furthermore, they indicate the most critical roles for the β1 and the collagen-binding α10 integrin subunits.

8.3.2 Integrin Function: In Vitro Studies

The functional role of integrins in cartilage differentiation was intensively studied in micromass culture of limb-bud mesenchymal cells, which mimics several steps of developmental chondrogenesis in vitro. Administration of β1 integrin-blocking antibodies to mouse micromass culture prevented the formation of cartilaginous nodules indicating a pivotal role of β1 integrins in chondrogenic differentiation of skeletogenic mesenchymal progenitor cells (Shakibaei 1998; Bang et al. 2000). β1 integrin-mediated tyrosine phosphorylation of FAK and fibronectin were required for precartilaginous condensation and chondrogenic differentiation of chick mesenchymal cells (Bang et al. 2000). Another study has demonstrated that matrix metalloprotease-2 (MMP-2) activity regulated by microRNA-488 impairs chondrogenic cell condensation by disrupting matrix adhesion sites and downregulating FAK-β1 integrin interaction (Jin et al. 2007). On the other hand, mouse fibroblasts lacking FAK could form cartilage nodules, and pharmacological inhibition of FAK/Scr signaling induces the expression of chondrogenic genes suggesting that FAK signaling may suppress chondrogenesis by regulating the shift from cell-ECM to cell-cell interactions (Pala et al. 2008). Collectively, these observations indicate that focal adhesion complexes containing fibronectin, β1 integrins, and FAK/Scr are important for the transition of mesenchymal precursors to chondrocytes in vitro.

Studies using chick whole sternum and isolated sternum chondrocytes revealed that inhibition of β1, α1, and α2 integrins by blocking antibodies results in growth retardation, increased apoptosis, and disorganization of the actin cytoskeleton indicating important regulatory roles of integrins for chondrocyte behavior (Hirsch et al. 1997). Further in vitro studies have demonstrated that proliferation of rabbit growth plate chondrocytes requires α5β1 and fibronectin (Enomoto-Iwamoto et al. 1997), whereas survival of chicken chondrocytes in suspension culture is mediated via β1 integrin-collagen interactions (Cao et al. 1999). The importance of integrins in hypertrophic chondrocyte differentiation was demonstrated by experiments showing that transglutaminase-induced hypertrophy of cultured articular cartilage chondrocytes requires signaling through α1β1 and α5β1 integrins (Johnson et al. 2008), and that β1 integrin-blocking antibodies impaired the deposition of the hypertrophic chondrocyte marker collagen X in chick sternal organ culture (Hirsch et al. 1997). Interaction of annexin V, a collagen II receptor and cytoplasmic signaling protein, with β5 integrin cytoplasmic tail in hypertrophic chondrocytes induced apoptosis of cultured growth plate chondrocytes, while siRNA-mediated silencing of annexin V increased chondrocyte survival (Wang and Kirsch 2006).

Cultured primary chondrocytes rapidly lose their chondrogenic phenotype on plastic monolayer characterized by changes in cell geometry (from spread to rounded), actin cytoskeleton organization (from cortical, sub-membrane localization to stress fiber formation throughout the cell), and gene expression (e.g., from cartilage-specific collagen II and aggrecan expression to fibroblast-specific collagen I and fibronectin expression). A unique and sensitive marker for monitoring chondrocyte phenotypic changes during cell culture expansion is the relative ratio of the α10 and α11 integrin subunits (Gouttenoire et al. 2010). Long-term culturing and

TGF-β1 treatment result in chondrocyte dedifferentiation associated with decreased α10 integrin and increased α11 integrin expression levels. BMP-2 supplementation, however, stabilizes the chondrogenic phenotype accompanied by high levels of α10 expression (Gouttenoire et al. 2010). Interestingly, chondrocytes cultured on plastic monolayer coated with type II collagen keep their rounded shape and express collagen II suggesting that collagen-binding integrins may stabilize the chondrogenic phenotype (Shakibaei et al. 1997). In line with this hypothesis, the formation of chondrocyte aggregates with a cartilage-like pericellular matrix environment in suspension culture requires α10β1 integrin-collagen II interaction (Gigout et al. 2008).

8.3.3 In Vivo Studies with Knockout Mice: The Essential Role of β1 Integrins in Cartilage Development

In order to investigate the in vivo function of β1 integrins during endochondral bone formation, the floxed β1 integrin gene (*β1fl/fl*) was conditionally inactivated in the entire cartilaginous skeleton using a transgene which drives expression of the cre recombinase in differentiated chondrocytes under the control of the mouse collagen II promoter (*Col2a1cre*) (Sakai et al. 2001). *β1fl/fl-Col2a1cre* mice display general chondrodysplasia during embryogenesis leading to lethality shortly after birth due to breathing difficulties (Aszodi et al. 2003). To overcome lethality, the β1 integrin gene was also deleted in limb-bud mesenchymal progenitor cells using the Prx1cre transgene (Raducanu et al. 2009; Logan et al. 2002). The *β1fl/fl-Prx1cre* mice survive after birth, but mesenchymal limb tissues such as the cartilage, tendon, and ligaments or the synovium completely or partially lack β1 integrins (Raducanu et al. 2009). The *β1fl/fl-Prx1cre* mutant mice show similar embryonic limb phenotype characterized by reduced growth of the cartilaginous skeletal elements as the *β1fl/fl-Col2a1cre* mice but develop postnatal abnormalities of synovial joint tissues (see later). Importantly, β1 integrin deficiency in mesenchymal progenitors does not influence chondrogenic differentiation implicating that β1 integrin-mediated cell-matrix interactions and signaling processes are not essential for early stages of endochondral bone formation (Raducanu et al. 2009). Supporting this conclusion, homozygous deletion of the floxed β1 integrin alleles at the time of mesenchymal condensation by the *Twist2-cre* transgene allows cartilage formation but severely impairs later stages of skeletal development (Shekaran et al. 2014).

The most dramatic phenotype of mice lacking β1 integrins in chondrocytes and hence lacking chondrocyte adhesion to collagen II, laminin, and partially fibronectin was observed in the growth plate (Fig. 8.2) (Aszodi et al. 2003). The normal columnar growth plate is a polarized structure characterized by specifically shaped and oriented cells in the proliferative zone (for details, see Chap. 4 in volume I (Aszódi 2016)). Proliferative chondrocytes display a strongly flattened, anisotropic geometry and align along the mediolateral (ML) axis of the growth plate, thus perpendicular to the proximo-distal (PD) direction of the longitudinal growth. The mitotic figures in these elongated chondrocytes lie also perpendicular to the PD axis, and cell divisions occur parallel with the columns (Dodds 1930). The two

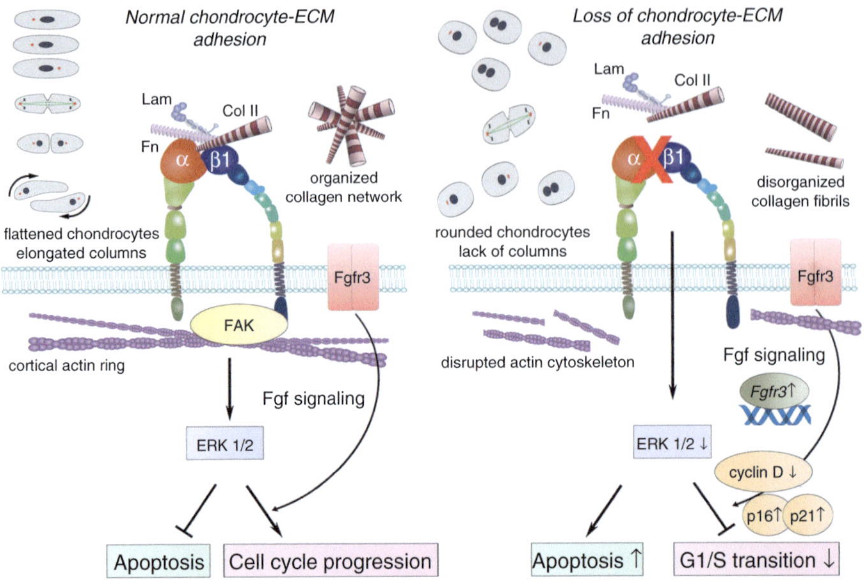

Fig. 8.2 The role of β1 integrins in the growth plate cartilage. Adhesion through multiple αβ1 heterodimers to the ECM proteins collagen II (*Col II*), fibronectin (*Fn*), and laminin (*Lam*) regulates the shape, oriented cell division, and intercalation movements of proliferative chondrocytes resulting in the formation of elongated growth plate columns (*left panel*). Signaling pathways including FAK and ERK1/2 activation control chondrocyte survival and, together with fibroblast growth factor (Fgf) signaling, cell cycle progression in the proliferative zone. When β1 integrin is ablated in chondrocytes (*right panel*), the loss of matrix attachment mediated via αβ1 integrins leads to cell rounding, impaired cytokinesis, diminished rotation movements, and the complete lack of the columnar structure of the growth plate. In the absence of β1 integrins, the link between the ECM and the actin cytoskeleton is disrupted leading to disorganized collagen (extracellularly) and actin (intracellularly) networks. Furthermore, ERK1/2 activation and Fgf receptor 3 (*Fgfr3*) mRNA expression are perturbed resulting in increased apoptosis and impaired G1/S transition due to abnormal expression of cyclin D and the cell cycle inhibitors p16 and p21

mediolaterally oriented, semicircular daughter cells subsequently undergo gradual flattening and extensive rotational or gliding movements around each other to arrange into the longitudinal column. Polarized cell migration and ML intercalation is characteristic for convergent extension, an essential morphogenetic process resulting in tissue narrowing along the ML axis and lengthening along the PD axis. This mechanism was completely disrupted in β1 integrin-deficient growth plates resulting in shortened and broadened long bones. The mutant chondrocytes in the proliferative zone were rounded up, failed to rotate, and showed random orientation of the mitotic spindle and the cleavage plain (Aszodi et al. 2003 and Aszodi, 2016 unpublished). Although an increasing body of evidence suggests that β1 integrins are actively involved in spindle and cell division positioning in various tissues including the skin (Lechler and Fuchs 2005) and mammary epithelium (Taddei et al. 2008), the mechanisms by which β1 integrins governs mitotic spindle orientation and gliding movements in cartilage are still not fully understood. In contrast to other cell types, normal proliferative chondrocytes do not round up during mitosis but remain flattened throughout the cell cycle. For such elongated cells Hertwig's laws can be applied postulating that (1) spindles are formed in the longest axis of the cells and (2) the division plane is always perpendicular to the long axis (reviewed in Shah 2010). In this model, simple geometric constraints provide the necessary cue for orientation of the mitotic spindle and direction of cell division. β1-null chondrocytes do not adhere to collagen II and laminin, and only partially to fibronectin (Aszodi et al. 2003), therefore tempting to speculate that such significant loss of cell anchorage to the cartilage ECM causes the shape change (rounding) of chondrocytes in the proliferative zone of the mutant growth plate. Consequently, the isotropic β1-deficient chondrocytes lack geometric and adhesive cues which guide oriented cell division and rotational movements resulting in disorganized growth plate.

In addition to the crucial role of β1 integrins for spindle and division positioning, *β1fl/fl-Col2a1cre* mice exhibited further cellular and ECM defects (Aszodi et al. 2003). Mutant chondrocytes were frequently binucleated and displayed actin abnormalities demonstrating that disrupted cell-ECM interactions may impair adhesion-generated pulling forces and actomyosin ring constriction at the cleavage furrow, which are required to complete cytokinesis. β1-null chondrocytes also showed abnormal G1/S transition caused by reduced cyclin D level, increased nuclear translocation of signal transducer, and activator of transcription 1 (Stat1) and Stat5a, and elevated expression of the cell cycle inhibitors p16 and p21. The defective cell cycle progression was associated with significantly elevated expression of fibroblast growth factor receptor 3 (*Fgfr3*) mRNA in proliferative growth plate chondrocytes suggesting a crosstalk of β1 integrin to Fgf signaling. Since FGFR3 activation mutations in chondrocytes activate Stat1/Stat5 and upregulate cell cycle inhibitors resulting in skeletal dysplasias both in human and in mice (Li et al. 1999; Su et al. 1997), it is likely that β1 integrin-Fgf signaling pathways are cooperating in the regulation of chondrocyte proliferation. Furthermore, an increased apoptosis rate of growth plate chondrocytes was observed in the mutant mice. These abnormalities were accompanied with reduced activation of FAK and Erk1/Erk2 suggesting that altered integrin signaling is likely contributing to the proliferation and survival defects. Indeed, cartilage-specific deletion of the gene encoding integrin-linked kinase (Ilk), an effector molecule interacting with the cytoplasmic domain of β1/β3 integrins, leads to similar but milder growth plate abnormalities including reduced proliferation, disorganization of the cytoskeleton, and chondrocyte shape change (Grashoff et al. 2003; Terpstra et al. 2003).

Finally, the *β1fl/fl-Col2a1cre* cartilage has abnormal ultrastructure characterized by thick and disordered collagen fibrils indicating that β1 integrins on chondrocytes are important for proper assembly of the collagenous fibrillar network. In fibroblast, a preformed fibronectin matrix is essential for the formation of collagen fibrillar network, and the collagen-binding α2β1 and α11β1 integrins strongly enhance this process (Velling et al. 2002). It has been shown that mice lacking fibronectin in cartilage have normal growth plate structure (Aszodi et al. 2003) predicting that one or more collagen-binding β1 integrins directly modulate the polymerization of collagen II and/or incorporation of the fibrils into the meshwork. In accordance, β1 integrins were shown to play a role in the arrangement of the ECM around chondrocytes by anchoring and bending type II collagen fibrils (Lee and Loeser 1999). The absence of β1 integrins on the cell surface and the ECM changes had profound effect on the diffusion and biomechanical properties of the cartilage (Bougault et al. 2013). Rheological measurements on intact costal cartilage isolated from 17.5-day-old embryos revealed increased compressive stiffness of *β1fl/fl-Col2a1cre* samples likely due to high cellular rigidity induced by uncoupling the cytoskeleton-integrin interaction. In contrast, the diffusivity of macromolecular fluorescent solutes was greatly reduced in close vicinity of the cell membrane indicating that the lack of β1 integrin-ECM interactions influences the organization and/or composition of the pericellular matrix compartment.

8.3.4 In Vivo Studies with Knockout Mice: Mild Role of Integrin α Subunits in Cartilage Morphogenesis

In contrast to the severe cartilage phenotype of the mice lacking all β1 integrins in chondrocytes, most mutant strains deficient for a particular α/β1 heterodimer such as α1β1, α2β1, or α6β1 do not show defects in endochondral bone formation (Bouvard et al. 2001). However, the α10 integrin-deficient mice partially recapitulate the growth plate abnormalities observed in the *β1fl/fl-Col2a1cre* and in *β1fl/fl-Prx1cre* mice. Constitutive deletion of the gene encoding α10 (*Itga10*) results in a mild, nonlethal chondrodysplasia characterized by mild chondrocyte shape change and disorganization of the growth plate (Bengtsson et al. 2005). Furthermore, the lack of α10 on chondrocytes impairs proliferation via the modulation of Stat1/Stat5a/p16 pathway and leads to reduced collagen fibrillar density in the cartilage matrix as seen in the β1 null growth plate. These findings imply that although α10β1 integrin is the most critical collagen-binding integrin, which modulate chondrocyte function, the absence of multiple β1 integrins is necessary for the most severe cartilage defects indicating partial compensation among β1 integrin-containing heterodimers. Interestingly, double knockout mice lacking both α2β1 and α11β1 integrin develop dwarfism which is, however, caused by decreased production of insulin-like growth factor-1 in the liver and does not result from growth plate dysfunction (Blumbach et al. 2012).

Heritable bone diseases are frequently caused by mutations in genes coding for ECM structural proteins or molecules associated with extracellular signaling (Briggs et al. 2015; Tosi and Warman 2015), suggesting the importance of cell-matrix interactions in proper development and normal function of the cartilaginous skeleton. However, no mutations in genes encoding integrin subunits have so far been identified in human chondrodysplasias (Warman et al. 2011). Therefore, interesting to note that a truncating mutation in the canine *ITGA10* gene has been recently correlated with the disproportionate short-limbed dwarfism phenotype of two breeds of Nordic hunting dogs (Kyostila et al. 2013). The autosomal recessive chondrodysplasia of Norwegian Elkhound and Karelian Bear Dog breeds is characterized by irregular and disorganized growth plates (Kyostila et al. 2013; Bingel and Sande 1982) and is associated with a nonsense mutation in exon 16 of *ITGA10* causing the complete lack of α10 protein in the cartilage of the affected dogs. As the canine and human *ITGA10* genes are highly homologous, these findings suggest that a similar, naturally occurring loss-of-function mutation in human *ITG10* is a potential candidate for chondrodysplasias with so far unidentified genetic background.

Recently, the in vivo role of the fibronectin-binding α5β1 integrin in endochondral bone formation was analyzed using an ex utero manipulation system (Inoue et al. 2014). Injection of RGDS (Arg-Gly-Asp-Ser) peptides or anti-α5β1 integrin antibody into the upper part of the forearms of 15.5-day-old embryos resulted in shortened humerus, just one day after the surgical manipulation. Decreased BrdU incorporation, apoptosis, and lower ratios of collagen X/collagen II RNA and protein were observed in the cartilage of the micro-manipulated embryos implicating that α5ββ1 integrin-mediated cell-matrix interactions through the RGD sequence may regulate proliferation, survival, and differentiation of growth plate chondrocytes.

8.3.5 Integrin-Associated Signaling Pathways Involved in Endochondral Bone Growth

Filamin B is an intracellular scaffolding protein that mediates signaling between cell membrane receptors and the actin cytoskeleton (Nakamura et al. 2011). Mutations in the gene coding for filamin B (*FLNB*) result in a range of human skeletal disorders. Loss-of-function mutations (null alleles) lead to recessive spondylocarpotarsal syndrome (SCT) characterized by short stature and bone fusions (Krakow et al. 2004). Autosomal dominant (gain of function) mutations cause a range of skeletal dysplasias including boomerang dysplasia and atelosteogenesis I and III (Bicknell et al. 2005; Farrington-Rock et al. 2006). A mouse model lacking *Flnb* mimics the human characteristic of SCT (Farrington-Rock et al. 2008) and exhibits reduced phosphorylation of β1 integrin at the residue Ser785 associated with reduced chondrocyte adhesion to ECM ligands, premature hypertrophic differentiation, and abnormal organization of the collagen fibrillar network in the cartilage (Lu et al. 2007; Hu et al. 2014). The interaction between filamin B and β1 integrin in cartilage also modulates chondrocyte proliferation through the regulation of Cdk1/Cyclin B, an effector complex required for G2/M progression of the cell cycle. Mechanistically, the Pi3k/Akt pathway is activated downstream of Flnb-β1 integrin to promote endochondral bone growth (Hu et al. 2014). Thus, disruption of the ECM-β1 integrin-filamin B axis may contribute to skeletal dysplasias both in mouse and human.

The CCN matricellular proteins are a group of extracellular modular molecules which bind to ECM components, integrins, and growth hormone receptors to regulate diverse cellular processes (Krupska et al. 2015). All the six members (CCN1-6) of the family are expressed in the developing cartilage (Kawaki et al. 2008), and CCN2/connective tissue growth factor (CTGF) deficiency in mice results in severe, perinatal lethal chondrodysplasia accompanied by diverse growth plate abnormalities including reduced chondrocyte proliferation, delayed hypertrophic differentiation, and impaired ECM production (Hall-Glenn et al. 2013; Ivkovic et al. 2003). Mutant chondrocytes exhibit enlarged endoplasmatic reticulum in vivo, a typical sign of cellular stress, and upregulate the stress responsive genes *BiP* and *Chop* in three-dimensional alginate culture upon chemically induced stress (Hall-Glenn et al. 2013). CCN2/CTGF is localized pericellularly, where it interacts with α5β1 integrin to modulate chondrocyte proliferation and ECM production (Nishida et al. 2007). In stress situation, binding of the C-terminal part of CCN2/CTGF to α5β1 decreases *Chop* mRNA levels, while α5 integrin-blocking antibodies prevent the antistress role of CCN2/CTGF in cultured chondrocytes (Hall-Glenn et al. 2013). These findings suggest that the protective role of CCN2/CTGF on chondrocytes under stress conditions could be, at least partially, mediated through α5β1 integrin.

Lysophosphatidic acid (LPA), small glycerophospholipid produced from lysophosphatidylcholine by the extracellular enzyme autotaxin (ATX), modulates diverse cellular functions in various tissues through the activation of LPA receptors (LPA_{1-6}) (Aikawa et al. 2015). A recent study has demonstrated through genetic studies in zebrafish and mice that the ATX-LPA signaling axis is pivotal for proper cartilage formation (Nishioka et al. 2016). *LPA1* knockout and *ATXflox*$^{/-}$ mice with significantly reduced ATX level in the serum show skeletal growth retardation associated with chondrocyte misalignment and proliferation deficit. Mechanistically, the ATX-LPA1 axis enhances proliferation by promoting S-phase entry of the cell cycle through activation of β1 integrin signaling and subsequent deposition the pericellular fibronectin matrix. The ATX-LPA1-induced, integrin-dependent fibronectin assembly may promote chondrocyte attachment to the ECM and support proliferation, both processes required for proper cartilage morphogenesis.

8.4 Synovial Joint and Articular Cartilage

Synovial joints unite different types of tissues to ensure load transmission within skeletal structures and coordinate their movements. The opposing bones are covered with articular cartilage (AC) and are separated by a joint cavity enclosed in a capsule linking the skeletal elements. The capsule is lined by a synovial membrane secreting synovial fluids, which lubricates the AC surface and provides nutrition to the components of joints.

Synovial joints develop through two main processes called joint specification and joint morphogenesis (Pacifici et al. 2005) (for details, see also Chap. 7 in Volume 1 (Wang et al. 2016)). In the step of joint specification, skeletogenic MSCs expressing the Sox trio are directed to the articular fate under the control of TGF-beta receptor 2 (Spagnoli et al. 2007), canonical Wnt (e.g., Wnt4 and Wnt9a/14) (Hartmann and Tabin 2001; Guo et al. 2004) and growth differentiation factor-5 (Gdf5) (Settle et al. 2003) signaling cascades, resulting in the formation of presumptive joint regions called interzones. Subsequently, a cavity forms in the interzone between the articulating skeletal elements driven by a not fully understood mechanism involving high-level synthesis of hyaluronan (Matsumoto et al. 2009), specific ECM composition (Pacifici et al. 1993; Choocheep et al. 2010), and skeletal movement (Kahn et al. 2009). During joint morphogenesis, specific articular cells differentiate and form the joint tissues. The continued expression of the Sox trio is pivotal for articular chondrocyte differentiation, while it seems that Wnt/beta-catenin signaling prevents chondrocyte differentiation at sites not destined for articular cartilage such as the synovial lining and ligaments.

8.4.1 Integrins and Joint Morphogenesis

The function of integrin-mediated cell-cell or cell-matrix interactions in joint determination and morphogenesis is relatively poorly investigated. Garciadiego-Cázares and colleagues reported that α5 integrin is downregulated in the interzones of developing mouse fingers suggesting a role of α5β1 integrin for patterning of the hands (Garciadiego-Cazares et al. 2004). Indeed, by microinjecting anti-β1 and anti-α5 monoclonal antibodies or RGD peptides into the wrist region of embryonic mouse forelimb induced ectopic joints formation at the distal part of the radius and the ulna. A narrow gap region expressing interzonal markers such as Gdf5 and Wnt9a/14 but not expressing the prehypertrophic marker Indian hedgehog (Ihh) had formed in between the proliferative and hypertrophic zones. In contrast, forced expression of human cDNAs coding for human α5 and β1 integrin subunits in embryonic chick legs had resulted in joint fusions of the phalanges accompanied by the differentiation of prehypertrophic chondrocytes expressing Ihh. Consequently, the authors proposed that α5β1 integrin-RGD ligand (fibronectin) interactions are pivotal in cell fate determination during the development of the appendicular skeleton by dictating either the differentiation of interzonal cells or prehypertrophic chondrocytes on the basis of α5β1 expression level in chondrocytes. Although this hypothesis is clearly attractive, genetic experiments with conditional knockout mice did not support a mechanistic role for α5β1-mediated matrix attachments in joint morphogenesis. To date, no in vivo experiments have been performed to investigate the role of α5 integrins in skeletogenesis. However, neither *β1fl/fl-Col2cre* mice nor *β1fl/fl-Prx1cre* mice display ectopic joint formation, and in both models, Ihh is expressed at the normal level in the prehypertrophic zone (Aszodi et al. 2003; Raducanu et al. 2009). The lack of any obvious phenotype in joint formation in the absence of β1 integrins on skeletogenic MSCs, perichondrial cells, and chondrocytes suggests that antibody perturbation might trigger nonspecific and/or nonphysiological responses (e.g., apoptosis), which cannot be seen upon genetic ablation.

8.4.2 Integrins in the Articular Cartilage

Articular cartilage (AC) is a permanent, specialized type of hyaline cartilage that functions as a load bearing tissue and can resist shearing forces and friction. The AC, similar to the growth plate, is characterized by a high degree of anisotropy and is divided into four zones with distinct organization of the matrix and cells: superficial (or tangential), intermediate (or transitional), deep (or radial), and calcified (Poole et al. 2001). The superficial zone is the thin, outermost layer of the AC, which contains elongated cells and densely packed collagen fibrils arranged parallel to the surface. This zone has low aggrecan content, but it is enriched in small proteoglycans such as decorin and biglycan (Poole et al. 1996). The intermediate zone contains individual spherical chondrocytes and a relatively disorganized collagen network, whereas in the deep zone, the cells tend to organize into columns with radially oriented collagen fibril bundles between them. The pericellular compartment of these zones has high collagen VI, decorin, and perlecan content, while aggrecan is more abundant in the territorial and interterritorial regions. The calcified zone is separated from the upper zones by the tidemark, and its main function is to anchor the AC to the subchondral bone. The hypertrophic chondrocyte marker collagen X is expressed in this zone (Gannon et al. 1991; Hoyland et al. 1991; von der Mark et al. 1992), but unlike in the growth plate, the calcified matrix normally resists vascular invasion and resorption. Cartilage homeostasis is carefully balanced by metabolic and catabolic pathways to maintain ECM structure and function throughout the articular cartilage zones. Mechanical overloading, age-associated changes in chondrocyte function, and disturbed cytokine activities may all initiate ECM degradation and contribute to the onset and progression of articular cartilage deterioration resulting in osteoarthritis (for further details, see also Chaps. 1 and 2). The upregulation of matrix-degrading enzymes such as matrix metalloproteinases (MMPs) and aggrecanases belonging to the ADAMTS (a disintegrin and metalloproteinase with thrombospondin motifs) family causes the degradation of the collagenous and proteoglycan networks which in turn weakens the biomechanical function of the articular cartilage (for details, see Chaps. 3 and 4). Matrix turnover in normal and osteoarthritic articular cartilage occurs more rapidly in the pericellular compartment than in the interterritorial area (Wu et al. 2002) suggesting a role of cell surface receptors like integrins in the control of matrix remodeling.

Normal articular cartilage chondrocytes expose various integrins on their surfaces with the most prominent expression of α1β1, α2β1, and α10β1 (for collagens); α5β1, αvβ3,and αvβ5 (for fibronectin); and α6b1 (for laminin) (Durr et al. 1993; Salter et al. 1995; Camper et al. 2001; Woods et al. 1994). In vitro studies have demonstrated that human articular cartilage chondrocytes predominantly use β1 integrins, including α5β1 and αvβ5 integrins, for attachment to the cartilage extracellular matrix (Kurtis et al. 2003). In articular cartilage, chondrocytes interact with a dense collagen network in an environment which has low oxygen tension and exposed to mechanical stimuli. Bovine articular chondrocytes cultured in three-dimensional collagen sponges express high levels of collagen II, aggrecan, and integrin α10 mRNAs (Cortial et al. 2006). Chondrocytes cultured in alginate beads respond to hypoxia by elevating the expression of β1 integrin mRNA (Grimshaw and Mason 2001). As a response for mechanical loading, articular cartilage chondrocytes elevate the expression levels of α5 subunit in bovine explants (Lucchinetti et al. 2004) and the expression of α5β1 integrin in monolayer culture (Wright et al. 1997). α5β1-ECM interaction was shown to promote survival of AC chondrocytes cultured on fibronectin-coated plastic surface or in alginate-suspension culture under serum-free conditions (Pulai et al. 2002). Blocking antibodies against the α5 subunit also inhibited insulin-like growth factor I-mediated chondrocyte survival emphasizing the role of α5β1 integrin-growth factor signaling crosstalk in preventing chondrocyte death (Pulai et al. 2002). Overexpression and siRNA-mediated silencing studies revealed that β1 integrins regulate articular chondrocyte proliferation and survival, at least partially, by modulating the expression of the β1 integrin downstream target molecule G-protein-coupled receptor kinase interacting protein-1 (GIT1) (Zhang et al. 2015). Collectively, these results imply that integrins are important sensors and mediators of chemical and mechanical signals of the extracellular matrix which regulate numerous physiological functions of articular cartilage chondrocytes.

8.4.3 The Role of Integrins in Mechanotransduction of Articular Cartilage Chondrocytes

Biomechanical factors play a pivotal role in the normal function and health of synovial joints. Under normal circumstances, physiological mechanical loading results in a variety of environmental alterations including ECM and chondrocyte deformity, changes in fluid flow, ion concentrations, and hydrostatic and osmotic pressures which in turn lead to controlled anabolic changes that maintain the integrity of the articular cartilage. In contrast, nonphysiological mechanical stress, due to overloading or injury, is promoting cartilage catabolism and predisposes to OA. Upon stimuli, chondrocytes transmit mechanical signals through membrane associates called mechanotransducers such as stretch-activated ion channels, receptor tyrosine kinases, the hyaluronan receptor CD44, and integrins (Millward-Sadler and Salter 2004; Bader et al. 2011; Vincent 2013). Under physiological loading, integrins are activated on the surface of isolated chondrocytes leading to recruitment of adaptor proteins in focal adhesion complexes and to lateral interactions with growth factor receptors and ion channels (Millward-Sadler and Salter 2004). Induction of the integrin mechanotransduction pathway by intermittent hydrostatic pressure in monolayer-cultured chondrocytes is accompanied by α5β1 integrin- and protein kinase C (PKC)-dependent membrane hypopolarization and the secretion of the anabolic cytokine interleukin-4 (IL-4), which in turn reduces the expression of MMP-3 and increases aggrecan synthesis and glycosaminoglycan (GAG) production (Millward-Sadler et al. 1999, 2000; Wright et al. 1997; Holledge et al. 2008). Mechanical load elevates the surface expression of the α5 subunit on chondrocytes in bovine articular cartilage explants, which may suggest the existence of a positive feedback loop in the regulation of integrin-associated mechanotransduction processes (Lucchinetti et al. 2004). The application of dynamic cyclic compression on human chondrocytes in the presence of TGF-β3 also promotes proliferation and proteoglycan synthesis, which can be reversed by addition of the α5β1 ligand GRGDSP to the system (Chowdhury et al. 2004). Mechanotransduction via α5β1 is modulated through multiple integrin-associated pathways including activation of the focal adhesion proteins FAK and paxillin (Lee et al. 2000), assembly of PKCα-RACK1-β1 integrin complexes (Lee et al. 2002), and the interaction of α5β1 with the integrin-associated protein (IAP) CD47 (Orazizadeh et al. 2008). In addition to α5β1, normal human articular chondrocytes also respond for cyclic pressure-induced strain by the activation of αvβ5 and αvβ3 integrin-mediated signaling, since blocking αvβ5 or αvβ3 abolishes mechanical stimulation-induced GAG and proteoglycan synthesis (Holledge et al. 2008; Chai et al. 2010). Similar to monolayer-cultured human chondrocytes, IL-4 expression is upregulated in three-dimensionally embedded, mechanically stimulated rat chondrocytes and regulates matrix synthesis through the activation of the p38 MAPK signaling pathway (Shioji et al. 2014). Further experiments with rat chondrocytes exposed to periodic mechanical stress revealed the β1 integrin-mediated activation of the FAK-ERK1/2 signaling pathway indicating a complex role of MAP kinases downstream of integrins in chondrocyte mechanotransduction (Liang et al. 2013).

In contrast to the protective role of physiological loading, nonphysiological mechanical stress disrupts the actin cytoskeleton via α5β1-mediated signaling that activates MAP kinases and NF-kappa B. Those, in turn, result in elevation of nitric oxide, prostaglandin E2, reactive oxygen species, proteases, and cytokines, which mediate articular cartilage degradation (Bader et al. 2011). Furthermore, impact injury-induced chondrocyte death was also found to be dependent on cell-matrix interactions and cytoskeletal rearrangements which require integrin activation and FAK signaling (Jang et al. 2014) (further details, see Chap. 2).

Compression of cartilage leads to pressurization and exudation of the interstitial water, while the retained negatively charged GAGs attract positive counterions resulting in changes in osmolarity and osmotic pressure in the vicinity of the chondrocyte surface. Chondrocytes respond for osmotic stress by transient increase of intracellular Ca^{2+}, mostly mediated by the nonspecific cation channel, transient receptor potential vanilloid-4 (TRPV4) (McNulty et al. 2015; Clark et al. 2010). A recent study has revealed that α1 integrin subunit-deficient chondrocytes exhibit impaired intracellular calcium transient in response to osmotic stress and the TRPV4 agonist GSK1016790A (Jablonski et al. 2014). These observations implicate that integrin α1β1 may play important, adhesion-independent role in the osmotransduction of chondrocytes.

8.4.4 Integrins and Osteoarthritis

Articular cartilage is highly prone to injury- and age-associated pathological deterioration. Once damaged, the AC is unable to initiate an efficient regenerative process owing to the loss of mitotic activity of adult AC chondrocytes, decreased responsiveness to anabolic growth factors, and synthesis of less functional matrix molecules. Osteoarthritis (OA) is a common degenerative disorder of the synovial joints that not only affects the articular cartilage but also involves changes in subchondral bone, synovium, ligaments, and muscle. OA is a complex disorder influenced by both environmental and genetic factors. Although the precise etiology is largely unknown, it is generally accepted that uncontrolled metabolism of skeletal tissues is critical for the pathophysiology of this disabling condition. In primary OA, the normal turnover of ECM molecules produced by chondrocytes is disturbed, and the equilibrium between anabolic and catabolic processes is shifted toward proteolytic activities, which eventually results in progressive, age-dependent articular cartilage destruction (Goldring and Goldring 2007) (for overview, see also Chap. 1). In addition, congenital abnormalities of the joints, as a typical result of impaired development of endochondral bones, are often associated with early-onset degeneration and loss of the articular cartilage (secondary OA). Physiological ECM remodeling of the articular cartilage occurs in a spatially and temporally controlled fashion and involves the activities of both proteinases and proteinase inhibitors, which are tightly regulated at multiple levels. The regulatory mechanisms are only partially understood and include, in addition to growth factor and cytokine signaling pathways, chemical and mechanical signals modulated by matrix-matrix and cell-matrix interaction (Knudson and Loeser 2002; Loeser 2014). There is a growing consensus that changes of ECM composition, e.g., by proteolytic degradation of matrix constituents, or alterations of the biomechanical microenvironment of chondrocytes caused by chronic stress or injury significantly increase the risk of OA through the perturbation of integrin-mediated signaling. The strongest evidence for the participation of integrins in OA has been emerged from experiments studying integrin expression and ECM fragments-induced signaling processes. In osteoarthritic monkey articular cartilage, increased immunostaining of α1, α3, and α5β1 integrins was reported compared with normal cartilage (Loeser et al. 1995). In human OA samples, the expression of β1 integrin was inversely correlated with the severity of the lesions suggesting that β1 integrin-mediated-ECM adhesion is decreasing with OA progression (Lapadula et al. 1998). In contrast, osteoarthritic human femoral head cartilage expressed the integrin subunits β2, α2, and α4 which were not observed in normal AC cartilage (Ostergaard et al. 1998).

Subsequently, numerous reports have shown that fibronectin (Fn) and collagen II degradation fragments are present in synovial fluid and cartilage of OA patients as a result of increased catabolic activity, and these fragments can further mediate joint destruction through the activation of integrin signaling which in turn elevates the gene expression of catabolic cytokines, matrix metalloproteinases (MMPs), and catabolic mediators such as nitric oxide (NO) and prostaglandin E_2 (PGE_2) (reviewed by (Sofat 2009; Yasuda 2006; Loeser 2014) (Fig. 8.3). Treatments of cultured rabbit synovial fibroblasts with the central 110–120 kDa cell-binding Fn fragment (Fn-f) carrying the RGD sequence or ligation of α5β1 integrin with a monoclonal antibody were shown to stimulate MMP-1 and MMP-3 expression (Werb et al. 1989). Similarly, the central Fn-f and α5β1/α2β1 activating antibodies were shown to upregulate MMP-13 synthesis accompanied by elevated expression of various interleukins (ILs) including IL-1β, IL-6, IL7, and IL-8 and the growth-related oncogene β (Groβ) in cultured human articular cartilage chondrocytes (Forsyth et al. 2002; Loeser et al. 2003; Pulai et al. 2005; Long et al. 2008). In addition to central Fn-f, the amino-terminal heparin binding 29 kDa fragment, which can interact with α5β1 in a non-RGD-dependent manner (Dzamba et al. 1994; Hocking et al. 1998),

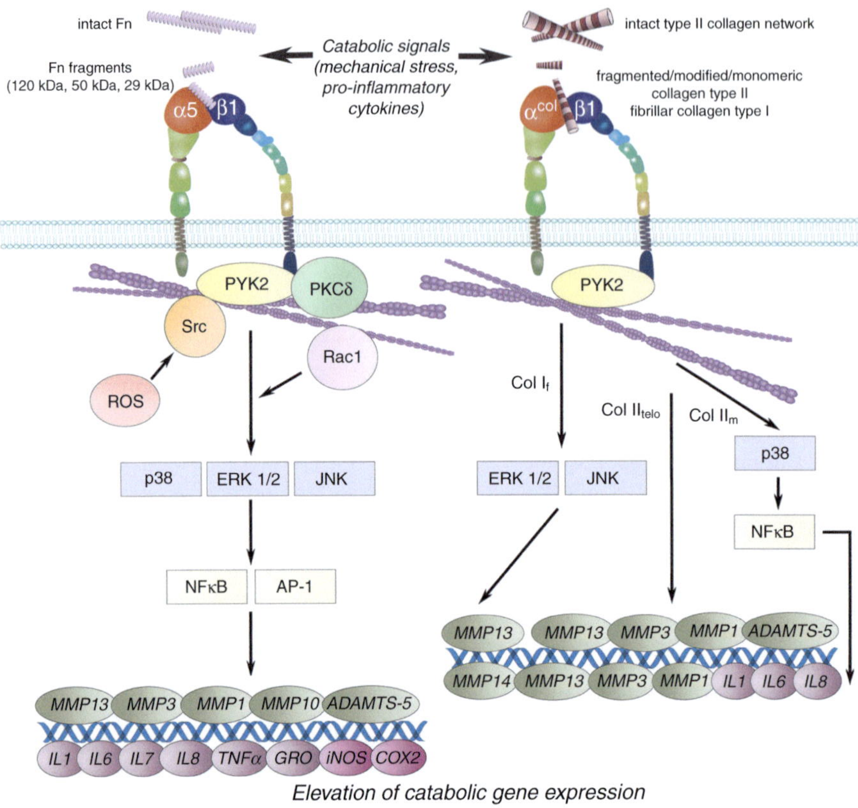

Fig. 8.3 Integrins-associated signaling pathways induced in articular cartilage chondrocytes by cartilage degradation fragments. Fibronectin (Fn) fragmentation products signals through α5β1 integrin to induce catabolic gene expression. Collagen-binding integrin heterodimers (such as α1β1 and α2β1) might activate alternative intracellular pathways for catabolic response to fibrillar type I collagen (*Col If*), type II collagen telopeptides (*Col IItelo*), or monomeric collagen type II (*Col IIm*)

and the gelatin-binding 45–50 kDa fragment are also chondrolytic and likely interfere with the native Fn/α5β1 signaling pathway (Homandberg and Hui 1994; Homandberg et al. 2002a, b). The 29 kDa Fn-f in micromolar concentration stimulated the expression of MMP-3 and the catabolic cytokines tumor necrosis factor α (TNFα), IL1 and IL-6; and neutralizing antibodies against these cytokines were able to reduce MMP-3 release in human articular cartilage explants (Homandberg et al. 1997). It has been reported that synthetic RGD peptides regulate the expression of MMP-3 via an IL-1 autocrine regulatory loop in cultured rabbit articular chondrocytes (Arner and Tortorella 1995); however, IL-1 is not required for the synthesis of MMP-13 induced by the 120 kDa Fn-f but did increase the activation of MMP-13 in human articular cartilage chondrocytes (Forsyth et al. 2002). The 45 kDa fragment, besides increasing MMP-3 and MMP-13 synthesis, was shown to elevate the levels of TEGE, an aggrecan degradation neoepitope generated by ADAMTS-4 and

ADAMTS-5 (see also Chap. 3), independently of IL-1 in porcine articular cartilage chondrocytes (Stanton et al. 2002). Mechanistically, the stimulation of α5β1 integrin with ligation antibodies or Fn-fs activates protein kinase C (PKC)-delta, which in turn phosphorylates the proline-rich tyrosine kinase-2 (PYK2) resulting in activation of MAPK (Erk-1/2, p38, and JNK) signaling cascades, increased phosphorylation of c-jun subunit of the AP-1 transcription factor, and the activation of the NFkB signaling (Loeser et al. 2003; Pulai et al. 2005; Ding et al. 2008, 2009).

Fibronectin fragment-induced upregulation of MMPs requires endogenous production of reactive oxygen species (ROS) (Del Carlo et al. 2007; Ding et al. 2009), the active form of the small GTPase Rac1 (Long et al. 2013), and Src, a focal adhesion tyrosine kinase activating FAK and PYK2 (Ding et al. 2009). In human chondrocytes, Src activity is apparently under the regulation of redox signaling-associated oxidative posttranslational modification, which mediates S-sulfenylation of Scr and required for Fn-f-induced MMP-13 production (Wood et al. 2016). In bovine cartilage explants, Fn-fs increased NO release and inducible NO synthase (iNOS) protein and mRNA levels, and iNOS inhibitors blocked proteoglycan depletion and repaired damaged cartilage (Pichika and Homandberg 2004). The physiological relevance of Fn-fs in cartilage destruction was evidenced by studies demonstrating that injection of Fn-fs into knee joint dramatically increases proteoglycan loss of rabbit articular cartilage (Homandberg et al. 1993). In pathological situations such as joint injury caused by mechanical impact, cartilage degradation is associated with increased Fn-f levels, the activation of p38 and ERK1/ERK2 MAP kinases, and the upregulation of MMP-3, MMP-13, TNF-α, and ADAMTS-5 (Ding et al. 2010, 2014). All together these observations implicate that mechanical or catabolic stress-induced fibronectin fragmentation converges with integrin and MAP kinase signaling in the control of matrix destruction in osteoarthritis.

Collagen II degradation fragments and synthetic N- and C-terminal telopeptides (NT and CT) have also been shown to upregulate MMPs and ADAMTS-5 in cultured articular cartilage chondrocytes or in cartilage explants implicating an important role of these collagen-derived degradation products in matrix homeostasis (Jennings et al. 2001; Fichter et al. 2006; Guo et al. 2009). Both NT and CT stimulate MMP-1, MMP-3, and MMP-13 expression and proteoglycan depletion in bovine cartilage explants at lesser extent and do not induce IL-1ß and TNFα expression compared with the 29 kDa Fn fragment (Guo et al. 2009). In contrast, monomeric, non-fibrillar type II collagen dose-dependently upregulates not only MMP-1, -3, -13, and -14 but also the expression of the catabolic cytokines IL-1ß, IL-6, and IL-8 in primary human articular chondrocytes, which may lead to collagen fragmentation and further MMPs and cytokines production (Klatt et al. 2009). Similar to Fn fragments, the expression of these matrix destructive molecules required the activation of the p38 MAPK and NFkB signaling pathways (Klatt et al. 2009). Collagen fragments and peptides likely interact with collagen-binding integrins on chondrocytes, since the CB10 cyanogen bromide fragment of type II collagen binds to α2β1 integrin (Tuckwell et al. 1994). It is important to note, however, that chondrocytes might utilize alternative non-integrin receptors for the modulation of collagen fragment-induced cartilage metabolism such as annexin V (Lucic et al. 2003) and discoidin domain receptor II (DDR2) (Xu et al. 2010).

The exact mechanism by which ECM degradation fragments accelerate cartilage catabolism is unclear. The most popular explanation supposes that FN and collagen degradation products disturb or even disrupt normal integrin clustering induced by native FN or collagen II, which in turn alters physiological outside-in signaling and cartilage metabolisms (Peters et al. 2002). However, beside degradation fragments, fibrillar collagen itself is able to induce MMP expression upon binding to α1β1 integrin. Collagen I is upregulated in OA cartilage (Adam and Deyl 1983; Miosge et al. 2004) and in the chondrogenic cell line MC615, and it stimulates the expression of MMP-13 via the activation of ERK and JNK MAP kinases. In the presence of α1 or β1 blocking antibodies, the induction of MMP-13 expression was inhibited (Ronziere et al. 2005). Furthermore, a recent report has suggested that collagen II fibrils modified by the lipid peroxidation end-product 4-hydroxynonenal, which is upregulated in synovial fluid of OA patients, induce cell death and catabolic and inflammatory responses of human AC chondrocytes, presumably through perturbation of α1β1 signaling (El-Bikai et al. 2010).

8.4.5 Genetic Models: The In Vivo Role of Integrins in the Articular Cartilage

Although the expression studies and in vitro experiments highlight the importance of integrin signaling in OA and suggest that inhibition of integrin pathways might be a potential approach to ameliorate ECM and ECM fragment-mediated articular cartilage destruction, the in vivo role of integrins in OA pathogenesis has just begun to be elucidated. Most knockout mouse lines lacking a particular α subunit, except the α1 integrin-null mice, develop an apparently normal AC without reported abnormalities. The collagen-binding α1β1 integrin normally is expressed at low level in the deep zone of healthy mouse AC, but it is significantly upregulated in the middle and upper zones of early osteoarthritic cartilage suggesting that this integrin heterodimer supports cartilage remodeling (Zemmyo et al. 2003). Knee joints of α1-null mice exhibit proteoglycan loss, cartilage erosion accompanied by increased MMP-2 and MMP-3 expression, and synovial hyperplasia in younger animals than wild-type controls (Zemmyo et al. 2003). It has been shown that microRNA-101 and the long noncoding RNA HOTTIP (Hoxa transcript at the distal tip) regulate the expression of α1 integrin during chondrogenic differentiation of mesenchymal stem cells and in osteoarthritic chondrocytes (Kim et al. 2013). Lentiviral-mediated overexpression of miR-101 or α1 integrin in the joint ameliorates osteoarthritis induced by surgical destabilization of the medial meniscus (DMM) in mice. Precocious development of knee OA was also observed in mice lacking the membrane-anchored disintegrin and metalloproteinase ADAM15 (Bohm et al. 2005). ADAM15 carries a RGD motif and has multiple pericellular functions including the modulation of outside-in signaling in chondrocyte-matrix interactions (Bohm et al. 2009), which in turn enhances chondrocyte adhesion to collagen II and promotes AC homeostasis. In contrast to α1 integrin, α2 integrin deficiency does not cause spontaneous OA but suppresses cartilage erosion and synovial inflammation associated with reduced MMP-3 expression in mouse models of rheumatoid arthritis (Peters et al. 2012). The ablation of the gene coding for the articular chondrocyte-abundant, collagen-binding α10 integrin in mice results no reported pathology of the articular cartilage up to 1 year of age (Bengtsson et al. 2005).

Similarly to the growth plate, β1 integrins are critical regulators of chondrocyte phenotype in the articular cartilage as well. *β1fl/fl-Prx1cre* mice display severe abnormalities of AC chondrocytes including rounding of the superficial zone cells, loss of columnar arrangement in the deep zone, reduced proliferation, increased cell death, frequent bi-nucleation, and disrupted actin cytoskeleton (Raducanu et al. 2009). These observations extend the view that in vivo β1 integrins are pivotal for maintaining the normal shape, organization, proliferation, and survival of chondrocytes independent of their location in the cartilaginous skeleton. Furthermore, β1 integrin deficiency of AC chondrocytes leads to prominent ECM defects including disorganized collagen II network in the interterritorial compartment and reduced pericellular deposition of collagen VI indicating that β1 integrins are important organizers throughout the ECM. Importantly, the ablation of β1 integrins differentially affects chondrocyte maturation in the growth plate and the articular cartilage. While in *β1fl/fl-Col2cre* and *β1fl/fl-Prx1cre* mice the hypertrophic differentiation of growth plate and epiphyseal chondrocytes is delayed, the AC in the *β1fl/fl-Prx1cre* mice exhibits reduced ratio of the noncalcified and calcified zones and contains collagen X expressing chondrocytes above the tidemark, suggesting accelerated hypertrophic differentiation (Aszodi et al. 2003; Raducanu et al. 2009). This may imply that β1 integrins have a spatially restricted, tissue context-dependent role in chondrocyte maturation.

Despite the osteoarthritis-like structural abnormalities, *β1fl/fl-Prx1cre* mice exhibit no obvious articular cartilage erosion or elevated expression of matrix-degrading proteases compared with controls (Raducanu et al. 2009). In contrast to the previous in vitro experiments, no evidence was found for disturbed activation of MAP kinases in mutant AC and in hip explants upon stimulation with a 40 kDA cell-binding FN fragment suggesting that the lack of β1 integrins might be efficiently compensated via signaling through β3/β5 integrins and/or growth factor receptors. In addition to the growth plate and articular cartilage phenotypes, *β1fl/fl-Prx1cre* mice develop ectopic cartilage and bone in various limb tissues including tendons, synovial ligaments, and vessels (Raducanu et al. 2009). These pathologies, in turn, result in significant immobilization of the legs and high mortality rate between 8 and 12 months of age due to vessel rupture. Since the onset and progression of OA are influenced by biomechanical factors, the articular cartilage of the *β1fl/fl-Prx1cre* mice mechanically might be less exposed providing a possible explanation for the lack of cartilage deterioration.

The secreted melanoma inhibiting activity/cartilage-derived retinoic acid-sensitive protein (MIA/CD-RAP)-deficient mice partially recapitulate the AC abnormalities seen in the *β1fl/fl-Prx1cre* mice. MIA/CD-RAP is expressed in chondrocytes (Bosserhoff et al. 1997; Dietz and Sandell 1996), and its absence in mice leads to mild ultrastructural abnormalities of the collagen fibrils in the developing growth plate cartilage (Moser et al. 2002). In 3-month-old mice, MIA/CD-RAP-deficient articular cartilage shows disturbed columnar arrangement of the chondrocytes and increased height of the calcified zone (Schubert et al. 2010). Based on in vitro studies with cultured chondrocytes, it was proposed that MIA/CD-RAP physically binds to and inhibits α5β1 integrin leading to reduced ERK signaling, stabilization of the chondrocyte phenotype, and delay of hypertrophy. This finding is consistent with a recent study demonstrating correlation between ERK activation and hypertrophic differentiation of human AC chondrocytes (Prasadam et al. 2010). Furthermore, an antibody activating α5β1 integrin/ERK signaling (Forsyth et al. 2002) was shown to abolish the negative effect of MIA/CD-RAP on the expression of the osteogenic marker osteocalcin in cultured AC chondrocytes (Schubert et al. 2010).

Taken together, in vitro and loss-of-function genetic studies in mice strongly suggest that integrins are important for proper differentiation and spatial organization of AC chondrocytes. Since the ablation of both β1 integrin and the α5β1 integrin-inhibitory MIA/CD-RAP results in similar phenotype, the exact nature of integrin-mediated signaling pathways on articular chondrocyte hypertrophy remains to be elucidated.

8.5 Articular Cartilage Repair

Human articular cartilage has a limited intrinsic capacity to repair due to low proliferation of chondrocytes, inefficient regeneration potential of resident chondroprogenitors, and the lack of vasculature to deliver mesenchymal stem cells (MSCs) into the damaged tissue. Current orthopedic therapeutic strategies include marrow stimulation-based, osteochondral transfer, and cell-based techniques (Richter 2009) with various advantages and disadvantages of each. Adult MSCs derived from the bone marrow or adipose tissues, for example, have better proliferative activity compared to autologous chondrocytes in vitro but require specific culture conditions to differentiate into chondrocytes. Cell-based regeneration of the articular cartilage requires the induction and stabilization of the chondrogenic phenotype, and the pivotal role of integrins in these processes is increasingly recognized.

In vitro, MSC differentiation into the chondrogenic, osteogenic, or adipogenic lineages can be modulated via growth factors, cell-matrix interactions, three-dimensional (3D) microenvironment, cytoskeletal tension, or mechanical stimulation (McBeath et al. 2004; Engler et al. 2006; Woods et al. 2007; Kelly and Jacobs 2010). Collagen II promotes chondrogenesis of bone marrow-derived MSCs (BMSCs) and adipose-derived MSCs (ASCs) in hydrogels (Lu et al. 2010; Bosnakovski et al. 2006) by inducing cell rounding through β1 integrin-mediated RhoA/Rock signaling (Lu et al. 2010). Culturing of human BMSCs in pellet together with type II collagen fibrils requires β1 and α2 integrins for chondrocyte differentiation and the expression of cartilaginous markers (Chang et al. 2007). In contrast, RGD peptides block chondrogenesis of BMSCs by inducing cell spreading through

$\alpha v\beta 3$ integrin and actin cytoskeletal reorganization (Connelly et al. 2007, 2008). Stem cell commitment toward the chondrogenic lineage can be further modulated through ECM remodeling. MT1-MMP (MMP-14), a membrane-bound pericellular collagenase, activates $\beta 1$ integrin-Rho GTPase signaling cascade in mesenchymal progenitors and triggers the nuclear translocation of YAP/TAZ transcriptional coactivators (Tang et al. 2013). An elegant study using MT1-MMP-deficient MSCs and mice lacking MT1-MMP in skeletogenic MSCs has demonstrated that in the absence of MT1-MMP, MSCs preferentially differentiate toward the chondrogenic and adipogenic lineages. Changing the internal stiffness of the hydrogels and external mechanical loading were also shown also influence MSC chondrogenesis (Huang et al. 2004; Pelaez et al. 2009; Meyer et al. 2011), which involves integrin-mediated cell-matrix interactions (Steward et al. 2012, 2014).

Many further studies have demonstrated the potential of integrin signaling modulation in articular cartilage repair and regeneration. The activation of FAK is increasing upon exogenous stimulation of alginate beads with collagen II, whereas FAK knockdown reduces chondrocyte redifferentiation in the same culture system (Kim and Lee 2009). MIA/CDRAP-deficient mice exhibit increased regeneration potential of the AC in a surgically induced OA model due to enhanced proliferation of the chondrocytes (Schmid et al. 2010). In contrast, the lack of tenascin C, a specific ligand for some $\beta 1$ and $\beta 3$ integrins, delays the repair of cartilage defects in mice (Okamura et al. 2010). Magnesium, a divalent cation required for activation of $\beta 1$ integrins, enhances the adherence and cartilaginous differentiation of human synovial MSCs in a rabbit osteochondral defect model (Shimaya et al. 2010).

Human BMSCs expanded in monolayer express both $\alpha 10\beta 1$ and $\alpha 11\beta 1$ integrins, and during chondrogenic differentiation in pellet culture, the expression of $\alpha 10\beta 1$ is increasing, while the expression of $\alpha 11\beta 1$ is decreasing. Simple administration of FGF-2 to the monolayer culture medium elevates the levels of $\alpha 10\beta 1$ integrin and the chondrogenic marker Sox9, which in turn results in better chondrogenic differentiation potential of MSCs (Varas et al. 2007). This observation raises the possibility that $\alpha 10\beta 1$ is a potent and unique biomarker for quality assurance of MSCs used for cartilage repair.

8.6 Perspectives

Chondrocyte integrins have indispensable roles in cartilage development, skeletal growth, and articular cartilage function. Acting as signaling hubs, integrins sense chemical and mechanical stimuli arising from the cartilage extracellular matrix, transmit them into the chondrocytes, and orchestrate a variety of cellular responses important for differentiation, survival, proliferation, polarity, matrix production, and assembly. Integrating and connecting ECM cues with the actin cytoskeleton through signaling pathways are the essential function of integrins to maintain normal physiology of cartilaginous tissues. The complexity of integrin signaling in the control of chondrocyte behavior is increasingly recognized through in vitro and in vivo models, and drawing a valid conclusion for integrin functions from the expanding data obtained from two- or three-dimensional culture systems and genetic animal models is a great challenge of the future research. Careful morphologic evaluation of mice carrying ablation for integrin subunits in combination with cutting-edge proteomics, genomics, and metabolomics techniques held a promise to dissect more precisely the role of integrin heterodimers and their associated signaling cascades in cartilage physiology. Furthermore, disturbed integrin-mediated adhesion or altered signaling in response to nonphysiological ECM signals is associated with pathological situations such as chondrodysplasia and osteoarthritis. Overwhelming body of evidence supports the view that integrin-mediated signaling play pivotal roles in matrix destruction, and interfering with the way how chondrocytes sense the environment through integrins may open novel paths of therapeutic approaches to slow down osteoarthritis or correct growth abnormalities. Finally, modulating integrin expression and/or signaling during chondrogenic differentiation of stem cells could help to design better tissue engineering constructs for cell-based regeneration of the damaged articular cartilage.

Acknowledgment The author is supported by the DFG grants AS 150/11-1 (FOR2407) and AS 150/7-1.

References

Adam M, Deyl Z (1983) Altered expression of collagen phenotype in osteoarthrosis. Clin Chim Acta: Int J Clin Chem 133(1):25–32

Aikawa S, Hashimoto T, Kano K, Aoki J (2015) Lysophosphatidic acid as a lipid mediator with multiple biological actions. J Biochem 157(2):81–89. doi:10.1093/jb/mvu077

Arner EC, Tortorella MD (1995) Signal transduction through chondrocyte integrin receptors induces matrix metalloproteinase synthesis and synergizes with interleukin-1. Arthritis Rheum 38(9):1304–1314

Aszódi A (2016) The cartilaginous growth plate. In: Grässel S, Aszódi A (eds) Cartilage: volume 1: Physiology and development. Springer International Publishing, Cham, pp 83–113. doi:10.1007/978-3-319-29568-8_4

Aszodi A, Bateman JF, Gustafsson E, Boot-Handford R, Fassler R (2000) Mammalian skeletogenesis and extracellular matrix: what can we learn from knockout mice? Cell Struct Funct 25(2):73–84

Aszodi A, Hunziker EB, Brakebusch C, Fassler R (2003) Beta1 integrins regulate chondrocyte rotation, G1 progression, and cytokinesis. Genes Dev 17(19):2465–2479. doi:10.1101/gad.277003

Bader DL, Salter DM, Chowdhury TT (2011) Biomechanical influence of cartilage homeostasis in health and disease. Arthritis 2011:979032. doi:10.1155/2011/979032

Bang OS, Kim EJ, Chung JG, Lee SR, Park TK, Kang SS (2000) Association of focal adhesion kinase with fibronectin and paxillin is required for precartilage condensation of chick mesenchymal cells. Biochem Biophys Res Commun 278(3):522–529. doi:10.1006/bbrc.2000.3831

Bengtsson T, Aszodi A, Nicolae C, Hunziker EB, Lundgren-Akerlund E, Fassler R (2005) Loss of alpha10beta1 integrin expression leads to moderate dysfunction of growth plate chondrocytes. J Cell Sci 118(Pt 5):929–936. doi:10.1242/jcs.01678

Bicknell LS, Morgan T, Bonafe L, Wessels MW, Bialer MG, Willems PJ, Cohn DH, Krakow D, Robertson SP (2005) Mutations in FLNB cause boomerang dysplasia. J Med Genet 42(7):e43. doi:10.1136/jmg.2004.029967

Bingel SA, Sande RD (1982) Chondrodysplasia in the Norwegian Elkhound. Am J Pathol 107(2):219–229

Blumbach K, Niehoff A, Belgardt BF, Ehlen HW, Schmitz M, Hallinger R, Schulz JN, Bruning JC, Krieg T, Schubert M, Gullberg D, Eckes B (2012) Dwarfism in mice lacking collagen-binding integrins alpha2beta1 and alpha11beta1 is caused by severely diminished IGF-1 levels. J Biol Chem 287(9):6431–6440. doi:10.1074/jbc.M111.283119

Bohm BB, Aigner T, Roy B, Brodie TA, Blobel CP, Burkhardt H (2005) Homeostatic effects of the metalloproteinase disintegrin ADAM15 in degenerative cartilage remodeling. Arthritis Rheum 52(4):1100–1109. doi:10.1002/art.20974

Bohm BB, Schirner A, Burkhardt H (2009) ADAM15 modulates outside-in signalling in chondrocyte-matrix interactions. J Cell Mol Med 13(8B):2634–2644. doi:10.1111/j.1582-4934.2008.00490.x

Bosnakovski D, Mizuno M, Kim G, Takagi S, Okumura M, Fujinaga T (2006) Chondrogenic differentiation of bovine bone marrow mesenchymal stem cells (MSCs) in different hydrogels: influence of collagen type II extracellular matrix on MSC chondrogenesis. Biotechnol Bioeng 93(6):1152–1163. doi:10.1002/bit.20828

Bosserhoff AK, Kondo S, Moser M, Dietz UH, Copeland NG, Gilbert DJ, Jenkins NA, Buettner R, Sandell LJ (1997) Mouse CD-RAP/MIA gene: structure, chromosomal localization, and expression in cartilage and chondrosarcoma. Dev Dyn 208(4):516–525. doi:10.1002/(sici)1097-0177(199704)208:4<516::aid-aja7>3.0.co;2-l

Bougault C, Cueru L, Bariller J, Malbouyres M, Paumier A, Aszodi A, Berthier Y, Mallein-Gerin F, Trunfio-Sfarghiu AM (2013) Alteration of cartilage mechanical properties in absence of beta1 integrins revealed by rheometry and FRAP analyses. J Biomech 46(10):1633–1640. doi:10.1016/j.jbiomech.2013.04.013

Bouvard D, Brakebusch C, Gustafsson E, Aszodi A, Bengtsson T, Berna A, Fassler R (2001) Functional consequences of integrin gene mutations in mice. Circ Res 89(3):211–223

Briggs MD, Bell PA, Wright MJ, Pirog KA (2015) New therapeutic targets in rare genetic skeletal diseases. Expert Opin Orphan Drugs 3(10):1137–1154. doi:10.1517/21678707.2015.1083853

Campbell ID, Humphries MJ (2011) Integrin structure, activation, and interactions. Cold Spring Harb Perspect Biol 3(3). doi:10.1101/cshperspect.a004994 a004994 [pii] cshperspect.a004994 [pii]

Camper L, Hellman U, Lundgren-Akerlund E (1998) Isolation, cloning, and sequence analysis of the integrin subunit alpha10, a beta1-associated collagen binding integrin expressed on chondrocytes. J Biol Chem 273(32):20383–20389

Camper L, Holmvall K, Wangnerud C, Aszodi A, Lundgren-Akerlund E (2001) Distribution of the collagen-binding integrin alpha10beta1 during mouse development. Cell Tissue Res 306(1):107–116

Cao L, Lee V, Adams ME, Kiani C, Zhang Y, Hu W, Yang BB (1999) beta-Integrin-collagen interaction reduces chondrocyte apoptosis. Matrix Biol: J Int Soc Matrix Biol 18(4):343–355

Chai DH, Arner EC, Griggs DW, Grodzinsky AJ (2010) Alphav and beta1 integrins regulate dynamic compression-induced proteoglycan synthesis in 3D gel culture by distinct complementary pathways. Osteoarthr Cartil/OARS Osteoarthritis Res Soc 18(2):249–256. doi:10.1016/j.joca.2009.09.002

Chang CF, Lee MW, Kuo PY, Wang YJ, Tu YH, Hung SC (2007) Three-dimensional collagen fiber remodeling by mesenchymal stem cells requires the integrin-matrix interaction. J Biomed Mater Res A 80(2):466–474. doi:10.1002/jbm.a.30963

Choocheep K, Hatano S, Takagi H, Watanabe H, Kimata K, Kongtawelert P, Watanabe H (2010) Versican facilitates chondrocyte differentiation and regulates joint morphogenesis. J Biol Chem 285(27):21114–21125. doi:10.1074/jbc.M109.096479

Chowdhury TT, Salter DM, Bader DL, Lee DA (2004) Integrin-mediated mechanotransduction processes in TGFbeta-stimulated monolayer-expanded chondrocytes. Biochem Biophys Res Commun 318(4):873–881. doi:10.1016/j.bbrc.2004.04.107

Clark AL, Votta BJ, Kumar S, Liedtke W, Guilak F (2010) Chondroprotective role of the osmotically sensitive ion channel transient receptor potential vanilloid 4: age- and sex-dependent progression of osteoarthritis in Trpv4-deficient mice. Arthritis Rheum 62(10):2973–2983. doi:10.1002/art.27624
Connelly JT, Garcia AJ, Levenston ME (2007) Inhibition of in vitro chondrogenesis in RGD-modified three-dimensional alginate gels. Biomaterials 28(6):1071–1083. doi:10.1016/j.biomaterials.2006.10.006
Connelly JT, Garcia AJ, Levenston ME (2008) Interactions between integrin ligand density and cytoskeletal integrity regulate BMSC chondrogenesis. J Cell Physiol 217(1):145–154. doi:10.1002/jcp.21484
Cortial D, Gouttenoire J, Rousseau CF, Ronziere MC, Piccardi N, Msika P, Herbage D, Mallein-Gerin F, Freyria AM (2006) Activation by IL-1 of bovine articular chondrocytes in culture within a 3D collagen-based scaffold. An in vitro model to address the effect of compounds with therapeutic potential in osteoarthritis. Osteoarthr Cartil/OARS Osteoarthritis Res Soc 14(7):631–640. doi:10.1016/j.joca.2006.01.008
Del Carlo M, Schwartz D, Erickson EA, Loeser RF (2007) Endogenous production of reactive oxygen species is required for stimulation of human articular chondrocyte matrix metalloproteinase production by fibronectin fragments. Free Radic Biol Med 42(9):1350–1358. doi:10.1016/j.freeradbiomed.2007.01.035
Dietz UH, Sandell LJ (1996) Cloning of a retinoic acid-sensitive mRNA expressed in cartilage and during chondrogenesis. J Biol Chem 271(6):3311–3316
Ding L, Guo D, Homandberg GA (2008) The cartilage chondrolytic mechanism of fibronectin fragments involves MAP kinases: comparison of three fragments and native fibronectin. Osteoarthr Cartil/OARS Osteoarthritis Res Soc 16(10):1253–1262. doi:10.1016/j.joca.2008.02.015
Ding L, Guo D, Homandberg GA (2009) Fibronectin fragments mediate matrix metalloproteinase upregulation and cartilage damage through proline rich tyrosine kinase 2, c-src, NF-kappaB and protein kinase Cdelta. Osteoarthr Cartil/OARS Osteoarthritis Res Soc 17(10):1385–1392. doi:10.1016/j.joca.2009.03.024
Ding L, Heying E, Nicholson N, Stroud NJ, Homandberg GA, Buckwalter JA, Guo D, Martin JA (2010) Mechanical impact induces cartilage degradation via mitogen activated protein kinases. Osteoarthr Cartil/OARS Osteoarthritis Res Soc 18(11):1509–1517. doi:10.1016/j.joca.2010.08.014
Ding L, Guo D, Homandberg GA, Buckwalter JA, Martin JA (2014) A single blunt impact on cartilage promotes fibronectin fragmentation and upregulates cartilage degrading stromelysin-1/matrix metalloproteinase-3 in a bovine ex vivo model. J Orthop Res 32(6):811–818. doi:10.1002/jor.22610
Docheva D, Popov C, Alberton P, Aszodi A (2014) Integrin signaling in skeletal development and function. Birth Defects Res C Embryol Today 102(1):13–36. doi:10.1002/bdrc.21059
Dodds, GS (1930) Row formation and other types of arrangement of cartilage cells in endochondral ossification. Anat Rec 46: 385–399. doi:10.1002/ar.1090460409
Durr J, Goodman S, Potocnik A, von der Mark H, von der Mark K (1993) Localization of beta 1-integrins in human cartilage and their role in chondrocyte adhesion to collagen and fibronectin. Exp Cell Res 207(2):235–244. doi:10.1006/excr.1993.1189
Durr J, Lammi P, Goodman SL, Aigner T, von der Mark K (1996) Identification and immunolocalization of laminin in cartilage. Exp Cell Res 222(1):225–233. doi:10.1006/excr.1996.0028
Dzamba BJ, Bultmann H, Akiyama SK, Peters DM (1994) Substrate-specific binding of the amino terminus of fibronectin to an integrin complex in focal adhesions. J Biol Chem 269(30):19646–19652
El-Bikai R, Welman M, Margaron Y, Cote JF, Macqueen L, Buschmann MD, Fahmi H, Shi Q, Maghni K, Fernandes JC, Benderdour M (2010) Perturbation of adhesion molecule-mediated chondrocyte-matrix interactions by 4-hydroxynonenal binding: implication in osteoarthritis pathogenesis. Arthritis Res Ther 12(5):R201. doi:10.1186/ar3173

Engler AJ, Sen S, Sweeney HL, Discher DE (2006) Matrix elasticity directs stem cell lineage specification. Cell 126(4):677–689. doi:10.1016/j.cell.2006.06.044

Enomoto-Iwamoto M, Iwamoto M, Nakashima K, Mukudai Y, Boettiger D, Pacifici M, Kurisu K, Suzuki F (1997) Involvement of alpha5beta1 integrin in matrix interactions and proliferation of chondrocytes. J Bone Miner Res 12(7):1124–1132. doi:10.1359/jbmr.1997.12.7.1124

Farrington-Rock C, Firestein MH, Bicknell LS, Superti-Furga A, Bacino CA, Cormier-Daire V, Le Merrer M, Baumann C, Roume J, Rump P, Verheij JB, Sweeney E, Rimoin DL, Lachman RS, Robertson SP, Cohn DH, Krakow D (2006) Mutations in two regions of FLNB result in atelosteogenesis I and III. Hum Mutat 27(7):705–710. doi:10.1002/humu.20348

Farrington-Rock C, Kirilova V, Dillard-Telm L, Borowsky AD, Chalk S, Rock MJ, Cohn DH, Krakow D (2008) Disruption of the Flnb gene in mice phenocopies the human disease spondylocarpotarsal synostosis syndrome. Hum Mol Genet 17(5):631–641. doi:10.1093/hmg/ddm188

Fassler R, Meyer M (1995) Consequences of lack of beta 1 integrin gene expression in mice. Genes Dev 9(15):1896–1908

Fichter M, Korner U, Schomburg J, Jennings L, Cole AA, Mollenhauer J (2006) Collagen degradation products modulate matrix metalloproteinase expression in cultured articular chondrocytes. J Orthop Res 24(1):63–70. doi:10.1002/jor.20001

Forsyth CB, Pulai J, Loeser RF (2002) Fibronectin fragments and blocking antibodies to alpha2beta1 and alpha5beta1 integrins stimulate mitogen-activated protein kinase signaling and increase collagenase 3 (matrix metalloproteinase 13) production by human articular chondrocytes. Arthritis Rheum 46(9):2368–2376. doi:10.1002/art.10502

Gannon JM, Walker G, Fischer M, Carpenter R, Thompson RC Jr, Oegema TR Jr (1991) Localization of type X collagen in canine growth plate and adult canine articular cartilage. J Orthop Res 9(4):485–494. doi:10.1002/jor.1100090404

Garciadiego-Cazares D, Rosales C, Katoh M, Chimal-Monroy J (2004) Coordination of chondrocyte differentiation and joint formation by alpha5beta1 integrin in the developing appendicular skeleton. Development 131(19):4735–4742. doi:10.1242/dev.01345

Geiger B, Spatz JP, Bershadsky AD (2009) Environmental sensing through focal adhesions. Nat Rev Mol Cell Biol 10(1):21–33. doi:10.1038/nrm2593

Gigout A, Jolicoeur M, Nelea M, Raynal N, Farndale R, Buschmann MD (2008) Chondrocyte aggregation in suspension culture is GFOGER-GPP- and beta1 integrin-dependent. J Biol Chem 283(46):31522–31530. doi:10.1074/jbc.M804234200

Goldring MB, Goldring SR (2007) Osteoarthritis. J Cell Physiol 213(3):626–634. doi:10.1002/jcp.21258

Gouttenoire J, Bougault C, Aubert-Foucher E, Perrier E, Ronziere MC, Sandell L, Lundgren-Akerlund E, Mallein-Gerin F (2010) BMP-2 and TGF-beta1 differentially control expression of type II procollagen and alpha 10 and alpha 11 integrins in mouse chondrocytes. Eur J Cell Biol 89(4):307–314. doi:10.1016/j.ejcb.2009.10.018

Grashoff C, Aszodi A, Sakai T, Hunziker EB, Fassler R (2003) Integrin-linked kinase regulates chondrocyte shape and proliferation. EMBO Rep 4(4):432–438. doi:10.1038/sj.embor.embor801

Grimshaw MJ, Mason RM (2001) Modulation of bovine articular chondrocyte gene expression in vitro by oxygen tension. Osteoarthr Cartil/OARS Osteoarthritis Res Soc 9(4):357–364. doi:10.1053/joca.2000.0396

Gullberg DE, Lundgren-Åkerlund E (2002) Collagen-binding I domain integrins – what do they do? Prog Histochem Cytochem 37(1):3–54, doi:http://dx.doi.org/10.1016/S0079-6336(02)80008-0

Guo X, Day TF, Jiang X, Garrett-Beal L, Topol L, Yang Y (2004) Wnt/beta-catenin signaling is sufficient and necessary for synovial joint formation. Genes Dev 18(19):2404–2417. doi:10.1101/gad.1230704

Guo D, Ding L, Homandberg GA (2009) Telopeptides of type II collagen upregulate proteinases and damage cartilage but are less effective than highly active fibronectin fragments. Inflamm Res: Off J Eur Histamine Res Soc 58(3):161–169. doi:10.1007/s00011-009-8090-5

Hall-Glenn F, Aivazi A, Akopyan L, Ong JR, Baxter RR, Benya PD, Goldschmeding R, van Nieuwenhoven FA, Hunziker EB, Lyons KM (2013) CCN2/CTGF is required for matrix organization and to protect growth plate chondrocytes from cellular stress. J Cell Commun Signal 7(3):219–230. doi:10.1007/s12079-013-0201-y

Hartmann C, Tabin CJ (2001) Wnt-14 plays a pivotal role in inducing synovial joint formation in the developing appendicular skeleton. Cell 104(3):341–351

Hausler G, Helmreich M, Marlovits S, Egerbacher M (2002) Integrins and extracellular matrix proteins in the human childhood and adolescent growth plate. Calcif Tissue Int 71(3):212–218. doi:10.1007/s00223-001-2083-x

Hirsch MS, Lunsford LE, Trinkaus-Randall V, Svoboda KK (1997) Chondrocyte survival and differentiation in situ are integrin mediated. Dev Dyn 210(3):249–263. doi:10.1002/(sici)1097-0177(199711)210:3<249::aid-aja6>3.0.co;2-g

Hocking DC, Sottile J, McKeown-Longo PJ (1998) Activation of distinct alpha5beta1-mediated signaling pathways by fibronectin's cell adhesion and matrix assembly domains. J Cell Biol 141(1):241–253

Holledge MM, Millward-Sadler SJ, Nuki G, Salter DM (2008) Mechanical regulation of proteoglycan synthesis in normal and osteoarthritic human articular chondrocytes – roles for alpha5 and alphaVbeta5 integrins. Biorheology 45(3–4):275–288

Homandberg GA, Hui F (1994) Arg-Gly-Asp-Ser peptide analogs suppress cartilage chondrolytic activities of integrin-binding and nonbinding fibronectin fragments. Arch Biochem Biophys 310(1):40–48. doi:10.1006/abbi.1994.1137

Homandberg GA, Meyers R, Williams JM (1993) Intraarticular injection of fibronectin fragments causes severe depletion of cartilage proteoglycans in vivo. J Rheumatol 20(8):1378–1382

Homandberg GA, Hui F, Wen C, Purple C, Bewsey K, Koepp H, Huch K, Harris A (1997) Fibronectin-fragment-induced cartilage chondrolysis is associated with release of catabolic cytokines. Biochem J 321(Pt 3):751–757

Homandberg GA, Costa V, Ummadi V, Pichika R (2002a) Antisense oligonucleotides to the integrin receptor subunit alpha(5) decrease fibronectin fragment mediated cartilage chondrolysis. Osteoarthr Cartil/OARS Osteoarthritis Res Soc 10(5):381–393. doi:10.1053/joca.2002.0524

Homandberg GA, Costa V, Wen C (2002b) Fibronectin fragments active in chondrocytic chondrolysis can be chemically cross-linked to the alpha5 integrin receptor subunit. Osteoarthr Cartil/OARS Osteoarthritis Res Soc 10(12):938–949

Hoyland JA, Thomas JT, Donn R, Marriott A, Ayad S, Boot-Handford RP, Grant ME, Freemont AJ (1991) Distribution of type X collagen mRNA in normal and osteoarthritic human cartilage. Bone Miner 15(2):151–163

Hu J, Lu J, Lian G, Zhang J, Hecht JL, Sheen VL (2014) Filamin B regulates chondrocyte proliferation and differentiation through Cdk1 signaling. PLoS One 9(2):e89352. doi:10.1371/journal.pone.0089352

Huang CY, Hagar KL, Frost LE, Sun Y, Cheung HS (2004) Effects of cyclic compressive loading on chondrogenesis of rabbit bone-marrow derived mesenchymal stem cells. Stem Cells 22(3):313–323. doi:10.1634/stemcells.22-3-313

Humphries JD, Byron A, Humphries MJ (2006) Integrin ligands at a glance. J Cell Sci 119(Pt 19):3901–3903, 119/19/3901 [pii]. doi:10.1242/jcs.03098

Inoue T, Hashimoto R, Matsumoto A, Jahan E, Rafiq AM, Udagawa J, Hatta T, Otani H (2014) In vivo analysis of Arg-Gly-Asp sequence/integrin alpha5beta1-mediated signal involvement in embryonic enchondral ossification by exo utero development system. J Bone Miner Res 29(7):1554–1563. doi:10.1002/jbmr.2166

Ivkovic S, Yoon BS, Popoff SN, Safadi FF, Libuda DE, Stephenson RC, Daluiski A, Lyons KM (2003) Connective tissue growth factor coordinates chondrogenesis and angiogenesis during skeletal development. Development 130(12):2779–2791

Jablonski CL, Ferguson S, Pozzi A, Clark AL (2014) Integrin alpha1beta1 participates in chondrocyte transduction of osmotic stress. Biochem Biophys Res Commun 445(1):184–190. doi:10.1016/j.bbrc.2014.01.157

Jang KW, Buckwalter JA, Martin JA (2014) Inhibition of cell-matrix adhesions prevents cartilage chondrocyte death following impact injury. J Orthop Res 32(3):448–454. doi:10.1002/jor.22523

Jennings L, Wu L, King KB, Hammerle H, Cs-Szabo G, Mollenhauer J (2001) The effects of collagen fragments on the extracellular matrix metabolism of bovine and human chondrocytes. Connect Tissue Res 42(1):71–86

Jin EJ, Choi YA, Kyun Park E, Bang OS, Kang SS (2007) MMP-2 functions as a negative regulator of chondrogenic cell condensation via down-regulation of the FAK-integrin beta1 interaction. Dev Biol 308(2):474–484. doi:10.1016/j.ydbio.2007.06.003

Johnson KA, Rose DM, Terkeltaub RA (2008) Factor XIIIA mobilizes transglutaminase 2 to induce chondrocyte hypertrophic differentiation. J Cell Sci 121(Pt 13):2256–2264. doi:10.1242/jcs.011262

Kahn J, Shwartz Y, Blitz E, Krief S, Sharir A, Breitel DA, Rattenbach R, Relaix F, Maire P, Rountree RB, Kingsley DM, Zelzer E (2009) Muscle contraction is necessary to maintain joint progenitor cell fate. Dev Cell 16(5):734–743. doi:10.1016/j.devcel.2009.04.013

Kawaki H, Kubota S, Suzuki A, Lazar N, Yamada T, Matsumura T, Ohgawara T, Maeda T, Perbal B, Lyons KM, Takigawa M (2008) Cooperative regulation of chondrocyte differentiation by CCN2 and CCN3 shown by a comprehensive analysis of the CCN family proteins in cartilage. J Bone Miner Res 23(11):1751–1764. doi:10.1359/jbmr.080615

Kelly DJ, Jacobs CR (2010) The role of mechanical signals in regulating chondrogenesis and osteogenesis of mesenchymal stem cells. Birth Defects Res C: Embryo Today 90(1):75–85. doi:10.1002/bdrc.20173

Kim YH, Lee JW (2009) Targeting of focal adhesion kinase by small interfering RNAs reduces chondrocyte redifferentiation capacity in alginate beads culture with type II collagen. J Cell Physiol 218(3):623–630. doi:10.1002/jcp.21637

Kim D, Song J, Han J, Kim Y, Chun CH, Jin EJ (2013) Two non-coding RNAs, MicroRNA-101 and HOTTIP contribute cartilage integrity by epigenetic and homeotic regulation of integrin-alpha1. Cell Signal 25(12):2878–2887. doi:10.1016/j.cellsig.2013.08.034

Klatt AR, Paul-Klausch B, Klinger G, Kuhn G, Renno JH, Banerjee M, Malchau G, Wielckens K (2009) A critical role for collagen II in cartilage matrix degradation: collagen II induces pro-inflammatory cytokines and MMPs in primary human chondrocytes. J Orthop Res 27(1):65–70. doi:10.1002/jor.20716

Knudson W, Loeser RF (2002) CD44 and integrin matrix receptors participate in cartilage homeostasis. Cell Mol Life Sci: CMLS 59(1):36–44

Krakow D (2015) Skeletal dysplasias. Clin Perinatol 42(2):301–319. doi:10.1016/j.clp.2015.03.003,viii

Krakow D, Rimoin DL (2010) The skeletal dysplasias. Genet Med: Off J Am Coll Med Genet 12(6):327–341. doi:10.1097/GIM.0b013e3181daae9b

Krakow D, Robertson SP, King LM, Morgan T, Sebald ET, Bertolotto C, Wachsmann-Hogiu S, Acuna D, Shapiro SS, Takafuta T, Aftimos S, Kim CA, Firth H, Steiner CE, Cormier-Daire V, Superti-Furga A, Bonafe L, Graham JM Jr, Grix A, Bacino CA, Allanson J, Bialer MG, Lachman RS, Rimoin DL, Cohn DH (2004) Mutations in the gene encoding filamin B disrupt vertebral segmentation, joint formation and skeletogenesis. Nat Genet 36(4):405–410. doi:10.1038/ng1319

Krupska I, Bruford EA, Chaqour B (2015) Eyeing the Cyr61/CTGF/NOV (CCN) group of genes in development and diseases: highlights of their structural likenesses and functional dissimilarities. Hum Genomics 9(1):24. doi:10.1186/s40246-015-0046-y

Kurtis MS, Schmidt TA, Bugbee WD, Loeser RF, Sah RL (2003) Integrin-mediated adhesion of human articular chondrocytes to cartilage. Arthritis Rheum 48(1):110–118. doi:10.1002/art.10704

Kyostila K, Lappalainen AK, Lohi H (2013) Canine chondrodysplasia caused by a truncating mutation in collagen-binding integrin alpha subunit 10. PLoS One 8(9):e75621. doi:10.1371/journal.pone.0075621

Lapadula G, Iannone F, Zuccaro C, Grattagliano V, Covelli M, Patella V, Lo Bianco G, Pipitone V (1998) Chondrocyte phenotyping in human osteoarthritis. Clin Rheumatol 17(2):99–104
Lechler T, Fuchs E (2005) Asymmetric cell divisions promote stratification and differentiation of mammalian skin. Nature 437(7056):275–280. doi:10.1038/nature03922
Lee GM, Loeser RF (1999) Cell surface receptors transmit sufficient force to bend collagen fibrils. Exp Cell Res 248(1):294–305. doi:10.1006/excr.1999.4418
Lee SK, Malpeli M, Cancedda R, Utani A, Yamada Y, Kleinman HK (1997) Laminin chain expression by chick chondrocytes and mouse cartilaginous tissues in vivo and in vitro. Exp Cell Res 236(1):212–222. doi:10.1006/excr.1997.3710
Lee HS, Millward-Sadler SJ, Wright MO, Nuki G, Salter DM (2000) Integrin and mechanosensitive ion channel-dependent tyrosine phosphorylation of focal adhesion proteins and beta-catenin in human articular chondrocytes after mechanical stimulation. J Bone Miner Res 15(8):1501–1509. doi:10.1359/jbmr.2000.15.8.1501
Lee HS, Millward-Sadler SJ, Wright MO, Nuki G, Al-Jamal R, Salter DM (2002) Activation of integrin-RACK1/PKCalpha signalling in human articular chondrocyte mechanotransduction. Osteoarthr Cartil/OARS Osteoarthritis Res Soc 10(11):890–897
Lefebvre V, Bhattaram P (2010) Vertebrate skeletogenesis. Curr Top Dev Biol 90:291–317. doi:10.1016/s0070-2153(10)90008-2
Legate KR, Wickstrom SA, Fassler R (2009) Genetic and cell biological analysis of integrin outside-in signaling. Genes Dev 23(4):397–418. doi:10.1101/gad.175870923/4/397[pii]
Li C, Chen L, Iwata T, Kitagawa M, Fu XY, Deng CX (1999) A Lys644Glu substitution in fibroblast growth factor receptor 3 (FGFR3) causes dwarfism in mice by activation of STATs and ink4 cell cycle inhibitors. Hum Mol Genet 8(1):35–44
Liang W, Ren K, Liu F, Cui W, Wang Q, Chen Z, Fan W (2013) Periodic mechanical stress stimulates the FAK mitogenic signal in rat chondrocytes through ERK1/2 activity. Cell Physiol Biochem: Int J Exp Cell Physiol Biochem Pharmacol 32(4):915–930. doi:10.1159/000354495
Loeser RF (2014) Integrins and chondrocyte-matrix interactions in articular cartilage. Matrix Biol: J Int Soc Matrix Biol 39:11–16. doi:10.1016/j.matbio.2014.08.007
Loeser RF, Carlson CS, McGee MP (1995) Expression of beta 1 integrins by cultured articular chondrocytes and in osteoarthritic cartilage. Exp Cell Res 217(2):248–257. doi:10.1006/excr.1995.1084
Loeser RF, Forsyth CB, Samarel AM, Im HJ (2003) Fibronectin fragment activation of proline-rich tyrosine kinase PYK2 mediates integrin signals regulating collagenase-3 expression by human chondrocytes through a protein kinase C-dependent pathway. J Biol Chem 278(27):24577–24585. doi:10.1074/jbc.M304530200
Logan M, Martin JF, Nagy A, Lobe C, Olson EN, Tabin CJ (2002) Expression of Cre Recombinase in the developing mouse limb bud driven by a Prxl enhancer. Genesis (New York, NY: 2000) 33(2):77–80. doi:10.1002/gene.10092
Long D, Blake S, Song XY, Lark M, Loeser RF (2008) Human articular chondrocytes produce IL-7 and respond to IL-7 with increased production of matrix metalloproteinase-13. Arthritis Res Ther 10(1):R23. doi:10.1186/ar2376
Long DL, Willey JS, Loeser RF (2013) Rac1 is required for matrix metalloproteinase 13 production by chondrocytes in response to fibronectin fragments. Arthritis Rheum 65(6):1561–1568. doi:10.1002/art.37922
Lu J, Lian G, Lenkinski R, De Grand A, Vaid RR, Bryce T, Stasenko M, Boskey A, Walsh C, Sheen V (2007) Filamin B mutations cause chondrocyte defects in skeletal development. Hum Mol Genet 16(14):1661–1675. doi:10.1093/hmg/ddm114
Lu Z, Doulabi BZ, Huang C, Bank RA, Helder MN (2010) Collagen type II enhances chondrogenesis in adipose tissue-derived stem cells by affecting cell shape. Tissue Eng A 16(1):81–90. doi:10.1089/ten.TEA.2009.0222
Lucchinetti E, Bhargava MM, Torzilli PA (2004) The effect of mechanical load on integrin subunits alpha5 and beta1 in chondrocytes from mature and immature cartilage explants. Cell Tissue Res 315(3):385–391. doi:10.1007/s00441-003-0836-8
Lucic D, Mollenhauer J, Kilpatrick KE, Cole AA (2003) N-telopeptide of type II collagen interacts with annexin V on human chondrocytes. Connect Tissue Res 44(5):225–239

Lundgren-Akerlund E, Aszodi A (2014) Integrin alpha10beta1: a collagen receptor critical in skeletal development. Adv Exp Med Biol 819:61–71. doi:10.1007/978-94-017-9153-3_4

Ma SU, Garciadiego-CÁZares D, Vargas C, Chimal-Monroy J, De LeÓN LD (1998) The spatial pattern of integrins during development of mouse articular cartilagea. Ann N Y Acad Sci 857(1):241–244. doi:10.1111/j.1749-6632.1998.tb10122.x

Matsumoto K, Li Y, Jakuba C, Sugiyama Y, Sayo T, Okuno M, Dealy CN, Toole BP, Takeda J, Yamaguchi Y, Kosher RA (2009) Conditional inactivation of Has2 reveals a crucial role for hyaluronan in skeletal growth, patterning, chondrocyte maturation and joint formation in the developing limb. Development 136(16):2825–2835. doi:10.1242/dev.038505

McBeath R, Pirone DM, Nelson CM, Bhadriraju K, Chen CS (2004) Cell shape, cytoskeletal tension, and RhoA regulate stem cell lineage commitment. Dev Cell 6(4):483–495

McNulty AL, Leddy HA, Liedtke W, Guilak F (2015) TRPV4 as a therapeutic target for joint diseases. Naunyn Schmiedeberg's Arch Pharmacol 388(4):437–450. doi:10.1007/s00210-014-1078-x

Meyer EG, Buckley CT, Steward AJ, Kelly DJ (2011) The effect of cyclic hydrostatic pressure on the functional development of cartilaginous tissues engineered using bone marrow derived mesenchymal stem cells. J Mech Behav Biomed Mater 4(7):1257–1265. doi:10.1016/j.jmbbm.2011.04.012

Millward-Sadler SJ, Salter DM (2004) Integrin-dependent signal cascades in chondrocyte mechanotransduction. Ann Biomed Eng 32(3):435–446

Millward-Sadler SJ, Wright MO, Lee H, Nishida K, Caldwell H, Nuki G, Salter DM (1999) Integrin-regulated secretion of interleukin 4: a novel pathway of mechanotransduction in human articular chondrocytes. J Cell Biol 145(1):183–189

Millward-Sadler SJ, Wright MO, Davies LW, Nuki G, Salter DM (2000) Mechanotransduction via integrins and interleukin-4 results in altered aggrecan and matrix metalloproteinase 3 gene expression in normal, but not osteoarthritic, human articular chondrocytes. Arthritis Rheum 43(9):2091–2099. doi:10.1002/1529-0131(200009)43:9<2091::aid-anr21>3.0.co;2-c

Miosge N, Hartmann M, Maelicke C, Herken R (2004) Expression of collagen type I and type II in consecutive stages of human osteoarthritis. Histochem Cell Biol 122(3):229–236. doi:10.1007/s00418-004-0697-6

Moser M, Bosserhoff AK, Hunziker EB, Sandell L, Fassler R, Buettner R (2002) Ultrastructural cartilage abnormalities in MIA/CD-RAP-deficient mice. Mol Cell Biol 22(5):1438–1445

Nakamura F, Stossel TP, Hartwig JH (2011) The filamins: organizers of cell structure and function. Cell Adhes Migr 5(2):160–169

Nishida T, Kawaki H, Baxter RM, Deyoung RA, Takigawa M, Lyons KM (2007) CCN2 (connective tissue growth factor) is essential for extracellular matrix production and integrin signaling in chondrocytes. J Cell Commun Signal 1(1):45–58. doi:10.1007/s12079-007-0005-z

Nishioka T, Arima N, Kano K, Hama K, Itai E, Yukiura H, Kise R, Inoue A, Kim SH, Solnica-Krezel L, Moolenaar WH, Chun J, Aoki J (2016) ATX-LPA1 axis contributes to proliferation of chondrocytes by regulating fibronectin assembly leading to proper cartilage formation. Sci Rep 6:23433. doi:10.1038/srep23433

Okamura N, Hasegawa M, Nakoshi Y, Iino T, Sudo A, Imanaka-Yoshida K, Yoshida T, Uchida A (2010) Deficiency of tenascin-C delays articular cartilage repair in mice. Osteoarthr Cartil/OARS Osteoarthritis Res Soc 18(6):839–848. doi:10.1016/j.joca.2009.08.013

Orazizadeh M, Lee HS, Groenendijk B, Sadler SJ, Wright MO, Lindberg FP, Salter DM (2008) CD47 associates with alpha 5 integrin and regulates responses of human articular chondrocytes to mechanical stimulation in an in vitro model. Arthritis Res Ther 10(1):R4. doi:10.1186/ar2350

Ostergaard K, Salter DM, Petersen J, Bendtzen K, Hvolris J, Andersen CB (1998) Expression of alpha and beta subunits of the integrin superfamily in articular cartilage from macroscopically normal and osteoarthritic human femoral heads. Ann Rheum Dis 57(5):303–308

Pacifici M, Iwamoto M, Golden EB, Leatherman JL, Lee YS, Chuong CM (1993) Tenascin is associated with articular cartilage development. Dev Dyn 198(2):123–134. doi:10.1002/aja.1001980206

Pacifici M, Koyama E, Iwamoto M (2005) Mechanisms of synovial joint and articular cartilage formation: recent advances, but many lingering mysteries. Birth Defects Res C: Embryo Today 75(3):237–248. doi:10.1002/bdrc.20050

Pala D, Kapoor M, Woods A, Kennedy L, Liu S, Chen S, Bursell L, Lyons KM, Carter DE, Beier F, Leask A (2008) Focal adhesion kinase/Src suppresses early chondrogenesis: central role of CCN2. J Biol Chem 283(14):9239–9247. doi:10.1074/jbc.M705175200

Pelaez D, Huang CY, Cheung HS (2009) Cyclic compression maintains viability and induces chondrogenesis of human mesenchymal stem cells in fibrin gel scaffolds. Stem Cells Dev 18(1):93–102. doi:10.1089/scd.2008.0030

Peters JH, Loredo GA, Benton HP (2002) Is osteoarthritis a 'fibronectin-integrin imbalance disorder'? Osteoarthritis and cartilage / OARS. Osteoarthr Res Soc 10(11):831–835

Peters MA, Wendholt D, Strietholt S, Frank S, Pundt N, Korb-Pap A, Joosten LA, van den Berg WB, Kollias G, Eckes B, Pap T (2012) The loss of alpha2beta1 integrin suppresses joint inflammation and cartilage destruction in mouse models of rheumatoid arthritis. Arthritis Rheum 64(5):1359–1368. doi:10.1002/art.33487

Pichika R, Homandberg GA (2004) Fibronectin fragments elevate nitric oxide (NO) and inducible NO synthetase (iNOS) levels in bovine cartilage and iNOS inhibitors block fibronectin fragment mediated damage and promote repair. Inflamm Res: Off J Eur Histamine Res Soc 53(8):405–412. doi:10.1007/s00011-004-1279-8

Poole AR, Rosenberg LC, Reiner A, Ionescu M, Bogoch E, Roughley PJ (1996) Contents and distributions of the proteoglycans decorin and biglycan in normal and osteoarthritic human articular cartilage. J Orthop Res 14(5):681–689. doi:10.1002/jor.1100140502

Poole AR, Kojima T, Yasuda T, Mwale F, Kobayashi M, Laverty S (2001) Composition and structure of articular cartilage: a template for tissue repair. Clin Orthop Relat Res (391 Suppl):S26–33

Prasadam I, van Gennip S, Friis T, Shi W, Crawford R, Xiao Y (2010) ERK-1/2 and p38 in the regulation of hypertrophic changes of normal articular cartilage chondrocytes induced by osteoarthritic subchondral osteoblasts. Arthritis Rheum 62(5):1349–1360. doi:10.1002/art.27397

Pulai JI, Jr. Del Carlo M, Loeser RF (2002) The alpha5beta1 integrin provides matrix survival signals for normal and osteoarthritic human articular chondrocytes in vitro. Arthritis Rheum 46(6):1528–1535. doi:10.1002/art.10334

Pulai JI, Chen H, Im HJ, Kumar S, Hanning C, Hegde PS, Loeser RF (2005) NF-kappa B mediates the stimulation of cytokine and chemokine expression by human articular chondrocytes in response to fibronectin fragments. J Immunol 174(9):5781–5788

Raducanu A, Hunziker EB, Drosse I, Aszodi A (2009) Beta1 integrin deficiency results in multiple abnormalities of the knee joint. J Biol Chem 284(35):23780–23792. doi:10.1074/jbc.M109.039347

Richter W (2009) Mesenchymal stem cells and cartilage in situ regeneration. J Intern Med 266(4):390–405. doi:10.1111/j.1365-2796.2009.02153.x

Ronziere MC, Aubert-Foucher E, Gouttenoire J, Bernaud J, Herbage D, Mallein-Gerin F (2005) Integrin alpha1beta1 mediates collagen induction of MMP-13 expression in MC615 chondrocytes. Biochim Biophys Acta 1746(1):55–64. doi:10.1016/j.bbamcr.2005.08.003

Sakai K, Hiripi L, Glumoff V, Brandau O, Eerola R, Vuorio E, Bosze Z, Fassler R, Aszodi A (2001) Stage-and tissue-specific expression of a Col2a1-Cre fusion gene in transgenic mice. Matrix Biol: J Int Soc Matrix Biol 19(8):761–767

Salter DM, Godolphin JL, Gourlay MS (1995) Chondrocyte heterogeneity: immunohistologically defined variation of integrin expression at different sites in human fetal knees. J Histochem Cytochem: Off J Histochem Soc 43(4):447–457

Schmid R, Schiffner S, Opolka A, Grassel S, Schubert T, Moser M, Bosserhoff AK (2010) Enhanced cartilage regeneration in MIA/CD-RAP deficient mice. Cell Death Dis 1:e97. doi:10.1038/cddis.2010.78

Schubert T, Schlegel J, Schmid R, Opolka A, Grassel S, Humphries M, Bosserhoff AK (2010) Modulation of cartilage differentiation by melanoma inhibiting activity/cartilage-derived retinoic acid-sensitive protein (MIA/CD-RAP). Exp Mol Med 42(3):166–174. doi:10.3858/emm.2010.42.3.017

Settle SH Jr, Rountree RB, Sinha A, Thacker A, Higgins K, Kingsley DM (2003) Multiple joint and skeletal patterning defects caused by single and double mutations in the mouse Gdf6 and Gdf5 genes. Dev Biol 254(1):116–130

Shah JV (2010) Cells in tight spaces: the role of cell shape in cell function. J Cell Biol 191(2):233–236. doi:10.1083/jcb.201009048

Shakibaei M (1998) Inhibition of chondrogenesis by integrin antibody in vitro. Exp Cell Res 240(1):95–106. doi:10.1006/excr.1998.3933

Shakibaei M, De Souza P, Merker HJ (1997) Integrin expression and collagen type II implicated in maintenance of chondrocyte shape in monolayer culture: an immunomorphological study. Cell Biol Int 21(2):115–125. doi:10.1006/cbir.1996.0118

Shekaran A, Shoemaker JT, Kavanaugh TE, Lin AS, LaPlaca MC, Fan Y, Guldberg RE, Garcia AJ (2014) The effect of conditional inactivation of beta 1 integrins using twist 2 Cre, Osterix Cre and osteocalcin Cre lines on skeletal phenotype. Bone 68:131–141. doi:10.1016/j.bone.2014.08.008

Shimaya M, Muneta T, Ichinose S, Tsuji K, Sekiya I (2010) Magnesium enhances adherence and cartilage formation of synovial mesenchymal stem cells through integrins. Osteoarthr Cartil/OARS Osteoarthritis Res Soc 18(10):1300–1309. doi:10.1016/j.joca.2010.06.005

Shioji S, Imai S, Ando K, Kumagai K, Matsusue Y (2014) Extracellular and intracellular mechanisms of mechanotransduction in three-dimensionally embedded rat chondrocytes. PLoS One 9(12):e114327. doi:10.1371/journal.pone.0114327

Sofat N (2009) Analysing the role of endogenous matrix molecules in the development of osteoarthritis. Int J Exp Pathol 90(5):463–479. doi:10.1111/j.1365-2613.2009.00676.x

Spagnoli A, O'Rear L, Chandler RL, Granero-Molto F, Mortlock DP, Gorska AE, Weis JA, Longobardi L, Chytil A, Shimer K, Moses HL (2007) TGF-beta signaling is essential for joint morphogenesis. J Cell Biol 177(6):1105–1117. doi:10.1083/jcb.200611031

Stanton H, Ung L, Fosang AJ (2002) The 45 kDa collagen-binding fragment of fibronectin induces matrix metalloproteinase-13 synthesis by chondrocytes and aggrecan degradation by aggrecanases. Biochem J 364(Pt 1):181–190

Steward AJ, Thorpe SD, Vinardell T, Buckley CT, Wagner DR, Kelly DJ (2012) Cell-matrix interactions regulate mesenchymal stem cell response to hydrostatic pressure. Acta Biomater 8(6):2153–2159. doi:10.1016/j.actbio.2012.03.016

Steward AJ, Wagner DR, Kelly DJ (2014) Exploring the roles of integrin binding and cytoskeletal reorganization during mesenchymal stem cell mechanotransduction in soft and stiff hydrogels subjected to dynamic compression. J Mech Behav Biomed Mater 38:174–182. doi:10.1016/j.jmbbm.2013.07.020

Su WC, Kitagawa M, Xue N, Xie B, Garofalo S, Cho J, Deng C, Horton WA, Fu XY (1997) Activation of Stat1 by mutant fibroblast growth-factor receptor in thanatophoric dysplasia type II dwarfism. Nature 386(6622):288–292. doi:10.1038/386288a0

Taddei I, Deugnier MA, Faraldo MM, Petit V, Bouvard D, Medina D, Fassler R, Thiery JP, Glukhova MA (2008) Beta1 integrin deletion from the basal compartment of the mammary epithelium affects stem cells. Nat Cell Biol 10(6):716–722. doi:10.1038/ncb1734

Tang Y, Rowe RG, Botvinick EL, Kurup A, Putnam AJ, Seiki M, Weaver VM, Keller ET, Goldstein S, Dai J, Begun D, Saunders T, Weiss SJ (2013) MT1-MMP-dependent control of skeletal stem cell commitment via a beta1-integrin/YAP/TAZ signaling axis. Dev Cell 25(4):402–416. doi:10.1016/j.devcel.2013.04.011

Terpstra L, Prud'homme J, Arabian A, Takeda S, Karsenty G, Dedhar S, St-Arnaud R (2003) Reduced chondrocyte proliferation and chondrodysplasia in mice lacking the integrin-linked kinase in chondrocytes. J Cell Biol 162(1):139–148. doi:10.1083/jcb.200302066

Tosi LL, Warman ML (2015) Mechanistic and therapeutic insights gained from studying rare skeletal diseases. Bone 76:67–75. doi:10.1016/j.bone.2015.03.016

Tuckwell DS, Ayad S, Grant ME, Takigawa M, Humphries MJ (1994) Conformation dependence of integrin-type II collagen binding. Inability of collagen peptides to support alpha 2 beta 1 binding, and mediation of adhesion to denatured collagen by a novel alpha 5 beta 1-fibronectin bridge. J Cell Sci 107(Pt 4):993–1005

Varas L, Ohlsson LB, Honeth G, Olsson A, Bengtsson T, Wiberg C, Bockermann R, Jarnum S, Richter J, Pennington D, Johnstone B, Lundgren-Akerlund E, Kjellman C (2007) Alpha10 integrin expression is up-regulated on fibroblast growth factor-2-treated mesenchymal stem cells with improved chondrogenic differentiation potential. Stem Cells Dev 16(6):965–978. doi:10.1089/scd.2007.0049

Velling T, Risteli J, Wennerberg K, Mosher DF, Johansson S (2002) Polymerization of type I and III collagens is dependent on fibronectin and enhanced by integrins alpha 11beta 1 and alpha 2beta 1. J Biol Chem 277(40):37377–37381. doi:10.1074/jbc.M206286200

Vincent TL (2013) Targeting mechanotransduction pathways in osteoarthritis: a focus on the pericellular matrix. Curr Opin Pharmacol 13(3):449–454. doi:10.1016/j.coph.2013.01.010

von der Mark K, Kirsch T, Nerlich A, Kuss A, Weseloh G, Gluckert K, Stoss H (1992) Type X collagen synthesis in human osteoarthritic cartilage. Indication of chondrocyte hypertrophy. Arthritis Rheum 35(7):806–811

Wang W, Kirsch T (2006) Annexin V/beta5 integrin interactions regulate apoptosis of growth plate chondrocytes. J Biol Chem 281(41):30848–30856. doi:10.1074/jbc.M605937200

Wang X, Chan WC, Chan D (2016) Joint development. In: Grässel S, Aszódi A (eds) Cartilage, vol 1: Physiology and development. Springer International Publishing, Cham, pp 169–189. doi:10.1007/978-3-319-29568-8_7

Warman ML, Cormier-Daire V, Hall C, Krakow D, Lachman R, LeMerrer M, Mortier G, Mundlos S, Nishimura G, Rimoin DL, Robertson S, Savarirayan R, Sillence D, Spranger J, Unger S, Zabel B, Superti-Furga A (2011) Nosology and classification of genetic skeletal disorders: 2010 revision. Am J Med Genet A 155A(5):943–968. doi:10.1002/ajmg.a.33909

Wegener KL, Campbell ID (2008) Transmembrane and cytoplasmic domains in integrin activation and protein-protein interactions (review). Mol Membr Biol 25(5):376–387. doi:10.1080/09687680802269886

Werb Z, Tremble PM, Behrendtsen O, Crowley E, Damsky CH (1989) Signal transduction through the fibronectin receptor induces collagenase and stromelysin gene expression. J Cell Biol 109(2):877–889

Wood ST, Long DL, Reisz JA, Yammani RR, Burke EA, Klomsiri C, Poole LB, Furdui CM, Loeser RF (2016) Cysteine-mediated redox regulation of cell signaling in chondrocytes stimulated with fibronectin fragments. Arthritis Rheumatol (Hoboken, NJ) 68(1):117–126. doi:10.1002/art.39326

Woods VL Jr, Schreck PJ, Gesink DS, Pacheco HO, Amiel D, Akeson WH, Lotz M (1994) Integrin expression by human articular chondrocytes. Arthritis Rheum 37(4):537–544

Woods A, Wang G, Beier F (2007) Regulation of chondrocyte differentiation by the actin cytoskeleton and adhesive interactions. J Cell Physiol 213(1):1–8. doi:10.1002/jcp.21110

Wright MO, Nishida K, Bavington C, Godolphin JL, Dunne E, Walmsley S, Jobanputra P, Nuki G, Salter DM (1997) Hyperpolarisation of cultured human chondrocytes following cyclical pressure-induced strain: evidence of a role for alpha 5 beta 1 integrin as a chondrocyte mechanoreceptor. J Orthop Res 15(5):742–747. doi:10.1002/jor.1100150517

Wu W, Billinghurst RC, Pidoux I, Antoniou J, Zukor D, Tanzer M, Poole AR (2002) Sites of collagenase cleavage and denaturation of type II collagen in aging and osteoarthritic articular cartilage and their relationship to the distribution of matrix metalloproteinase 1 and matrix metalloproteinase 13. Arthritis Rheum 46(8):2087–2094. doi:10.1002/art.10428

Xu L, Servais J, Polur I, Kim D, Lee PL, Chung K, Li Y (2010) Attenuation of osteoarthritis progression by reduction of discoidin domain receptor 2 in mice. Arthritis Rheum 62(9):2736–2744. doi:10.1002/art.27582

Yasuda T (2006) Cartilage destruction by matrix degradation products. Mod Rheumatol/Jpn Rheum Assoc 16(4):197–205. doi:10.1007/s10165-006-0490-6

Yeung Tsang K, Wa Tsang S, Chan D, Cheah KS (2014) The chondrocytic journey in endochondral bone growth and skeletal dysplasia. Birth Defects Res C Embryo Today 102(1):52–73. doi:10.1002/bdrc.21060

Zemmyo M, Meharra EJ, Kuhn K, Creighton-Achermann L, Lotz M (2003) Accelerated, aging-dependent development of osteoarthritis in alpha1 integrin-deficient mice. Arthritis Rheum 48(10):2873–2880. doi:10.1002/art.11246

Zhang LQ, Zhao GZ, Xu XY, Fang J, Chen JM, Li JW, Gao XJ, Hao LJ, Chen YZ (2015) Integrin-beta1 regulates chondrocyte proliferation and apoptosis through the upregulation of GIT1 expression. Int J Mol Med 35(4):1074–1080. doi:10.3892/ijmm.2015.2114

The Sensory and Sympathetic Nervous System in Cartilage Physiology and Pathophysiology

9

Susanne Grässel, Rainer H. Straub, and Zsuzsa Jenei-Lanzl

Abstract

The peripheral nervous system is critically involved in metabolism of joint tissue and intervertebral disks (IVD). Nerve fibers of sympathetic and sensory origin innervate synovial tissue and subchondral bone of diarthrodial joints. In pathophysiological situations as in osteoarthritis (OA), rheumatoid arthritis (RA), and IVD degeneration, innervation patterns of sympathetic and sensory nerve fibers are partly altered in joint tissue and IVD.

Various resident cell types of the musculoskeletal system express receptors for sensory and sympathetic neurotransmitters allowing response to neuronal stimuli. Among them are mesenchymal stem cells, synovial fibroblasts, bone cells, and different types of chondrocytes, which express distinct subtypes of adrenoceptors, receptors for vasoactive intestinal peptide (VIP), for substance P (SP), and calcitonin gene-related peptide (CGRP). Some of these cell types even synthesize neuropeptides such as SP, and they are positive for tyrosine hydroxylase (TH), the rate limiting enzyme for biosynthesis of catecholamines. During

S. Grässel (✉)
Experimental Orthopaedics, Department of Orthopaedic Surgery, University of Regensburg, ZMB/BioPark 1, Am Biopark, 93053 Regensburg, Germany
e-mail: susanne.graessel@ukr.de

R.H. Straub
Laboratory of Experimental Rheumatology and Neuroendocrine Immunology, Department of Internal Medicine I, University of Regensburg, BioPark 1, Am Biopark 9, 93053 Regensburg, Germany
e-mail: rainer.straub@ukr.de

Z. Jenei-Lanzl
Dr. Rolf M. Schwiete Research Unit for Osteoarthritis, Orthopaedic University Hospital Friedrichsheim gGMBH Johann Wolfgang Goethe-University, Marienburgstr. 2, 60528 Frankfurt a.M., Germany
e-mail: zsuzsa.jenei-lanzl@friedrichsheim.de

S. Grässel, A. Aszódi (eds.), *Cartilage*, DOI 10.1007/978-3-319-45803-8_9

endochondral ossification in embryonic limb development, sensory and sympathetic neurotransmitters modulate osteo-chondrogenic differentiation of mesenchymal progenitor cells, vascularization, and matrix differentiation indicating a distinct role in skeletal growth and possible limb regeneration processes. In adults, sensory and sympathetic neurotransmitters are involved in pathology of inflammatory diseases as rheumatoid arthritis which manifests mainly in joints. In addition, they might play a role in pathogenesis of a priori degenerative joint disorders, as osteoarthritis and intervertebral disk degeneration.

Altogether it became evident that sensory and sympathetic neurotransmitters have crucial trophic effects which are critical for proper limb formation during embryonic skeletal growth. In adults, they modulate articular cartilage, subchondral bone and synovial tissue homeostasis, and physiological and pathophysiological conditions, in addition to their classical neurological features.

9.1 Introduction

Clinical observations demonstrated the importance of the peripheral nervous system for maintaining body homeostasis and mediating organogenesis and tissue repair. On one hand, disorders of nerves (central or peripheral) can have substantial influence on bone health, repair, and regeneration. On the other hand, dramatic alterations in density and distribution of sympathetic and sensory nerve fibers are reported in some musculoskeletal pathophysiologies. Changes in the density of sympathetic nerve fibers, which are characterized by tyrosine hydroxylase and/or neuropeptide Y (NPY), in synovial tissue contribute to rheumatoid arthritis (RA) (Dirmeier et al. 2008). Loss of sensory joint innervation during aging is suggested to accelerate degenerative cartilage alterations, which adds to development of spontaneous osteoarthritis (OA) in mice (Salo et al. 2002). Nerve fibers of sympathetic and sensory origin innervate frequently the trabecular bone, periosteum, and fracture callus (Hukkanen et al. 1992; Madsen et al. 1998). They are involved in controlling vascularization and matrix differentiation during endochondral ossification in embryonic limb development (Gajda et al. 2000) indicating a distinct role in modulating skeletal growth and limb formation processes.

Sensory nerves contain in general four different nociceptive neuropeptide families: the tachykinins (Zhang et al. 2000), calcitonin-gene-related peptides (CGRP), glutamine, and galanin. Effects of all tachykinins are mediated by three neurokinin receptors with varying affinities for the individual ligands (Maggi 1995) (Table 9.1). Classically, tachykinin substance P (SP) is known as a mediator of nociception and of inflammation (Hartung and Toyka 1983). CGRP is encoded together with calcitonin and is generated by alternative splicing (Rosenfeld et al. 1983) and signals through a complex family of receptor proteins (Poyner et al. 2002) (Table 9.1). The most important neurotransmitter of the catecholaminergic sympathetic nervous system is norepinephrine (NE), which signals through α- and ß-adrenoceptors (AR) depending on concentration (Molinoff 1984) (Table 9.1; Fig. 9.1). However,

Table 9.1 Mammalian sensory and sympathetic neurotransmitters and their receptors

Sensory neurotransmitters	Receptor (highest affinity)	Neuropeptide genes	Neuropeptide receptor genes
Substance P (SP)	Neurokinin (NK)$_1$ receptor	Tachykinin (TAC) 1 or PPT-A or PPT-I	Tachykinin receptor (TACR)1
Neurokinin A (NKA) Alternatively spliced forms are neuropeptides (NP) K and NPγ	NK$_2$ receptor	TAC1	TACR2
Neurokinin B (NKB)	NK$_3$ receptor	TAC3 or PPT-B or PPT-II	TACR3
Hemokinin 1 (HK1) alternatively spliced forms are endokinins (EK) A, B, C, D	NK$_1$ receptor, hemokinin (HK)-1 receptor?	TAC4 or PPT-C	TACR1
Alpha-calcitonin-related peptide (αCGRP)	CL-receptor (CRLR)/ receptor activity modifying protein (RAMP-1)	CALCA	CALCR/RAMP1
Beta-calcitonin-related peptide (ßCGRP)	CRLR/RAMP-1	CALCB	CALCR/RAMP1
Sympathetic neurotransmitters	*Receptors*	*Neuropeptide genes*	*Neurotransmitter receptor genes*
Catecholaminergic			
Norepinephrine (NE) or noradrenaline (NA)	≤10^{-8}M: alpha-1a-, b-, d-adrenoceptors		ADRA1A, ADRA1B, ADRA1D
	≤10^{-8}M: alpha-2a-, b-, c-adrenoceptors		ADRA2A, ADRA2B, ADRA2C
	≥10^{-7}M: ß1-, ß2-, ß3-adrenoceptors		ADRB1, ADRB2, ADRB3
Peptidergic			
Vasoactive intestinal peptide (VIP)	VIP-1, VIP-2, and VIP/PACAP receptors	VIP	VIPR1, VIPR2
Neuropeptide Y (NPY)	NPY1-, 2-, and 5- receptor	NPY	NPY1R, NPY2R, NPY5R, PPYR1
	NPY4-receptor or pancreatic-polypeptide receptor-1 (PPR1)		

Modified from Grassel (2014)
Information on glutamine and galanin were not included

sympathetic nerves release not only NE but also different co-transmitters such as adenosine triphosphate (ATP), NPY, vasoactive intestinal peptide (VIP), and nitric oxide (Pongratz and Straub 2014; Straub 2015). In addition, neurotransmitters can be locally converted to differentially active neurotransmitters when respective converting enzymes are present in the tissue. Relevant for cartilage (patho)physiology

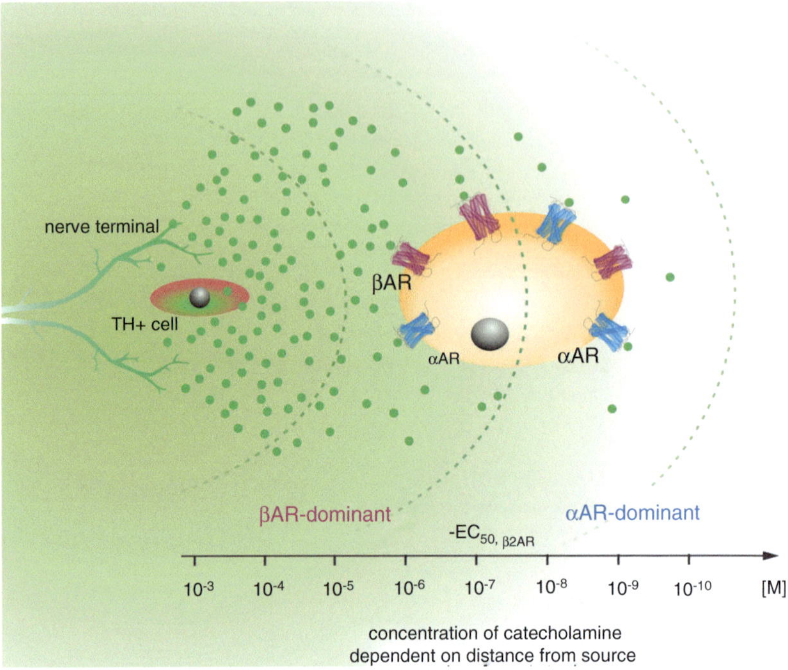

Fig. 9.1 Concentration-dependent influence of norepinephrine on ß- and α-adrenergic receptors. At high concentrations of 10^{-5} to 10^{-6} mol/l, norepinephrine (NE) binds to ß- and α-adrenergic receptors, which typically elicits an increase in cyclic AMP with anti-inflammatory effects (e.g., TNF inhibition). At low concentration of 10^{-8} to 10^{-9} mol/l, NE inhibits formation of cyclic AMP which is a pro-inflammatory signal. Thus, nerve fiber density and nerve terminal distance to target cells plays an important role in neuroimmuno-modulation. EC50 of NE is 10^{-7} mol/l

is the conversion of co-released ATP to adenosine by ectonucleotidases (CD39, CD73) located on the surface of cells (Romio et al. 2011; Van Linden and Eltzschig 2007). NPY is one of the most abundant peptides in the central (CNS) and peripheral (PNS) nervous systems and other peripheral tissue. NPY receptors are identical for all members of the NPY family and belong to class A G-protein-coupled receptors. Results from several studies suggest that the NPY system is linked to the aging process (for review, see Botelho and Cavadas 2015) (Table 9.1). VIP belongs to a family of structurally related peptides including secretin, glucagon, gastric inhibitory peptide, growth hormone-releasing factor, and PACAP and signals through three different subtypes of VIP receptors (reviewed in Lerner and Persson 2008) (Table 9.1).

A further important point to understand the effects of sympathetic neurotransmitters is that they do not have 0-1 effects. NE, for instance, binds to α2-AR below the concentration of 10^{-7} mol/l and decreases cAMP via inhibitory G-protein (G_i) signaling, whereas at concentrations higher than 10^{-7} to 10^{-8} mol/l, NE binds to β-AR leading to cAMP increase via stimulatory G-protein (G_s) signaling (Fig. 9.1).

This phenomenon explains why these neurotransmitters show U-shaped and bell-shaped dose-response curves (reviewed in detail in Straub 2015).

Unlike other musculoskeletal connective tissue as bone, periosteum, and synovium, cartilage does not contain blood vessels and is not innervated by nerve fibers indicating that cartilage for some reason might be a hostile environment for spreading of nerve fibers. However, despite lack of nervous innervation, cartilage metabolism is modulated and influenced by neurotransmitters released either from nerve fibers located in neighboring tissue or directly from chondrocytes (for review see Grassel 2014).

This chapter will review recent literature describing effects of sensory and sympathetic nerve fibers and their neurotransmitters on joint physiology and pathophysiology including articular cartilage, growth cartilage, intervertebral disk cartilage, synovium, and subchondral bone.

9.2 Sensory and Sympathetic Nerve Fibers in Cartilage Physiology

9.2.1 Synovial Joints

There is some evidence that sensory nerve fibers may come in contact with those chondrocytes located in growth cartilage and at the surface of articular cartilage. In rat knee joints, CGRP-positive fibers, which originate from the periosteum and near insertion regions of muscle and tendons, penetrate for up to 25 μm into the outer layer of articular and meniscal cartilage. These fibers are located between single chondrocytes indicating a local effector function. However, there are subpopulations of SP-positive axons in perichondrium and periosteum which for unknown reasons do not innervate the cartilage (Schwab and Funk 1998). CGRP- and SP-positive nerve fibers precede the development of cartilage canals, which are formed during skeletal growth shortly after birth. These fibers were detected when penetrating into canals of growth cartilage in the epiphysis of young rats, thereby, coming into close contact to chondrocytes (Edoff et al. 2000; Hara-Irie et al. 1996). The presence of cartilage canals carrying sensory nerve fibers precedes the development of the secondary ossification center, and it is speculated that these nerve fibers modulate the formation of synovial joints through trophic effects (Oliva et al. 2005).

These observations imply important functions of sensory nerve fibers for regulating chondrogenic differentiation during limb growth in embryonic development. Along this line, Edoff et al. reported that articular chondrocytes respond specifically to application of CGRP by increasing cAMP level (Edoff et al. 2000; Edoff and Hildebrand 2003). They assume that dorsal root ganglion (DRG) neurons, which project to growth cartilage, may influence chondrocyte differentiation via CGRP. It is described that increased levels of cAMP as response to CGRP suppresses terminal differentiation of chondrocytes and matrix mineralization (Jikko et al. 1996). This makes it likely that locally released CGRP can delay chondrocyte hypertrophy and subsequent terminal differentiation through modulating cAMP level (Fig. 9.2).

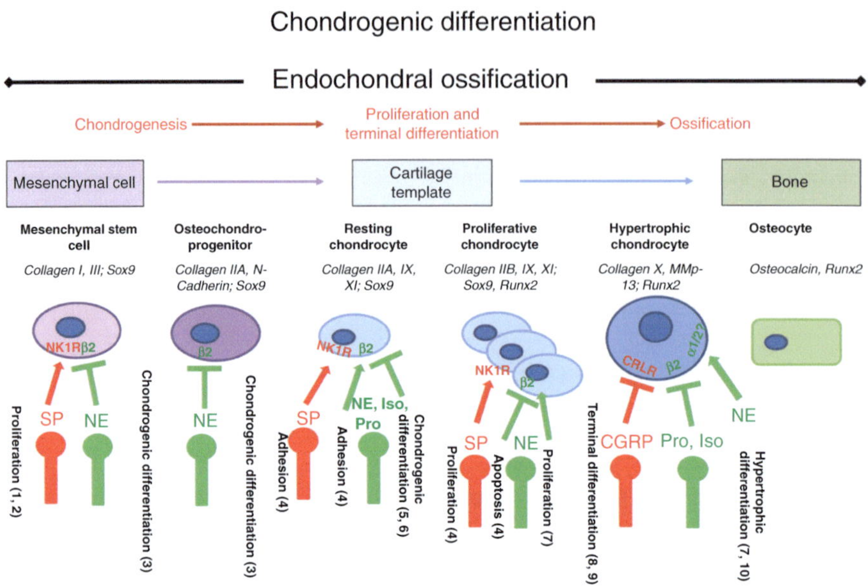

Fig. 9.2 Role of sensory and sympathetic neurotransmitters and their receptors in chondrogenic differentiation. Sensory (SP and CGRP) and sympathetic (NE) neurotransmitters and antagonists/agonists (isoproterenol, propranolol) affect chondrogenic differentiation and metabolisms of chondroprogenitor cells and bone marrow-derived stem cells (BMSC). These neurotropic effects are mediated through specific receptors for sensory neuropeptides, NK1R and CRLR/RAMP-1, and mainly the sympathetic ß2-AR. A *line with an arrow* indicates stimulation and a *line with a bar* indicates inhibition. The *red* (*green*) nerve ending represents sensory (sympathetic) nerve fibers (Modified from Grassel 2014). *SP* substance P, *CGRP* calcitonin gene-related peptide, *NE* norepinephrine, *Iso* isoproterenol, *Pro* propranolol, *Nk1* neurokinin 1 receptor, *CRLR* CGRP receptor, *ß2* ß2-adrenoceptor. *Numbers* indicate references according to bibliography: *1* Fortier and Nixon (1997), *2* Goto et al. (2007), *3* Jenei-Lanzl et al. (2014), *4* Opolka et al. (2012), *5* Mitchell et al. (2011), *6* Takahata et al. (2009), *7* Lai and Mitchell (2008), *8* Edoff et al. (2000), *9* Edoff and Hildebrand (2003), *10* Mauro et al. (2010)

Similarly, to sensory nerve fibers, sympathetic nerve fibers are abundantly present in subchondral bone marrow and synovial joint tissue (Maestroni 2000; Pongratz and Straub 2013) (Fig. 9.3). Under physiological conditions, sympathetic nerve fibers control the blood flow in adult joint vessels via vasoconstriction. Ferrell and colleagues described that these effects are mediated mainly through α1-AR but not through β-AR (Ferrell and Khoshbaten 1990, 1989). Since hyaline cartilage is an aneural tissue, there are very few studies that addressed the potential link between sympathetic nerve fibers or neurotransmitters and cartilage or chondrocytes, respectively.

The only existing report in this context observed type C nerve fibers (absence of myelin or Schwann cells) with sensory or sympathetic function in cartilage canals in human fetal epiphyseal uncalcified cartilage (from 22nd gestational week to 13 months after birth), which serves as bone template during endochondral ossification (Strange-Vognsen and Laursen 1997). These nerve fibers may come into close

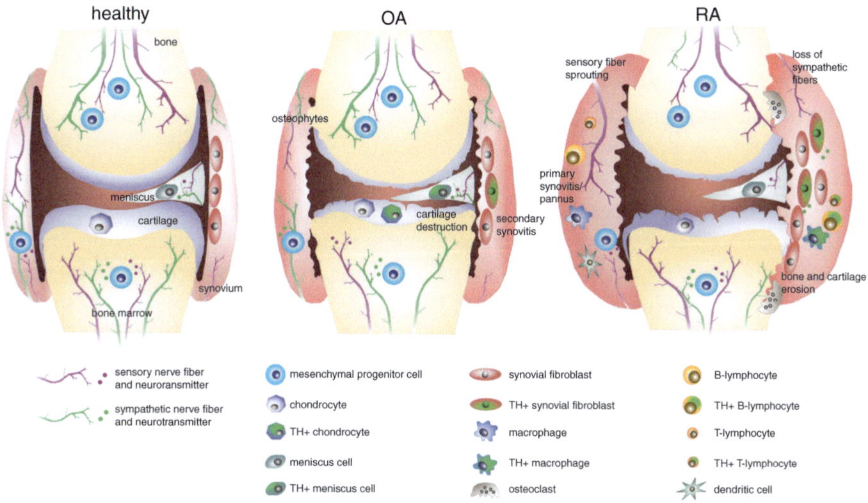

Fig. 9.3 Neuronal changes from healthy conditions to osteoarthritic (*OA*) and rheumatoid arthritic (*RA*) joints. In the healthy situation, nerve fiber densities of sensory and sympathetic nerve fibers are similar, and the ratio of nerve fiber density is 1:1. This situation is more or less the same in OA, but it markedly changes in RA, where sensory nerve fibers sprout into the tissue while sympathetic nerve fibers get lost. The nerve fiber ratio (sensory: sympathetic) changes from 1:1 to 10:1. This is accompanied by a profound thickening of the synovial tissue, called pannus formation. Since the loss of nerve fibers is specific for the sympathetic branch, it is hypothesized that specific repellent factors of sympathetic nerve fibers play a major role

contact to chondrocytes and regulate the developmental process of long bone formation as described for sensory nerves (see above) (Fig. 9.2).

For other components of the adult skeletal system, i.e., bone, the importance of the sympathetic nervous system has been shown. Best characterized are the effects of leptin-induced activation of sympathetic nerve fibers, which leads to changes in bone formation and remodeling (Elefteriou and Karsenty 2004; Elefteriou et al. 2005; Takeda et al. 2002). These studies focused mainly on β2-adrenergic induction of bone loss through upregulation of RANKL by osteoblasts and activation of osteoclasts.

So far, no data exist for sympathetic neurotransmitter concentrations in permanent cartilage or growth cartilage during skeletal growth. One can speculate that these neurotransmitters contribute to or work against the effects of sensory neurotransmitters via regulation of intracellular cAMP levels. In rabbit growth plate chondrocyte cultures, dibutyryl cAMP (membrane permeable analog of cyclic AMP) abrogated the increase in chondrocyte size and the expression of hypertrophic markers such as alkaline phosphatase (ALP) and collagen X, affirming that cAMP plays a crucial role in suppressing terminal differentiation of chondrocytes and cartilage-matrix calcification during skeletal development (Jikko et al. 1996).

Furthermore, Lai et al. observed that β2-AR-dependent stimulation of adenylyl cyclase by isoproterenol treatment leads to increased chondrocyte proliferation and decreased expression of hypertrophy markers such as collagen X in murine embryonic growth plate chondrocytes (Lai and Mitchell 2008). Although this study uses a non-selective β-AR agonist and does not provide information on growth plate sympathetic neurotransmitter influence in vivo, the data clearly show that the sympathetic nervous system probably is an important regulator of embryonic cartilage development (Fig. 9.2).

Sympathetic neurotransmitters secreted from adult joint tissue, e.g., subchondral bone marrow or synovial tissue, might influence chondrocyte physiology. Since a while, it is known that sympathetic nerve fibers are present in these joint structures (Maestroni 2000; Pongratz and Straub 2013) (Fig. 9.3). In synovial fluid of traumatic knee injury patients, norepinephrine was detected at a concentration between 10^{-8} and 10^{-9} mol/l (Jenei-Lanzl et al. 2014), but levels in cartilage tissue are not known.

The only report on the effect of chronic immobilization stress on cartilage and subchondral bone in murine temporomandibular joints demonstrated increased subchondral bone NE levels and concomitantly promoted condylar subchondral bone loss and cartilage degradation (Jiao et al. 2015). These authors concluded that increased activation of the sympathetic nervous system by stress might be a factor that is involved in degenerative changes of condylar subchondral bone.

A further illuminating point in the context of cartilage physiology and sympathetic neurotransmitters is the role of local anesthetics for diagnostic and trauma-therapeutic purposes (Rao et al. 2014). Rao and colleagues showed that more than 60 % of human chondrocytes in culture were apoptotic after addition of 1:1,000 epinephrine (~5 μM, thus β-AR-mediated) for only 90 min. Similarly, epinephrine caused a significant increase in cell death of synovial fibroblasts resulting in a higher release of matrix metalloproteinases (MMPs) and subsequent damage of the proximate cartilage extracellular matrix (ECM) (Braun et al. 2013).

Taken together, it is abundantly clear that despite the a priori aneural nature of cartilage tissue, the sensory and sympathetic nervous system influence cartilage physiology by releasing neurotransmitters from surrounding innervated joint tissue or by influencing cells of surrounding tissue which then in turn synthesize factors, i.e., growth factors, cytokines, proteases, etc., to influence chondrocyte function.

9.2.2 Intervertebral Disks

It is common knowledge that the normal nucleus pulposus (NP), the inner annular zones, and the cartilaginous endplate (CED) are devoid of nerves. Anatomical and experimental studies during the last decades have provided many details of lumbar spine innervation (see review Edgar 2007; Raj 2008 and Chap. 10). The normal intervertebral disk (IVD) is considered as an organ that is poorly innervated supplied only by sensory and sympathetic perivascular nerve fibers. Most of the studies performed in different animal species, including human, have demonstrated that nerve fibers in IVDs are found mostly in the periphery of the annulus fibrosus (AF) and are typically very fine in diameter presumably being C-fibers. The sensory fibers that innervate the IVD are mainly nociceptive and, to a lesser extent, proprioceptive. In the human healthy IVD, nerve fibers which innervate the outer layers of the AF are positive for SP, CGRP, VIP, and NPY, among others (see review Garcia-Cosamalon et al. 2010), whereas the NP is not innervated. The IVD is primarily innervated by small DRG neurons, which are classified into peptidergic nerve growth factor (NGF)-dependent and non-peptidergic glial cell line-derived neurotrophic factor (GDNF)-dependent neurons that are functionally different (Stucky and Lewin 1999). The NGF-dependent neurons – positive for SP and CGRP – are critical to hyperalgesic responses induced by inflammation (Mantyh et al. 1997), whereas the GNDF-dependent neurons are important in neuropathic pain (Malmberg et al. 1997). A number of repulsive factors that exist in the healthy IVD presumably prevent nerve and endothelial cell ingrowth. These factors are aggrecan (Johnson et al. 2006), chondromodulin (Takao et al. 2000), and the semaphorins (Tolofari et al. 2010).

Growth factors such as NGF, brain-derived neurotrophic factor (BNDF), NT3, and NT4/5, which are members of the neurotrophin (NT) family, and their signaling receptors (NTRs) have been detected in healthy IVDs in innervated regions and also in regions lacking sensory or sympathetic innervation (Gruber et al. 2008). Upon binding to their receptors (TRKs A, B, C), NTs initiate a variety of signals that lead to regulation of cell proliferation, differentiation, and survival. NGF and its receptor TrkA are expressed in AF cells and BNDF and its receptor TrkB are found predominantly in the outer AF region and to some extend in the NP cells.

Notably, NTs produced in the IVD can be retrogradely transported to DGR sensory neuron bodies where they regulate expression of SP and CGRP. In addition, the expression of NTs and NTRs in AF and NP cells suggests autocrine and paracrine roles in regulating the physiology of IVD (see review Garcia-Cosamalon et al. 2010). NTs also play a role in inflammatory responses and in pain transmission. In particular, NGF can act as a cytokine (Levi-Montalcini et al. 1996) influencing development, differentiation, chemotaxis, and mediator release of inflammatory cells, as well as fibroblast activation, through complex networks that involve other pro-inflammatory cytokines (Bonini et al. 2003). Furthermore, NTs play key roles in immunocompetent cells by participating in inflammatory responses (Tessarollo 1998; Sariola 2001; Vega et al. 2003) and in sensing pain, because NT signaling increases expression of pain-related peptides.

9.3 Sensory and Sympathetic Nerve Fibers in Cartilage Pathophysiology

9.3.1 Osteoarthritis

9.3.1.1 Sensory Nerve Fibers and Neuropeptides

Suri et al. have localized both sensory (SP and CGRP positive) and sympathetic nerve fibers (NPY positive) in similar distributions to the articular cartilage in human tibiofemoral osteoarthritis (OA) (Suri et al. 2007) (Figs. 9.3 and 9.4). These

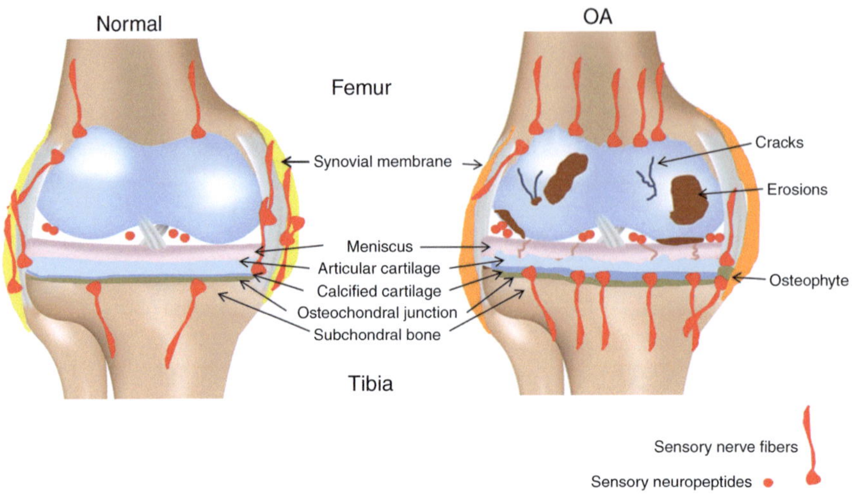

Fig. 9.4 Proposed alteration of sensory nervous innervation in osteoarthritis. It is discussed that there is an increase of sensory nerve fiber innervation which reaches into the calcified cartilage zone from the subchondral bone or which comes into close contact to the articular cartilage. Contrary, synovial sensory innervation decreases in an early OA stage and presumably also in later OA stages. In addition, concentration of sensory neuropeptides in synovial fluid increases with increasing OA severity

nerve fibers were present within vascular channels in both mild and severe OA. For the pathogenesis of OA, it is described that fine unmyelinated nerves grow into joint structures through vascular channels mainly from subchondral bone breaching through the tidemark.

The exclusively perivascular localization of nerves within the articular cartilage implies that vascularization is the driving force behind its innervation. Notably, vascularization of the non-calcified cartilage was found throughout a wide range of histological OA stages and was not restricted to end-stage OA. Possibly, scoring of nervous innervation and vascularization of cartilage might be exploited as a measure for severity of degradative changes in OA pathogenesis. The authors suggest that vascularization and accompanying innervation of the articular cartilage is a potential source of pain in patients with knee OA. They and others have also shown sympathetic and perivascular sensory innervation of tibial osteophytes, latter localized to the base of osteophytes (Suri et al. 2007; Wojtys et al. 1990). Sensory innervation of osteophytes may explain why radiological grading of osteophytosis is associated with reported pain severity (Fig. 9.4).

However, despite the clinical importance of pain in OA, pain mechanisms are poorly understood. It is unclear which structures in the joint give rise to OA pain, and the nature of OA pain (nociceptive versus neuropathic) is a matter of ongoing discussion (Schaible 2012). A tissue compartment, which can be a source of pain in OA, is the osteochondral junction. Interestingly, angiogenesis seems to predominate in both early and late OA, although, besides angiogenic factors, also anti-angiogenic factors have been shown to be upregulated in both OA stages as observed in rats (Mapp et al. 2008; Mapp and Walsh 2012).

During pathogenesis of OA, osteochondral angiogenesis was increased associated with increased NGF expression in subchondral spaces, vascular channels, and by chondrocytes themselves (Walsh et al. 2010). This NGF immunoreactivity was co-localized with CGRP-immunoreactive nerve fibers in the same vascular channels. Increased NGF production may thus contribute to OA pain, both structurally (increased aberrant innervation at the osteochondral junction) and through peripheral sensitization. This was confirmed in studies with human OA material showing that increased vascular penetration and density of perivascular CGRP-positive sensory nerve profiles in the meniscus is a potential source of pain in knee OA and leads in addition to further inflammation and tissue damage (Ashraf et al. 2011a, b).

In a murine OA model, it was suggested that in some soft joint tissue, CGRP- and SP-positive nerve fibers disappeared 5 weeks after induction of OA with intra-articular injection of collagenase (Buma et al. 1992). This observation was confirmed by a study of Murakami et al. who report that the densities of PGP 9.5- and CGRP-positive nerve fibers in the synovium were drastically decreased 1 week after the collagenase injection. However, by week 4, the density of PGP 9.5- and CGRP-positive nerve fibers had recovered to 84 % and 79 %, respectively, of their normal levels. Despite the poor correlation between the synovitis score and the density of CGRP- or SP-positive nerve fibers in the synovium, the ossification rate of chondrophytes in chondro/osteophyte lesions correlated strongly with the density of CGRP-positive nerve fibers (Murakami et al. 2015). This quite aggressive method of OA induction leads to profound degenerative alterations in synovial joint tissue (among other soft tissue) after a few weeks. Contrary, some studies of human OA have reported an increase in SP- and CGRP-positive nerve fibers in the synovial tissue of knee joints (Saito 1992; Saito and Koshino 2000; Saxler et al. 2007), whereas others have described a decrease in these nerve fibers in the synovium (Gronblad et al. 1988). Because most of these reports are from studies of the synovium in chronic OA, there is little information about the association between the pathological changes in the knee and changes in these nociceptive fiber innervation patterns during the acute phase of the disease. The mechanisms that possibly destroy the normal vascular and neural network in early or late OA stages are not identified yet, and, similarly, it is unclear if nerves are destroyed as a consequence of OA or if pathogenesis of OA is facilitated due to nerve disappearance.

A study by Salo et al. indicated that loss of SP- and CGRP-positive innervation from the whole joint always preceded histological changes of cartilage degeneration (Salo et al. 2002). They used a mouse model, where mice usually developed a mild form of OA, but surgical ablation of joint innervation caused the development of severe patellofemoral OA. Their findings would be consistent with the hypothesis that an age-related loss of sensory joint innervation may contribute to the development of OA. However, in human synovial tissue of late-stage OA patients, undergoing knee joint replacement surgery, the density of SP-positive sensory nerve fibers was markedly lower as compared to RA patients and trauma controls (Weidler et al. 2005) (Fig. 9.3). Whether sensory nerve fibers are lost, remain unaltered, increase or change tissue distribution as a prerequisite of OA pathogenesis in humans remains to be determined. Notably, CGRP and SP synovial fluid levels are increased in developmental dysplasia of the hip. CGRP concentration in synovial fluid is increased in knee OA patients and clearly correlated with increased Kellgren-Lawrence scores (indicating OA severity) (Dong et al. 2015; Wang et al. 2015; Li et al. 2015). Further insight into these mechanisms and relations will require reproducible OA-animal models resembling slow proceeding pathogenesis of human OA and enabling longitudinal studies from early onset of the disease to late stages.

9.3.1.2 Sympathetic Nerve Fibers and Neurotransmitters

The presence of specific neurotransmitter receptors is a prerequisite for neurotransmitter action. In healthy human cartilage, dopaminergic and β-adrenergic receptors have been detected and, interestingly, the expression of these receptors was significantly increased in OA cartilage (Vignon et al. 1990) suggesting the contribution of the sympathetic nervous system to OA manifestation.

Recent studies on human chondroprogenitor cell chondrogenesis demonstrated that the sympathetic neurotransmitter NE might influence the self-regeneration capacity of articular cartilage via β2-AR signaling. The deposition of cartilage-specific collagen II was significantly inhibited and concomitantly, collagen X expressed in hypertrophic chondrocytes, was significantly increased after NE treatment (Jenei-Lanzl et al. 2014). Thus, NE might play a pivotal role in the development and manifestation of OA. However, the exact effects of sympathetic neurotransmitters on cartilage tissue homeostasis, on its capacity for spontaneous regeneration, and, thus, on OA development are largely unclear.

Another interesting finding in long-standing OA is a higher density of sympathetic nerve fibers in synovial tissue in relation to sensory nerve fibers when compared to trauma controls (Fig. 9.3). This phenomenon might lead to unfavorable elevated levels of sympathetic neurotransmitters in relation to growth-promoting sensory neurotransmitters (Weidler et al. 2005).

NPY, a co-transmitter of NE, is increased in OA synovial fluid, which is interpreted as a sign of secondary inflammation. Wang et al. observed a direct correlation between NPY concentration in synovial fluid of OA patients and pain indicating that NPY may act as a putative regulator of pain transmission and perception in OA (Wang et al. 2014). High NPY concentrations lead to further degradation of cartilage, because NPY potentiates inflammation by modulating key immune cell functions via Y1/Y2 receptors (Dimitrijevic et al. 2008). However, higher levels of NPY might also demonstrate that the generally higher release of sympathetic neurotransmitters, and NE signaling via α2-AR can induce pain.

Contradictory to presence of sympathetic nerve fibers in OA synovium and cartilage channels, OA patients have lower VIP concentration in synovial fluid compared to trauma patients (Jiang et al. 2006). The reason for this negative association of VIP with progressive joint damage in OA is not known at present (Delgado et al. 2001). Since VIP has been shown to predominantly possess anti-inflammatory effects (Jiang et al. 2006), its insufficient expression might contribute to the characteristic secondary inflammation emerging in late OA which is very important to consider, because over 90 % of OA patients develop synovitis (Roemer et al. 2011). Since VIP is the neurotransmitter of the cholinergic subtype of sympathetic nerve fibers, it might also show that this particular sympathetic nerve fiber type decreases in OA joints.

Of note, sympathetic neurotransmitters have a limited chemical stability, which is not only caused by converting enzymes but also by degrading enzymes, e.g., monoaminooxidase (MAO) and catechol-o-methyltransferase (COMT) for catecholamines (Jiang et al. 2006; Napolitano et al. 1995) or adenosine deaminase (ADA) for adenosine (Antonioli et al. 2012; Cortes et al. 2015). Taken together, local availability of neurotransmitters depends on synthesizing, converting, and degrading enzymes.

Chambers et al. observed an abnormal localization of MAO in murine osteoarthritic cartilage, which is normally tightly bound to the outer membrane of mitochondria in healthy tissue cells. In OA cartilage, MAO was abnormally located extending beyond the chondrocytes into the surrounding matrix (Chambers et al. 1992). Under these conditions, catecholamines cannot be metabolized intracellularly, and they accumulate pericellularly and might exhibit unfavorable effects on chondrocyte function.

COMT, another catecholamine-degrading enzyme, has functional polymorphisms linked to different pain levels in OA patients (Stolk et al. 2007). A single-nucleotide polymorphism results in a valine to methionine mutation at position 158 (Val158Met). This variant results in three- to fourfold lower enzyme activity compared to the valine variant. It has been shown to be associated with higher pain levels and increased radiographic damage of the hip. Therefore, Stolk and colleagues propose that patients carrying the methionine allele could be treated more effectively with drugs affecting neurotransmitter function (Stolk et al. 2007).

Furthermore, it is important to consider adenosine, a co-transmitter of catecholamines, in OA pathophysiology. The enzyme ADA converts adenosine to inosine; thus, ADA deficiency leads to adenosine accumulation. Children with ADA deficiency due to genetic mutation have elevated adenosine levels in extracellular fluids. They exhibit marked changes in the chondro-osseous tissue of vertebrae, and the costochondral junction has increased numbers of necrotic chondrocytes and has large amounts of cellular debris (Mediero and Cronstein 2013). A similar phenomenon was observed by Mistry et al. in murine cartilage tissue showing that increased adenosine concentrations induce chondrocyte apoptosis and, therefore, might play a role in OA pathogenesis (Mistry et al. 2006). The adenosine effect at high concentrations might be induced by elevated intracellular cAMP levels because adenosine would bind to the A2A-/B-adenosine receptors.

9.3.2 Rheumatoid Arthritis

Although rheumatoid arthritis (RA) is a primary inflammatory systemic disease with autoimmune characteristics, in contrast to the mainly injury- or wear- and bear-driven OA, there are some analogies regarding disease outbreak and manifestation. Similar to OA development, pathogenesis of RA is multifactorial and influenced by the nervous system. Besides effects of the CNS and endocrine hormones (Del Rey et al. 2010; Straub et al. 2013) (Straub RH), the peripheral nervous system has been shown to influence development and progression of RA (Pongratz and Straub 2013; Levine et al. 1987; Lorton et al. 2013). The first proof of this involvement provided the observation that patients with hemiplegia develop RA symptoms only in the non-paralyzed hand, whereas the hemiplegic hand shows no RA-specific signs (Keyszer et al. 2004). Similar results of hemiplegic patients have been reported several times. In a recent study, this phenomenon was impressively confirmed in murine experimental arthritis, where denervation of limbs protected the animals from the disease (Stangenberg et al. 2014).

The elements of the peripheral nervous system have various functions and influence the inflammatory response differently. First, the sensory nervous system transmits pain and danger signals from the periphery to the brain, thereby, acting mainly pro-inflammatory. This function of the sensory nervous system is called the efferent function leading to release of local pro-inflammatory SP. Second, the sympathetic nervous system, which influences blood flow, vascular permeability, and plays a dual role in inflammation depending on the stage of arthritis (Lubahn et al. 2004; Harle et al. 2005), is both pro-inflammatory or anti-inflammatory. Third, the parasympathetic nervous system conveys mainly anti-inflammatory effects; however, its exact role in arthritis is still unclear (Pongratz and Straub 2013; Schaible and Straub 2014).

As soon as tissue homeostasis is disturbed, the CNS receives this information from activated peripheral nociceptive sensory nerve fibers or by circulating cytokines released locally due to tissue damage, e.g., IL-6 (Besedovsky and del Rey 1996; Pongratz and Straub 2013). In turn, the CNS initiates the activation of the hypothalamic–pituitary–adrenal (HPA) axis and the sympathetic nervous system both reacting locally at the site of inflammation or in secondary lymphoid organs. This first neuroendocrine response within hours is pro-inflammatory and leads to activation of the immune system (Pongratz and Straub 2013).

In addition, sensory (SP and αCGRP) neurotransmitters have strong vasodilatory properties, and sympathetic neurotransmitters (NE and NPY) are chemotactic, both leading to the recruitment of immune cells (Ruff et al. 1985; Smith et al. 1993; Straub et al. 2000; Wheway et al. 2007). In inflamed tissue, peripheral sympathetic and sensory neurons talk to each other leading to sensitization of sensory afferents via α2-AR signaling (Niissalo et al. 2002; Chen et al. 1996). In adjuvant-induced arthritis, β2-agonists and selective blocking of β2-adrenergic receptors and NE depletion aggravate arthritis symptoms (Niissalo et al. 2002; Lubahn et al. 2004). This confirms the pro-inflammatory nature of the sympathetic nervous system in an early immune encounter (Schaible and Straub 2014). This first pro-inflammatory response of the body is critical in order to fight the target antigen.

During pathogenesis of RA, the original balance between SP-positive sensory nerve fibers and TH-positive sympathetic nerve fibers is disturbed at the site of inflammation. Miller et al. observed the loss of TH-positive sympathetic nerve fibers in chronic inflamed synovial tissue, whereas SP-positive nerve fiber density increased (Miller et al. 2000) (Fig. 9.3). In contrast to SP-positive fibers, CGRP-positive nerve fibers are lost from sensory neurons (Imai et al. 1997; Kimberly and Byers 1988; Dirmeier et al. 2008). SP and CGPR exhibit opposite effects in RA. SP induces the release of pro-inflammatory factors such as TNF, IL-1, IL-6, IL-8, IL-12, IFN-γ, and superoxide anion and stimulates chemotaxis of immune cells (Dirmeier et al. 2008). In contrast, CGRP has anti-inflammatory effects, e.g., the inhibition of macrophage, monocyte, and human peripheral blood mononuclear functions. The reasons for the loss of CGRP from sensory nerve fibers are not known at present; however, it seems to be clear that this preponderance of SP-positive nerve fibers over CGRP-positive fibers contributes to the maintenance of the pro-inflammatory situation in RA (Dirmeier et al. 2008).

It has been proposed that the loss of TH-positive nerves is caused by the action of elevated nerve-repellent factor levels such as semaphorins 3C and 3F, which are specific for noradrenergic sympathetic nerve fibers (Fassold et al. 2009; Miller et al. 2004) (for details see Sect. 10.5.2.). The local loss of sympathetic nerve fibers has been observed also in other inflammatory diseases, e.g., in Crohn's disease (Straub et al. 2008) and in chronic Charcot foot (Koeck et al. 2009). Similarly, sprouting of SP-positive nerve fibers was detected in other inflammatory diseases, e.g., gastritis (Sipos et al. 2008), in esophagitis (Matthews et al. 2004), and in psoriatic skin disease (Naukkarinen et al. 1989) leading to the maintenance of a pro-inflammatory situation.

Of note, the loss of sympathetic nerves would lead to the local decrease of neurotransmitter concentration. It is important to consider that besides the first pro-inflammatory response mentioned above, sympathetic neurotransmitters at high concentrations have anti-inflammatory properties via Gαs-coupled catecholamine receptors (Straub 2015). NE and its co-transmitter adenosine exhibit anti-inflammatory effects mediated by β2-AR and adenosine A2A-/B-adenosine receptors, respectively (Sanders and Straub 2002; Sitkovsky 2003). In addition, the conversion of ATP to anti-inflammatory acting adenosine is elevated in RA, because the ectonucleotidases (CD39 and CD73), which convert purine precursor neurotransmitters such as ATP to adenosine, are increased in inflammation (Straub 2015).

The early repulsion of sympathetic nerve fibers from inflamed tissue has been positively selected to overcome acute wounding, because an anti-inflammatory influence does not make sense before bacteria are cleared. Thus, after repulsion, lower concentrations of NE are present locally and receptor signaling changes from β2-AR to a predominantly α-AR signaling pathway, which is Gi driven and exhibits mainly pro-inflammatory effects (Pongratz and Straub 2013).

Concomitant to TH-positive nerve fiber repulsion, TH-positive single cells appear in the human and murine arthritic synovial tissue (Capellino et al. 2010, 2012) (Fig. 9.3). These cells synthesize and release catecholamines, possibly in order to compensate the local loss of sympathetic nerve fibers (Jenei-Lanzl et al. 2015a; Miller et al. 2000, 2002). Interestingly, these recently described anti-inflammatory TH-positive cells also appear in lymphoid organs at the same time (Capellino et al. 2012). This appearance and function of TH-positive cells would explain the time-dependent dual effect of sympathectomy observed in a murine model of collagen-induced arthritis (CIA). In this model, the sympathetic nervous system is pro-inflammatory in the early phase and anti-inflammatory in the late chronic phase, where only TH-positive cells but no sympathetic nerves are present in synovium (Capellino et al. 2012; Harle et al. 2005; Jenei-Lanzl et al. 2015a). Cells of the inflamed synovium try to compensate sympathetic nerve fiber loss also by upregulation of receptors for D1/5 dopamine receptors known for Gαs-mediated signaling via cAMP (Capellino et al. 2014). However, even together with all these neurotransmitters locally produced by cells, neurotransmitter levels are probably still too low to exert anti-inflammatory effects.

Apart from NE and its co-transmitters, sympathetic nerve fibers after cholinergic transition can secrete VIP, which is normally released by cholinergic parasympathetic neurons (Steele et al. 1996). VIP-containing nerve fibers are present in human and murine synovial tissue (Bjurholm et al. 1990). Similar to NE, VIP acts via Gαs-coupled seven-transmembrane receptors, VIP receptor 1 and 2, by activating adenylyl cyclase (Straub 2015). Thus, it is not surprising that VIP treatment of CIA mice significantly reduced the incidence and severity of the disease and also destruction of bone and cartilage (Delgado et al. 2001).

In addition, anti-inflammatory effects of VIP have been reported in patients with RA (Gonzalez-Rey et al. 2006) and in animal experiments after CIA induction (Muschter et al. 2015a, b). Here, it was shown that VIP reduced proliferation of macrophages and has mostly protective effects on bone matrix by inhibition of cathepsin K activity and decrease of osteoclast numbers, an observation probably attributed to increased cAMP signaling. Therefore, the loss of sympathetic nerve fibers results in the loss of anti-inflammatory neurotransmitters.

In general, locally produced catecholamines, VIP, or opioids are not powerful enough to overcome the pro-inflammatory domination in chronic arthritis. Moreover, in a most recent study, an unexpected pro-inflammatory switch from Gαs to Gαi signaling depending on PDE4/β-arrestin interaction has been described in human arthritic synovial tissue. This phenomenon is most probably responsible for reduced efficacy of PDE4 inhibitors and Gαs agonistic neurotransmitters under these circumstances (Jenei-Lanzl et al. 2015b).

Multiple evidences confirmed that NPY influences the pathogenesis of RA (Bedoui et al. 2004; Harle et al. 2006). Interestingly, NPY concentration is elevated in RA joints, although the loss of NPY-containing nerve fibers from the superficial synovium was reported assuming that these nerve fibers become restricted to deeper zones of the synovium (Mapp et al. 1990; Larsson et al. 1991). One can speculate that, similar to TH-positive single cells, also NPY-producing cells appear in the synovium as a compensatory mechanism to nerve fiber loss (Capellino et al. 2012; Jenei-Lanzl et al. 2015a; Miller et al. 2000). NPY potentiates inflammation by manipulating immune cells via Y1/Y2 receptor signaling (Dimitrijevic et al. 2008) or by boosting the effects of NE (Bedoui et al. 2004).

In addition, NPY has angiogenic activities in several tissues, however, only at the healthy border of inflamed tissue because in the inner highly inflamed zone, the sympathetic innervation, including NPY, is lost. However, sensory nerves are still present, and in vivo experiments showed that SP induces angiogenesis in arthritic synovium enabling the further circulation of immune cells leading to maintenance of a pro-inflammatory condition (Ribatti et al. 2007).

This dominant and long-lasting pro-inflammatory situation in the synovium under strong neuronal influence most probably affects the neighboring cartilage tissue. Proliferating synovial pannus penetrates articular cartilage (Otero and Goldring 2007), and different cells of the pannus, e.g., fibroblasts, macrophages, and leukocytes, release a broad spectrum of factors causing cartilage-matrix destruction and bone erosion (Meyer et al. 2006; Knedla et al. 2007). Similarly, the action of invasive synoviocytes and the contact of cartilage with the synovial fluid results in destructive processes (Dodge and Poole 1989). In contrast to OA, where proteoglycan and collagen loss occurs mainly in the cartilage superficial zone, in RA, also middle and deeper zones are affected suggesting that the chondrocyte itself critically participates in destroying its own matrix by releasing ECM degrading enzymes (Dodge et al. 1991; Mitchell and Shepard 1978; Kane et al. 2004).

Apart from synovial tissue and chondrocytes, also adipose tissue surrounding the inflamed synovium secretes both pro-inflammatory and anti-inflammatory factors including leptin, IL-1RA, and IL-10 playing a role in cartilage physiology or pathophysiology (Dayer and Burger 2004; Loeser 2003; Otero et al. 2006). Leptin, one link between the neuroendocrine and immune system, is increased during inflammation and has been shown to exhibit mainly pro-inflammatory properties (Popa et al. 2005).

During skeletal development, mesenchymal progenitor cells in the growth plate differentiate to chondrocytes providing the templates for the developing bones (Goldring et al. 2006). A chondrogenesis-like process might also be important for cartilage repair in adults since mesenchymal progenitors are present in articular cartilage (Dowthwaite et al. 2004; Williams et al. 2010; O'Sullivan et al. 2011) and their number is increased in OA and RA (Pretzel et al. 2011; Fickert et al. 2004; Grogan et al. 2009). As reviewed by Kondo et al., most pro-inflammatory cytokines inhibit chondrogenic differentiation (Kondo et al. 2014). Thus, an inflammatory situation, where sympathetic nerve fibers are lost and Gαi-mediated effects are predominant or Gαs-mediated effects are not powerful enough, will have negative influence on progenitors and consequently on cartilage defect repair. In addition, catecholamines, released by sympathetic nerve fibers or by immune cells, have been shown to enhance mesenchymal stem cell recruitment from the bone marrow (Mignini et al. 2003; Stanojevic et al. 2013), given that these cells express a variety of adrenergic receptors (Silva et al. 2003; Jenei-Lanzl et al. 2014) (Fig. 9.1).

9.3.3 Intervertebral Disk Degeneration

The etiology of intervertebral disk degeneration (IVDD) is still unclear as is the cause of low back pain; however, a correlation of both has been documented, and the current consensus is that it is multifactorial (for review see (Vergroesen et al. 2015 and Chap. 10). Many changes in disk morphology and physiology have been described but, still, no widely accepted disease model has been developed. However, consensus exists that upon the degeneration process, IVDs become densely innervated even in regions that normally lack innervation. This increased innervation which is considered as part of a repair process concomitant with neovascularization has been associated with discogenic pain (for review see Garcia-Cosamalon et al. 2010; Benneker et al. 2014).

Members of the neurotrophin family are likely involved in IVD hyperinnervation (for details see Sect. 9.2.2). IVDs with symptoms of degeneration and those from patients suffering from discogenic pain express increased levels of neurotrophins and their receptors implying a positive correlation between levels of neurotrophins and density of IVD innervation (Purmessur et al. 2008).

In human and animal models of IVDD, nociceptive fibers grow into usually aneural parts of the AF and into the NP. These neurons are NGF sensitive and positive for SP and CGRP. Notably, the profile of these sensory neurons innervating pathological IVDs remains identical to normal conditions; however, the nerve fiber quantity is markedly increased constituting the phenomenon of sensory nerve fiber sprouting (Takahashi et al. 2009). Freemont et al. report that levels of NGF are higher in painful disks than in asymptomatic disks indicating that NGF may play a role in sensitizing the painful disk (Freemont et al. 2002). Of note, there are controversial reports that NGF expressing endothelial cells enter the IVD first followed by sensory nerve fibers tracking alongside.

A recent study by Binch et al. observed nerve fiber ingrowth into the IVD also in the absence of blood vessels challenging the hypothesis that neoinnervation requires neoangiogenesis first (Binch et al. 2015). In addition to sensory nerve fibers, there is growing evidence that afferents following sympathetic nerve fibers are also increased in degenerating IVDs and that they play a significant role in lower back pain (Takebayashi et al. 2006).

Binch et al. demonstrated that NP cells endogenously produce SP and CGRP which become upregulated upon cytokine stimulation (Binch et al. 2014). In painful IVDs, the levels of pro-inflammatory mediators such as TNFα, IL-1ß, and IL-8 are higher as in asymptomatic IVDs which in turn upregulates NGF production and secretion in cultured human IVD cells (Ahn et al. 2002; Lee et al. 2009). Notably, cells isolated from human NP increase production of SP after stimulation with TNFα, and levels of NGF together with SP were increased in human painful degenerated IVDs compared to non-degenerated NP cells (Purmessur et al. 2008; Richardson et al. 2009). An association of TNFα with SP expression and IVD degeneration is supported by the work of Chubinskaya and colleagues, who have demonstrated downregulation of both genes in an anti-catabolic disk model (Chubinskaya et al. 2007). Possibly, SP-containing nerve fibers contribute to nociceptive processes within the degenerate IVD since expression of SP and its receptor NK1R have been identified in both disk cells and nerve fibers growing into the diseased IVD (Ohtori et al. 2006).

To this end, Kepler et al. reported that AF and NP cells expressed SP at low levels, and expression did not change significantly with SP treatment but was significantly upregulated after treatment with IL-1β/TNF-α (Kepler et al. 2013). The same group detected expression of NK1R, neurokinin receptor 2 (NK2R), and neurokinin receptor 3 (NK3R) in AF and NP cells (Kepler et al. 2015). Treatment of disk cells with a specific NK1R antagonist was able to suppress expression of IL-1β, IL-6, and IL-8 in a dose-dependent manner. Their study suggests that NK1R is responsible for the pro-inflammatory effect of SP on IVD cells, and this effect can be blocked by preventing binding of SP to NK1R. It became clear that SP mediates signaling in disk cells through NK1R activating the pro-inflammatory p38-MAPK and ERK1/2 pathways.

9.4 Sensory and Sympathetic Neurotransmitters and Their Receptors in Chondrocytes

9.4.1 Sensory Neurotransmitters

Besides their classical function in nociception, SP and CGRP appear to have extra functions in the musculoskeletal system. Lately, it has been reported that newborn murine costal and adult human articular chondrocytes endogenously produce SP and its receptor NK1R (Opolka et al. 2012; Millward-Sadler et al. 2003). Prior, SP was immunolocalized to articular cartilage of dog shoulder joints (Karahan et al. 2002). Expression and localization were increased in chondrocytes and within the cartilaginous ECM after low-impact regimented exercise indicating a role in signaling pathways through which chondrocytes respond to mechanical stimulation. This was demonstrated by Millward-Sadler et al. who suggested that SP is involved in mechanotransduction via the NK1R (Millward-Sadler et al. 2000). In these experiments, SP was necessary for a hyperpolarization response of the cell membrane, and concomitant changes in gene expression as response to mechanical stimulation indicate a role of SP in maintenance of articular cartilage-matrix integrity and function in the presence of mechanical stress. The same group demonstrated that normal and OA chondrocytes reacted differently to mechanical stimulation in that OA chondrocytes upregulated gene expression of the SP encoding gene, Tachykinin 1, whereas non-OA chondrocytes did not show this phenomenon (Howard et al. 2008).

In addition, we recently demonstrated that proliferation was increased in costal chondrocytes from newborn mice when stimulated with SP, which also stimulated cell–matrix adherence by inducing formation of focal adhesion contacts. These effects are mediated specifically through the NK1R (Opolka et al. 2012) which is expressed mainly in middle and deep zones in articular cartilage chondrocytes obtained from OA patients (Fig. 9.5). Our observation implies that SP might modulate the proliferation rate of growth plate chondrocytes and consequently terminal

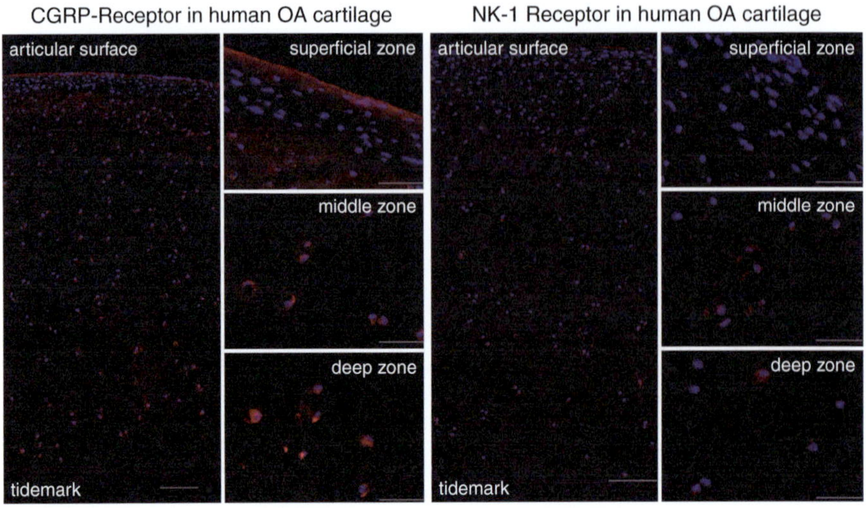

Fig. 9.5 Immunofluorescence staining of cryo-sections from human OA cartilage for NK1R and CRLR. In OA cartilage, chondrocytes in the middle and deep zones are immunoreactive for a NK1R antibody, whereas chondrocytes of the superficial zones remain mainly negative. In OA cartilage, chondrocytes in the middle and deep zones were immune-positive for a CRLR antibody, whereas chondrocytes of the superficial zone remain mainly negative. *Bars* (overview pictures) = 200 μm, *bars* (enlarged pictures) = 50 μm

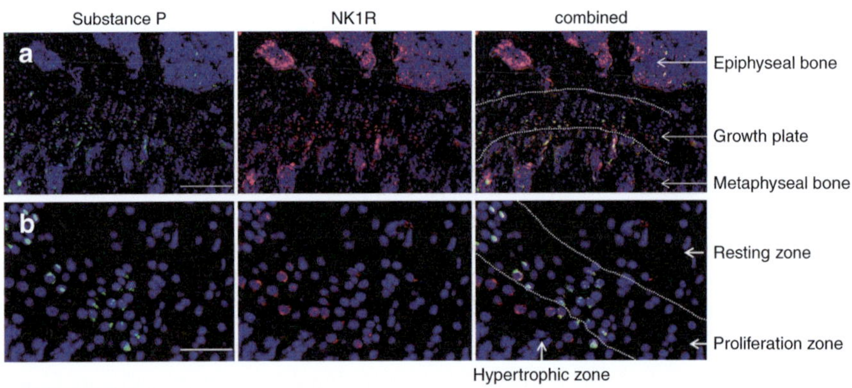

Fig. 9.6 SP and NK1R immunofluorescence staining on epiphyseal sections from 14-week-old wild-type mice. Proliferating and hypertrophic chondrocytes of the growth plate are positive for both NK1R (*red* fluorescence) and SP (*green* fluorescence). Additional, there are NK1R- and SP-positive regions in the epiphyseal and metaphyseal bone. (**a**) *Bars* = 200 μm, (**b**) *Bars* = 50 μm

differentiation during endochondral ossification. Unpublished data from our group demonstrate expression of SP and NK1R in growth plate chondrocytes (Fig. 9.6). It is thus conceivable that in chondrocyte physiology and in chondrogenic differentiation during skeletal growth, endogenous SP acts as a trophic, anabolic factor and does not function as a classical neuropeptide.

However, in adults, the detection of higher levels of SP in synovial fluid from patients with RA and OA, and increased expression of NK1R, indicates possibly catabolic effects of SP on articular cartilage (Inoue et al. 2001). In addition, transforming growth factor (TGF)-β and basic fibroblast growth factor (bFGF) play an important role as inductor or promoter for production of SP in synovial fibroblasts. These data are supported by Im et al. who elegantly demonstrated that SP induces IL-1ß release (Im et al. 2008). The authors propose a mechanism by which bFGF together with SP reduces proteoglycan deposition and stimulates production and release of matrix metalloprotease (MMP)-13 in human articular chondrocytes and thus accelerates catabolic processes in cartilage.

Altogether, these observations suggest that SP has autocrine functions and differentially modulates chondrocyte metabolism and cartilage homeostasis during skeletal growth and in pathophysiology. Like in synovial cells, where SP is described as potent mediator of inflammation by promoting secretion of PGE2, several MMPs (Lotz et al. 1987), reactive oxygen species (Tanabe et al. 1996), IL-1β and TNFα (Lotz et al. 1988), and SP might also act in a catabolic way on chondrocytes and to promote cartilage degradation.

Up to date, there are no reports listed in PubMed with respect to production of CGRP and its receptors in cartilage. However, our unpublished data demonstrate expression of CRLR/RAMP-1 in articular cartilage chondrocytes obtained from OA patients. Expression was detected mainly in middle and deep zones similar to the NK1R (Fig. 9.5). Since in bone metabolism CGRP is described as an anabolic factor by stimulating osteoblast activity and, thus, bone formation (Elefteriou 2005; Schinke et al. 2004), one might hypothesize that CGRP has similar anabolic effects in cartilage physiology. However, under pathophysiological conditions as in osteoarthritis, inhibition of CGPR effects by blocking its receptor with an antagonist, subchondral bone sclerosis was attenuated in a murine surgical OA model (Destabilization of the Medial Meniscus = DMM) (Nakasa et al. 2015). Consequently, articular cartilage erosion and degeneration were delayed in the early stage in this OA model. Of note, these effects were abrogated in a later stage of the disease (8 weeks after OA induction) indicating that other factors affect bone turnover compensating for CGRP effects.

9.4.2 Sympathetic Neurotransmitters

Given the fact that sympathetic nerve fibers and/or neurotransmitter-producing cells are present in joint tissue, except in articular cartilage and the avascular zone of the meniscus (Jenei-Lanzl et al. 2014; Miller et al. 2000), neurotransmitters released into the synovial fluid can influence cartilage tissue provided that specific neurotransmitter receptors are present on chondrocytes (Fig. 9.1).

Takarada and colleagues detected β2-AR and α2a-AR gene expression at different developmental stages in mice. In neonatal mouse tibial sections, β2-AR and α2a-AR mRNA expression was found in chondrocytes by in situ hybridization (Takarada et al. 2009). All other adrenergic receptors were not present. Similarly, Opolka et al. described that β2-AR and α2a-AR mRNA but also α1b-AR and

α1d-AR were expressed in newborn murine costal chondrocyte cultures (Opolka et al. 2012). A study performed by Lai et al. confirmed only the expression of β2-AR on growth plate chondrocytes from ribs of embryonic E18.5 mice (Lai and Mitchell 2008) (Fig. 9.2).

Studies on cartilage tissue explants derived from OA patients after knee replacement surgery and on chondrogenic progenitor cells isolated from human OA cartilage showed that also under degenerative conditions β2-AR and α2-AR are present (Jenei-Lanzl et al. 2014; Lorenz et al. 2016). Regarding alterations in biomechanical pathways in OA pathogenesis, it is important to consider that β2-AR might contribute to disease development related to overloading of cartilage, because β2-adrenergic drugs have been shown to influence mechanical events in bone tissue (Elefteriou et al. 2005, 2014).

In addition, high expression of the A2a and A2b adenosine receptors has been detected in OA chondrocytes (Koolpe et al. 1999). Both of these A2 adenosine receptor subtypes play a role in attenuation of inflammatory processes elicited by other cell types. It is known that NPY receptors play an important role in bone homeostasis (Lundberg et al. 2007; Baldock et al. 2002). However, no data exists about chondrocytes expressing NPY receptors, and only one study confirmed the existence of the Y1 receptor on bone marrow stromal cells being able to contribute to cartilage repair processes (Lundberg et al. 2007). Similarly, no studies described the presence of VIP receptors on chondrocytes until now.

So far, no studies exist comparing different sympathetic receptor expression levels between healthy and pathological conditions. In fact, sympathetic neurotransmitters influence both chondrogenic progenitor cell differentiation and mature chondrocyte function. Catecholamines released by sympathetic neurons or by immune cells influence mesenchymal progenitor cell migration from the bone marrow. There is growing evidence that G-CSF plays a major role in MSC recruitment driven by nervous stimuli (Dygai et al. 2012; Katayama et al. 2006; Saba et al. 2013). Interestingly, Maestroni and colleagues demonstrated that besides sympathetic nerves in bone marrow, also bone marrow cells can synthesize substantial amounts of catecholamines (Maestroni 2000; Maestroni et al. 1998). Moreover, NE and dopamine release is subject to a circadian rhythm with early morning peaks. Thus, increased sympathetic influences during stress or in an inflammatory situation might critically influence progenitor function in bone marrow.

Furthermore, Takarada and colleagues treated murine pre-chondrogenic cells with epinephrine and observed β2-AR-dependent inhibition of Sox9 and consequently of chondrogenic gene expression via classical Gαs-cAMP signaling (Takarada et al. 2009). Similar inhibitory effects of NE on BMSC and on OA cartilage-derived progenitor cell chondrogenesis were reported (Jenei-Lanzl et al. 2014). NE in high concentrations inhibited collagen II and glycosaminoglycan deposition via β2-AR signaling, and, in addition, it accelerated the expression of hypertrophic markers like collagen X and MMP-13. NE in low concentrations, acting preferentially via α-AR, had no effects.

In contrast to findings in progenitor cell chondrogenesis studies, Lai et al. reported that the specific β2-AR agonist isoproterenol inhibited Indian hedgehog (Ihh) and collagen X expression via cAMP and ERK1/2 activation in the murine growth plate (Lai and Mitchell 2008) (Fig. 9.2). Furthermore isoproterenol stimulated the proliferation of chondrocytes. In a later work, these authors elaborated on the involvement of the transcription factor Jun-B, activated by β2-AR in chondrogenesis, inhibiting the expression of Sox-6 and collagen II (Mitchell et al. 2011) similar as in MSCs (Jenei-Lanzl et al. 2014). In costal chondrocytes isolated from newborn mice, apoptosis decreased after NE stimulation; however, ECM formation was not influenced by NE (Opolka et al. 2012).

Jikko and colleagues confirmed the cAMP-driven suppression of chondrocyte hypertrophy by measuring alkaline phosphatase (ALP) activity and collagen X synthesis after dibutyryl-cAMP treatment in rabbit growth plate chondrocyte cultures (Jikko et al. 1996). Interestingly, Mauro et al. described completely contrary effects in the chicken growth plate. NE did not affect collagen II expression, but enhanced collagen X mRNA levels (Mauro et al. 2010) (Fig. 9.2). A further deviation from the studies mentioned above is the involvement of α-adrenergic receptors in their data. There is no explanation at present for these contrary findings regarding sympathetic influences on growth plate chondrocytes other than possible switch of G-protein-dependent signaling. High agonist concentrations, followed by permanently high intracellular cAMP concentrations, might lead to switching of stimulatory G-protein-mediated signaling to inhibitory G-protein-dependent signaling (Sun et al. 2007; Jenei-Lanzl et al. 2015b; Baillie et al. 2003).

NE influences the function of human OA chondrocytes as well. In OA, especially in late stages, inflammation is a hallmark. However, not much is known about the effect of inflammation on chondrocyte function in OA. For instance, Lorenz and colleagues investigated the effects of NE together with inflammatory stimuli like IL-1ß, simulating a late OA microenvironment (Lorenz et al. 2016). NE in high concentrations decreased chondrocyte proliferation via β2-AR signaling. In contrast, NE in low concentration increased the proliferation of OA chondrocytes via α1-AR signaling suggesting that NE might exhibit dual effects on chondrocyte proliferation in OA depending on sympathetic activity. In addition, NE reversed IL-1ß-mediated suppression of collagen II and glycosaminoglycan synthesis at high concentrations. Furthermore, Lorenz et al. showed the first time that TH+ chondrocytes are present OA cartilage. This indicates that the presence of sympathetic nerve fibers (in contrast to RA) and a high sympathetic activity would be beneficial in OA. However, no in vivo evidence exists at present confirming this hypothesis (Fig. 9.1).

It is important to think about further sympathetic neurotransmitters that signal via Gαs-coupled receptors and cAMP, e.g., adenosine. There is no evidence at present that adenosine influences progenitor cell chondrogenesis or mature chondrocyte function. However, one can speculate about similar effects compared to NE and epinephrine, given the fact that different adenosine receptors signaling via Gαs protein are detected on MSCs and on chondrocytes (Carroll and Ravid 2013).

Taken together, all those reports discussed above are providing different data which do not allow a clear prognosis in which way sympathetic neurotransmitters contribute to chondrogenesis of progenitor cells and in mature chondrocytes to articular cartilage repair in both a non-inflammatory and an inflammatory microenvironment.

9.5 Nerve-Repellent Factors in Articular Cartilage Pathophysiology

Why is it that healthy and also diseased articular cartilage is mostly not innervated? This is still not sufficiently clear. Possibly, the lack of blood vessels prevents innervation or vice versa. Putative underlying molecular mechanisms for this unique phenomenon might be found in specific axon guidance cues or nerve-repellent factors that are expressed in cartilage. Under physiological conditions, repellent factors regulate cell–matrix interactions and cell–cell interactions of migrating cells during embryonic development. This group consists of semaphorins and their receptors; netrins and their receptors, deleted in colon carcinoma (DCC) and UNCs; slits and roundabout receptors (Robo); and ephrins and the eph receptors (Hinck 2004; Miller et al. 2004; Serafini et al. 1994). Notably, most of the members of these families are expressed in cartilage already during embryonic limp development (Behar et al. 1996; Holmes and Niswander 2001; Wada et al. 2003; Kitsukawa et al. 1995).

9.5.1 Sensory Nerve-Repellent Factors

A selective repellent factor of SP- and CGRP-positive sensory nerve fibers is semaphorin (Sema) 3A which is present in neuronal tissue but also in developing cartilage and bone (Gomez et al. 2005), in intervertebral disks (Tolofari et al. 2010), and in adult human articular cartilage (Okubo et al. 2011). Gomez et al. showed convincingly that the semaphorin signaling system consisting of Sema3A and its receptors plexin (Plx) A3/neuropilin (NPN)-1 is expressed in resting, pre-hypertrophic, and hypertrophic chondrocytes in growth cartilage before onset of neurovascular invasion during endochondral ossification (Gomez et al. 2005). They suggest that the Sema3A/Plx-A3/NPN-1 pathway would inhibit neuro-vascularization of the cartilage anlage early in skeletal development. Another group reported that gene and protein expression of Sema3A, and its receptor Plx-A3/NPN-1, was significantly elevated in chondrocytes from OA cartilage as compared to chondrocytes from normal cartilage (Okubo et al. 2011). Sema3A expression is closely correlated with chondrocyte cloning, which is a characteristic feature of OA cartilage. The authors imply the possibility that Sema3A plays a role in the pathogenesis of chondrocyte cloning through antagonizing and inhibition of vascular endothelial growth factor (VEGF)-mediated cell migration.

Our group described a role of the receptor DCC in chondrocyte migration (Schubert et al. 2009b). We discovered strong induction of DCC in OA chondrocytes possibly induced by Sox9 and AP-1 suggesting that DCC mediates directed chondrocyte migration in vitro in response to netrin-1, its ligand. To this end, these findings may highlight an altered migratory phenotype of chondrocytes or misdirected migration during OA, which has important consequences for cartilage remodeling.

The induction of MMP1 and MMP3 via DCC in chondrocytes is an additional indication of a possible functional relevance of DCC in the pathophysiology of

OA. Notably, treating the synovial fibroblasts (SF) with recombinant netrin-1 resulted in inhibition of migration of RA-SFs and OA-SFs, whereas control cells were not affected (Schubert et al. 2009a). The stronger expression of UNC5B and UNC5C receptors might contribute to the disordered phenotype of RA-SFs and OA-SFs which in turn affect progression of cartilage degradation and severity during pathophysiology of these diseases.

9.5.2 Nerve Growth Factors and Sympathetic Nerve-Repellent Factors

Sympathetic nerve fiber density is regulated on one hand by NGF, which is an unspecific growth factor for different nerve fiber types. On the other hand, sympathetic nerve fiber density is regulated by specific nerve-repellent factors. In an inflammatory situation, i.e., in late OA and in RA, activated synovial fibroblasts and recruited macrophages produce NGF stimulating both sympathetic and sensory nerve fiber sprouting (Aloe et al. 1992, 1993). These cell types also release semaphorin 3C and semaphorin 3F resulting in the repulsion of sympathetic nerve fibers exclusively (Fassold et al. 2009; Miller et al. 2004). This phenomenon might be responsible for the preponderance of SP over sympathetic neurotransmitters in RA (Miller et al. 2000). Apart from NGF, other peripherally acting neurotrophins, e.g., brain-derived neurotrophic factor (BDNF), have been discovered. The positive effect on nerve fiber sprouting is similarly unspecific, because BDNF receptor expression has been confirmed on both sympathetic and sensory nerve fibers (Weidler et al. 2005).

Simao and colleagues did not find any correlation between plasma and synovial fluid BDNF concentrations and knee OA severity, although mean plasma BDNF levels of knee OA patients were significantly higher than in plasma of healthy controls (Simao et al. 2014). Instead, BDNF seems to play a role in pain sensitivity in knee OA. Given the fact that BDNF concentration is higher in RA synovium as well (Weidler et al. 2005), an additional function besides stimulation of nerve fiber sprouting would be possible. TH-positive single cells appear in both OA and RA synovial tissue when inflammation starts. BDNF is known for its ability to induce TH as shown, e.g., in dopaminergic differentiating stem cells (Trzaska et al. 2009). Thus, BDNF might contribute to the appearance of TH-positive single cells in inflamed synovium.

9.6 Perspectives

Sensory and sympathetic nerve fibers and their neurotransmitters are important neuronal effectors regulating cartilage physiology and playing decisive roles in musculoskeletal pathophysiologies (as OA and RA). Notably, many resident cells of the osteoarticular system have receptors for sympathetic and sensory neurotransmitters, and, thus, they can respond to these stimuli. Both types of neurotransmitters also modulate embryonic limb growth and postnatal long bone growth by targeting chondrocytes in the growth plate during endochondral ossification. More and more evidence accumulates that neuronal signaling critically influences tissue regeneration not only in joint disorders but also after bone and skin traumata.

In the meantime, many details with respect to sympathetic neuronal effects on synovial tissue inflammatory responses in RA pathogenesis are known. However, how changes in sensory and sympathetic joint innervation and their respective neurotransmitters contribute to abnormal subchondral bone remodeling, cartilage degradation, and osteophyte formation during the pathogenesis of OA is largely unknown yet.

One can speculate that nerve-repellent factors, next to their known axon guidance role, contribute to joint pathophysiology because several studies assign novel functions to semaphorins and other repellent factors in chondrogenic differentiation during embryonic development, cartilage degeneration, and synovial inflammation in adults.

Taken together, it becomes more and more evident that sensory and sympathetic nerve fibers and their neurotransmitters critically influence cartilage, synovial, and other joint tissue function and homeostasis. Doubtless the peripheral nervous system is crucially involved in the pathogenesis of musculoskeletal disorders such as OA, RA, and IVD degeneration.

Acknowledgments The authors are supported by the DFG grants to SG (GR 1301-19) and to ZJL (JE 642/4-1), both subprojects of FOR 2407-1.

References

Ahn SH, Cho YW, Ahn MW, Jang SH, Sohn YK, Kim HS (2002) mRNA expression of cytokines and chemokines in herniated lumbar intervertebral discs. Spine (Phila Pa 1976) 27(9):911–917

Aloe L, Tuveri MA, Carcassi U, Levi-Montalcini R (1992) Nerve growth factor in the synovial fluid of patients with chronic arthritis. Arthritis Rheum 35(3):351–355

Aloe L, Probert L, Kollias G, Bracci-Laudiero L, Spillantini MG, Levi-Montalcini R (1993) The synovium of transgenic arthritic mice expressing human tumor necrosis factor contains a high level of nerve growth factor. Growth Factors 9(2):149–155

Antonioli L, Colucci R, La Motta C, Tuccori M, Awwad O, Da Settimo F, Blandizzi C, Fornai M (2012) Adenosine deaminase in the modulation of immune system and its potential as a novel target for treatment of inflammatory disorders. Curr Drug Targets 13(6):842–862

Ashraf S, Mapp PI, Walsh DA (2011a) Contributions of angiogenesis to inflammation, joint damage, and pain in a rat model of osteoarthritis. Arthritis Rheum 63(9):2700–2710. doi:10.1002/art.30422

Ashraf S, Wibberley H, Mapp PI, Hill R, Wilson D, Walsh DA (2011b) Increased vascular penetration and nerve growth in the meniscus: a potential source of pain in osteoarthritis. Ann Rheum Dis 70(3):523–529. doi:10.1136/ard.2010.137844

Baillie GS, Sood A, McPhee I, Gall I, Perry SJ, Lefkowitz RJ, Houslay MD (2003) Beta-arrestin-mediated PDE4 cAMP phosphodiesterase recruitment regulates beta-adrenoceptor switching from Gs to Gi. Proc Natl Acad Sci U S A 100(3):940–945. doi:10.1073/pnas.262787199

Baldock PA, Sainsbury A, Couzens M, Enriquez RF, Thomas GP, Gardiner EM, Herzog H (2002) Hypothalamic Y2 receptors regulate bone formation. J Clin Invest 109(7):915–921. doi:10.1172/JCI14588

Bedoui S, Miyake S, Straub RH, von Horsten S, Yamamura T (2004) More sympathy for autoimmunity with neuropeptide Y? Trends Immunol 25(10):508–512. doi:10.1016/j.it.2004.08.005

Behar O, Golden JA, Mashimo H, Schoen FJ, Fishman MC (1996) Semaphorin III is needed for normal patterning and growth of nerves, bones and heart. Nature 383(6600):525–528. doi:10.1038/383525a0

Benneker LM, Andersson G, Iatridis JC, Sakai D, Hartl R, Ito K, Grad S (2014) Cell therapy for intervertebral disc repair: advancing cell therapy from bench to clinics. Eur Cell Mater 27:5–11

Besedovsky HO, del Rey A (1996) Immune-neuro-endocrine interactions: facts and hypotheses. Endocr Rev 17(1):64–102

Binch AL, Cole AA, Breakwell LM, Michael AL, Chiverton N, Cross AK, Le Maitre CL (2014) Expression and regulation of neurotrophic and angiogenic factors during human intervertebral disc degeneration. Arthritis Res Ther 16(5):416. doi:10.1186/s13075-014-0416-1

Binch AL, Cole AA, Breakwell LM, Michael AL, Chiverton N, Creemers LB, Cross AK, Le Maitre CL (2015) Nerves are more abundant than blood vessels in the degenerate human intervertebral disc. Arthritis Res Ther 17(1):370. doi:10.1186/s13075-015-0889-6

Bjurholm A, Kreicbergs A, Dahlberg L, Schultzberg M (1990) The occurrence of neuropeptides at different stages of DBM-induced heterotopic bone formation. Bone Miner 10(2):95–107

Bonini S, Rasi G, Bracci-Laudiero ML, Procoli A, Aloe L (2003) Nerve growth factor: neurotrophin or cytokine? Int Arch Allergy Immunol 131(2):80–84, doi:70922

Botelho M, Cavadas C (2015) Neuropeptide Y: an anti-aging player? Trends Neurosci 38(11):701–711. doi:10.1016/j.tins.2015.08.012

Braun HJ, Busfield BT, Kim HJ, Scuderi GJ, Dragoo JL (2013) The effect of local anaesthetics on synoviocytes: a possible indirect mechanism of chondrolysis. Knee Surg Sports Traumatol Arthrosc Official J ESSKA 21(6):1468–1474. doi:10.1007/s00167-012-2104-5

Buma P, Verschuren C, Versleyen D, Van der KP, Oestreicher AB (1992) Calcitonin gene-related peptide, substance P and GAP-43/B-50 immunoreactivity in the normal and arthrotic knee joint of the mouse. Histochemistry 98(5):327–339

Capellino S, Cosentino M, Wolff C, Schmidt M, Grifka J, Straub RH (2010) Catecholamine-producing cells in the synovial tissue during arthritis: modulation of sympathetic neurotransmitters as new therapeutic target. Ann Rheum Dis 69(10):1853–1860. doi:10.1136/ard.2009.119701, ard.2009.119701 [pii]

Capellino S, Weber K, Gelder M, Harle P, Straub RH (2012) First appearance and location of catecholaminergic cells during experimental arthritis and elimination by chemical sympathectomy. Arthritis Rheum 64(4):1110–1118. doi:10.1002/art.33431

Capellino S, Cosentino M, Luini A, Bombelli R, Lowin T, Cutolo M, Marino F, Straub RH (2014) Increased expression of dopamine receptors in synovial fibroblasts from patients with rheumatoid arthritis: inhibitory effects of dopamine on interleukin-8 and interleukin-6. Arthritis Rheumatol 66(10):2685–2693. doi:10.1002/art.38746

Carroll SH, Ravid K (2013) Differentiation of mesenchymal stem cells to osteoblasts and chondrocytes: a focus on adenosine receptors. Expert Rev Mol Med 15:e1. doi:10.1017/erm.2013.2

Chambers MG, Jasani MK, Bitensky L, Chayen J (1992) Chondrocytic monoamine oxidase activity in the development of natural murine osteoarthritis. Int J Exp Pathol 73(2):115–123

Chen Y, Michaelis M, Janig W, Devor M (1996) Adrenoreceptor subtype mediating sympathetic-sensory coupling in injured sensory neurons. J Neurophysiol 76(6):3721–3730

Chubinskaya S, Kawakami M, Rappoport L, Matsumoto T, Migita N, Rueger DC (2007) Anti-catabolic effect of OP-1 in chronically compressed intervertebral discs. J Orthop Res Off Publ Orthop Res Soc 25(4):517–530. doi:10.1002/jor.20339

Cortes A, Gracia E, Moreno E, Mallol J, Lluis C, Canela EI, Casado V (2015) Moonlighting adenosine deaminase: a target protein for drug development. Med Res Rev 35(1):85–125. doi:10.1002/med.21324

Dayer JM, Burger D (2004) Cell-cell interactions and tissue damage in rheumatoid arthritis. Autoimmun Rev 3(Suppl 1):S14–S16

Del Rey A, Wolff C, Wildmann J, Randolf A, Straub RH, Besedovsky HO (2010) When immune-neuro-endocrine interactions are disrupted: experimentally induced arthritis as an example. Neuroimmunomodulation 17(3):165–168. doi:10.1159/000258714

Delgado M, Abad C, Martinez C, Leceta J, Gomariz RP (2001) Vasoactive intestinal peptide prevents experimental arthritis by downregulating both autoimmune and inflammatory components of the disease. Nat Med 7(5):563–568. doi:10.1038/87887

Dimitrijevic M, Stanojevic S, Mitic K, Kustrimovic N, Vujic V, Miletic T, Kovacevic-Jovanovic V (2008) The anti-inflammatory effect of neuropeptide Y (NPY) in rats is dependent on dipeptidyl peptidase 4 (DP4) activity and age. Peptides 29(12):2179–2187. doi:10.1016/j.peptides.2008.08.017

Dirmeier M, Capellino S, Schubert T, Angele P, Anders S, Straub RH (2008) Lower density of synovial nerve fibres positive for calcitonin gene-related peptide relative to substance P in rheumatoid arthritis but not in osteoarthritis. Rheumatology (Oxford) 47(1):36–40. doi:10.1093/rheumatology/kem301, doi:47/1/36 [pii]

Dodge GR, Poole AR (1989) Immunohistochemical detection and immunochemical analysis of type II collagen degradation in human normal, rheumatoid, and osteoarthritic articular cartilages and in explants of bovine articular cartilage cultured with interleukin 1. J Clin Invest 83(2):647–661. doi:10.1172/JCI113929

Dodge GR, Pidoux I, Poole AR (1991) The degradation of type II collagen in rheumatoid arthritis: an immunoelectron microscopic study. Matrix 11(5):330–338

Dong T, Chang H, Zhang F, Chen W, Zhu Y, Wu T, Zhang Y (2015) Calcitonin gene-related peptide can be selected as a predictive biomarker on progression and prognosis of knee osteoarthritis. Int Orthop 39(6):1237–1243. doi:10.1007/s00264-015-2744-4

Dowthwaite GP, Bishop JC, Redman SN, Khan IM, Rooney P, Evans DJ, Haughton L, Bayram Z, Boyer S, Thomson B, Wolfe MS, Archer CW (2004) The surface of articular cartilage contains a progenitor cell population. J Cell Sci 117(Pt 6):889–897

Dygai AM, Khmelevskaya ES, Skurikhin EG, Pershina OV, Andreeva TV, Ermakova NN (2012) Catecholamine regulation of stromal precursors and hemopoietic stem cells in cytostatic myelosuppression. Bull Exp Biol Med 152(6):723–727

Edgar MA (2007) The nerve supply of the lumbar intervertebral disc. J Bone Joint Surg (Br) 89(9):1135–1139. doi:10.1302/0301-620X.89B9.18939

Edoff K, Hildebrand C (2003) Neuropeptide effects on rat chondrocytes and perichondrial cells in vitro. Neuropeptides 37(5):316–318

Edoff K, Grenegard M, Hildebrand C (2000) Retrograde tracing and neuropeptide immunohistochemistry of sensory neurones projecting to the cartilaginous distal femoral epiphysis of young rats. Cell Tissue Res 299(2):193–200

Elefteriou F, Takeda S, Ebihara K, Magre J, Patano N, Kim CA, Ogawa Y, Liu X, Ware SM, Craigen WJ, Robert JJ, Vinson C, Nakao K, Capeau J, Karsenty G (2004) Serum leptin level is a regulator of bone mass. Proc Natl Acad Sci U S A. 2004;101(9):3258–63. Epub 2004 Feb 20.

Elefteriou F (2005) Neuronal signaling and the regulation of bone remodeling. Cell Mol Life Sci CMLS 62(19–20):2339–2349. doi:10.1007/s00018-005-5175-3

Elefteriou F, Ahn JD, Takeda S, Starbuck M, Yang X, Liu X, Kondo H, Richards WG, Bannon TW, Noda M, Clement K, Vaisse C, Karsenty G (2005) Leptin regulation of bone resorption by the sympathetic nervous system and CART. Nature 434(7032):514–520

Elefteriou F, Campbell P, Ma Y (2014) Control of bone remodeling by the peripheral sympathetic nervous system. Calcif Tissue Int 94(1):140–151. doi:10.1007/s00223-013-9752-4

Fassold A, Falk W, Anders S, Hirsch T, Mirsky VM, Straub RH (2009) Soluble neuropilin-2, a nerve repellent receptor, is increased in rheumatoid arthritis synovium and aggravates sympathetic fiber repulsion and arthritis. Arthritis Rheum 60(10):2892–2901. doi:10.1002/art.24860

Ferrell WR, Khoshbaten A (1989) Adrenoceptor profile of blood vessels in the knee joint of the rabbit. J Physiol 414:377–383

Ferrell WR, Khoshbaten A (1990) Responses of blood vessels in the rabbit knee to electrical stimulation of the joint capsule. J Physiol 423:569–578

Fickert S, Fiedler J, Brenner RE (2004) Identification of subpopulations with characteristics of mesenchymal progenitor cells from human osteoarthritic cartilage using triple staining for cell surface markers. Arthritis Res Ther 6(5):R422–R432. doi:10.1186/ar1210

Fortier LA, Nixon AJ (1997) Distributional changes in substance P nociceptive fiber patterns in naturally osteoarthritic articulations. J Rheumatol 24(3):524–530

Freemont AJ, Watkins A, Le Maitre C, Baird P, Jeziorska M, Knight MT, Ross ER, O'Brien JP, Hoyland JA (2002) Nerve growth factor expression and innervation of the painful intervertebral disc. J Pathol 197(3):286–292. doi:10.1002/path.1108

Gajda M, Adriaensen D, Cichocki T (2000) Development of the innervation of long bones: expression of the growth-associated protein 43. Folia Histochem Cytobiol 38(3):103–110

Garcia-Cosamalon J, del Valle ME, Calavia MG, Garcia-Suarez O, Lopez-Muniz A, Otero J, Vega JA (2010) Intervertebral disc, sensory nerves and neurotrophins: who is who in discogenic pain? J Anat 217(1):1–15. doi:10.1111/j.1469-7580.2010.01227.x

Goldring MB, Tsuchimochi K, Ijiri K (2006) The control of chondrogenesis. J Cell Biochem 97(1):33–44

Gomez C, Burt-Pichat B, Mallein-Gerin F, Merle B, Delmas PD, Skerry TM, Vico L, Malaval L, Chenu C (2005) Expression of semaphorin-3A and its receptors in endochondral ossification: potential role in skeletal development and innervation. Dev Dyn Off Publ Am Assoc Anat 234(2):393–403. doi:10.1002/dvdy.20512

Gonzalez-Rey E, Fernandez-Martin A, Chorny A, Delgado M (2006) Vasoactive intestinal peptide induces CD4+, CD25+ T regulatory cells with therapeutic effect in collagen-induced arthritis. Arthritis Rheum 54(3):864–876. doi:10.1002/art.21652

Goto T, Nakao K, Gunjigake KK, Kido MA, Kobayashi S, Tanaka T (2007) Substance P stimulates late-stage rat osteoblastic bone formation through neurokinin-1 receptors. Neuropeptides 41(1):25–31

Grassel SG (2014) The role of peripheral nerve fibers and their neurotransmitters in cartilage and bone physiology and pathophysiology. Arthritis Res Ther 16(6):485

Grogan SP, Miyaki S, Asahara H, D'Lima DD, Lotz MK (2009) Mesenchymal progenitor cell markers in human articular cartilage: normal distribution and changes in osteoarthritis. Arthritis Res Ther 11(3):R85

Gronblad M, Konttinen YT, Korkala O, Liesi P, Hukkanen M, Polak JM (1988) Neuropeptides in synovium of patients with rheumatoid arthritis and osteoarthritis. J Rheumatol 15(12):1807–1810

Gruber HE, Ingram JA, Hoelscher G, Zinchenko N, Norton HJ, Hanley EN Jr (2008) Brain-derived neurotrophic factor and its receptor in the human and the sand rat intervertebral disc. Arthritis Res Ther 10(4):R82. doi:10.1186/ar2456

Hara-Irie F, Amizuka N, Ozawa H (1996) Immunohistochemical and ultrastructural localization of CGRP-positive nerve fibers at the epiphyseal trabecules facing the growth plate of rat femurs. Bone 18(1):29–39

Harle P, Mobius D, Carr DJ, Scholmerich J, Straub RH (2005) An opposing time-dependent immune-modulating effect of the sympathetic nervous system conferred by altering the cytokine profile in the local lymph nodes and spleen of mice with type II collagen-induced arthritis. Arthritis Rheum 52(4):1305–1313

Harle P, Straub RH, Wiest R, Mayer A, Scholmerich J, Atzeni F, Carrabba M, Cutolo M, Sarzi-Puttini P (2006) Increase of sympathetic outflow measured by neuropeptide Y and decrease of the hypothalamic-pituitary-adrenal axis tone in patients with systemic lupus erythematosus and rheumatoid arthritis: another example of uncoupling of response systems. Ann Rheum Dis 65(1):51–56. doi:10.1136/ard.2005.038059

Hartung HP, Toyka KV (1983) Activation of macrophages by substance P: induction of oxidative burst and thromboxane release. Eur J Pharmacol 89(3–4):301–305

Hinck L (2004) The versatile roles of "axon guidance" cues in tissue morphogenesis. Dev Cell 7(6):783–793. doi:10.1016/j.devcel.2004.11.002

Holmes G, Niswander L (2001) Expression of slit-2 and slit-3 during chick development. Dev Dyn Off Publ Am Assoc Anat 222(2):301–307. doi:10.1002/dvdy.1182

Howard MR, Millward-Sadler SJ, Vasilliou AS, Salter DM, Quinn JP (2008) Mechanical stimulation induces preprotachykinin gene expression in osteoarthritic chondrocytes which is correlated with modulation of the transcription factor neuron restrictive silence factor. Neuropeptides 42(5–6):681–686. doi:10.1016/j.npep.2008.09.004

Hukkanen M, Konttinen YT, Rees RG, Santavirta S, Terenghi G, Polak JM (1992) Distribution of nerve endings and sensory neuropeptides in rat synovium, meniscus and bone. Int J Tissue React 14(1):1–10

Im HJ, Li X, Muddasani P, Kim GH, Davis F, Rangan J, Forsyth CB, Ellman M, Thonar EJ (2008) Basic fibroblast growth factor accelerates matrix degradation via a neuro-endocrine pathway in human adult articular chondrocytes. J Cell Physiol 215(2):452–463

Imai S, Tokunaga Y, Konttinen YT, Maeda T, Hukuda S, Santavirta S (1997) Ultrastructure of the synovial sensory peptidergic fibers is distinctively altered in different phases of adjuvant induced arthritis in rats: ultramorphological characterization combined with morphometric and immunohistochemical study for substance P, calcitonin gene related peptide, and protein gene product 9.5. J Rheumatol 24(11):2177–2187

Inoue H, Shimoyama Y, Hirabayashi K, Kajigaya H, Yamamoto S, Oda H, Koshihara Y (2001) Production of neuropeptide substance P by synovial fibroblasts from patients with rheumatoid arthritis and osteoarthritis. Neurosci Lett 303(3):149–152

Jenei-Lanzl Z, Grassel S, Pongratz G, Kees F, Miosge N, Angele P, Straub RH (2014) Norepinephrine inhibits mesenchymal stem cell and chondrogenic progenitor cell chondrogenesis and accelerates chondrogenic hypertrophy. Arthritis Rheumatol. doi:10.1002/art.38695

Jenei-Lanzl Z, Capellino S, Kees F, Fleck M, Lowin T, Straub RH (2015a) Anti-inflammatory effects of cell-based therapy with tyrosine hydroxylase-positive catecholaminergic cells in experimental arthritis. Ann Rheum Dis 74(2):444–451. doi:10.1136/annrheumdis-2013-203925

Jenei-Lanzl Z, Zwingenberg J, Lowin T, Anders S, Straub RH (2015b) Proinflammatory receptor switch from galphas to galphai signaling by beta-arrestin-mediated PDE4 recruitment in mixed RA synovial cells. Brain Behav Immun 50:266–274. doi:10.1016/j.bbi.2015.07.020

Jiang JL, Qiu YH, Peng YP, Wang JJ (2006) Immunoregulatory role of endogenous catecholamines synthesized by immune cells. Sheng Li Xue Bao 58(4):309–317

Jiao K, Niu L, Xu X, Liu Y, Li X, Tay FR, Wang M (2015) Norepinephrine regulates condylar bone loss via comorbid factors. J Dent Res 94(6):813–820. doi:10.1177/0022034515577677

Jikko A, Murakami H, Yan W, Nakashima K, Ohya Y, Satakeda H, Noshiro M, Kawamoto T, Nakamura S, Okada Y, Suzuki F, Kato Y (1996) Effects of cyclic adenosine 3′,5′-monophosphate on chondrocyte terminal differentiation and cartilage-matrix calcification. Endocrinology 137(1):122–128. doi:10.1210/endo.137.1.8536602

Johnson WE, Sivan S, Wright KT, Eisenstein SM, Maroudas A, Roberts S (2006) Human intervertebral disc cells promote nerve growth over substrata of human intervertebral disc aggrecan. Spine (Phila Pa 1976) 31(11):1187–1193. doi:10.1097/01.brs.0000217669.04903.61

Kane D, Jensen LE, Grehan S, Whitehead AS, Bresnihan B, Fitzgerald O (2004) Quantitation of metalloproteinase gene expression in rheumatoid and psoriatic arthritis synovial tissue distal and proximal to the cartilage-pannus junction. J Rheumatol 31(7):1274–1280

Karahan S, Kincaid SA, Baird AN, Kammermann JR (2002) Distribution of beta-endorphin and substance P in the shoulder joint of the dog before and after a low impact exercise programme. Anat Histol Embryol 31(2):72–77

Katayama Y, Battista M, Kao WM, Hidalgo A, Peired AJ, Thomas SA, Frenette PS (2006) Signals from the sympathetic nervous system regulate hematopoietic stem cell egress from bone marrow. Cell 124(2):407–421. doi:10.1016/j.cell.2005.10.041

Kepler CK, Markova DZ, Hilibrand AS, Vaccaro AR, Risbud MV, Albert TJ, Anderson DG (2013) Substance P stimulates production of inflammatory cytokines in human disc cells. Spine (Phila Pa 1976) 38(21):E1291–E1299. doi:10.1097/BRS.0b013e3182a42bc2

Kepler CK, Markova DZ, Koerner JD, Mendelis J, Chen CM, Vaccaro AR, Risbud MV, Albert TJ, Anderson DG (2015) Substance P receptor antagonist suppresses inflammatory cytokine expression in human disc cells. Spine (Phila Pa 1976) 40(16):1261–1269. doi:10.1097/BRS.0000000000000954

Keyszer G, Langer T, Kornhuber M, Taute B, Horneff G (2004) Neurovascular mechanisms as a possible cause of remission of rheumatoid arthritis in hemiparetic limbs. Ann Rheum Dis 63(10):1349–1351. doi:10.1136/ard.2003.016410

Kimberly CL, Byers MR (1988) Inflammation of rat molar pulp and periodontium causes increased calcitonin gene-related peptide and axonal sprouting. Anat Rec 222(3):289–300. doi:10.1002/ar.1092220310

Kitsukawa T, Shimono A, Kawakami A, Kondoh H, Fujisawa H (1995) Overexpression of a membrane protein, neuropilin, in chimeric mice causes anomalies in the cardiovascular system, nervous system and limbs. Development 121(12):4309–4318

Knedla A, Neumann E, Muller-Ladner U (2007) Developments in the synovial biology field 2006. Arthritis Res Ther 9(2):209. doi:10.1186/ar2140

Koeck FX, Bobrik V, Fassold A, Grifka J, Kessler S, Straub RH (2009) Marked loss of sympathetic nerve fibers in chronic Charcot foot of diabetic origin compared to ankle joint osteoarthritis. J Orthop Res 27(6):736–741. doi:10.1002/jor.20807

Kondo M, Yamaoka K, Tanaka Y (2014) Acquiring chondrocyte phenotype from human mesenchymal stem cells under inflammatory conditions. Int J Mol Sci 15(11):21270–21285. doi:10.3390/ijms151121270

Koolpe M, Pearson D, Benton HP (1999) Expression of both P1 and P2 purine receptor genes by human articular chondrocytes and profile of ligand-mediated prostaglandin E2 release. Arthritis Rheum 42(2):258–267. doi:10.1002/1529-0131(199902)42:2<258::AID-ANR7>3.0.CO;2-O

Lai LP, Mitchell J (2008) Beta2-adrenergic receptors expressed on murine chondrocytes stimulate cellular growth and inhibit the expression of Indian hedgehog and collagen type X. J Cell Biochem 104(2):545–553

Larsson J, Ekblom A, Henriksson K, Lundeberg T, Theodorsson E (1991) Concentration of substance P, neurokinin A, calcitonin gene-related peptide, neuropeptide Y and vasoactive intestinal polypeptide in synovial fluid from knee joints in patients suffering from rheumatoid arthritis. Scand J Rheumatol 20(5):326–335

Lee S, Moon CS, Sul D, Lee J, Bae M, Hong Y, Lee M, Choi S, Derby R, Kim BJ, Kim J, Yoon JS, Wolfer L, Kim J, Wang J, Hwang SW, Lee SH (2009) Comparison of growth factor and cytokine expression in patients with degenerated disc disease and herniated nucleus pulposus. Clin Biochem 42(15):1504–1511. doi:10.1016/j.clinbiochem.2009.06.017

Lerner UH, Persson E (2008) Osteotropic effects by the neuropeptides calcitonin gene-related peptide, substance P and vasoactive intestinal peptide. J Musculoskelet Neuronal Interact 8(2):154–165

Levi-Montalcini R, Skaper SD, Dal Toso R, Petrelli L, Leon A (1996) Nerve growth factor: from neurotrophin to neurokine. Trends Neurosci 19(11):514–520. doi:10.1016/S0166-2236(96)10058-8

Levine JD, Goetzl EJ, Basbaum AI (1987) Contribution of the nervous system to the pathophysiology of rheumatoid arthritis and other polyarthritides. Rheum Dis Clin N Am 13(2):369–383

Li W, Long X, Jiang S, Li Y, Fang W (2015) Histamine and substance P in synovial fluid of patients with temporomandibular disorders. J Oral Rehabil 42(5):363–369. doi:10.1111/joor.12265

Loeser RF (2003) Systemic and local regulation of articular cartilage metabolism: where does leptin fit in the puzzle? Arthritis Rheum 48(11):3009–3012. doi:10.1002/art.11315

Lorenz J, Schafer N, Bauer R, Jenei-Lanzl Z, Springorum RH, Grassel S (2016) Norepinephrine modulates osteoarthritic chondrocyte metabolism and inflammatory responses. Osteoarthr Cartil OARS Osteoarthr Res Soc 24(2):325–334. doi:10.1016/j.joca.2015.08.007

Lorton D, Bellinger DL, Schaller JA, Shewmaker E, Osredkar T, Lubahn C (2013) Altered sympathetic-to-immune cell signaling via beta(2)-adrenergic receptors in adjuvant arthritis. Clin Dev Immunol 2013:764395. doi:10.1155/2013/764395

Lotz M, Carson DA, Vaughan JH (1987) Substance P activation of rheumatoid synoviocytes: neural pathway in pathogenesis of arthritis. Science 235(4791):893–895
Lotz M, Vaughan JH, Carson DA (1988) Effect of neuropeptides on production of inflammatory cytokines by human monocytes. Science 241(4870):1218–1221
Lubahn CL, Schaller JA, Bellinger DL, Sweeney S, Lorton D (2004) The importance of timing of adrenergic drug delivery in relation to the induction and onset of adjuvant-induced arthritis. Brain Behav Immun 18(6):563–571. doi:10.1016/j.bbi.2004.02.004
Lundberg P, Allison SJ, Lee NJ, Baldock PA, Brouard N, Rost S, Enriquez RF, Sainsbury A, Lamghari M, Simmons P, Eisman JA, Gardiner EM, Herzog H (2007) Greater bone formation of Y2 knockout mice is associated with increased osteoprogenitor numbers and altered Y1 receptor expression. J Biol Chem 282(26):19082–19091. doi:10.1074/jbc.M609629200
Madsen JE, Hukkanen M, Aune AK, Basran I, Moller JF, Polak JM, Nordsletten L (1998) Fracture healing and callus innervation after peripheral nerve resection in rats. Clin Orthop Relat Res 351:230–240
Maestroni GJ (2000) Neurohormones and catecholamines as functional components of the bone marrow microenvironment. Ann N Y Acad Sci 917:29–37
Maestroni GJ, Cosentino M, Marino F, Togni M, Conti A, Lecchini S, Frigo G (1998) Neural and endogenous catecholamines in the bone marrow. Circadian association of norepinephrine with hematopoiesis? Exp Hematol 26(12):1172–1177
Maggi CA (1995) The mammalian tachykinin receptors. Gen Pharmacol 26(5):911–944
Malmberg AB, Chen C, Tonegawa S, Basbaum AI (1997) Preserved acute pain and reduced neuropathic pain in mice lacking PKCgamma. Science 278(5336):279–283
Mantyh PW, Rogers SD, Honore P, Allen BJ, Ghilardi JR, Li J, Daughters RS, Lappi DA, Wiley RG, Simone DA (1997) Inhibition of hyperalgesia by ablation of lamina I spinal neurons expressing the substance P receptor. Science 278(5336):275–279
Mapp PI, Avery PS, McWilliams DF, Bowyer J, Day C, Moores S, Webster R, Walsh DA (2008) Angiogenesis in two animal models of osteoarthritis. Osteoarthritis Cartilage 16(1):61–9. Epub 2007 Jul 30.
Mapp PI, Kidd BL, Gibson SJ, Terry JM, Revell PA, Ibrahim NB, Blake DR, Polak JM (1990) Substance P-, calcitonin gene-related peptide- and C-flanking peptide of neuropeptide Y-immunoreactive fibres are present in normal synovium but depleted in patients with rheumatoid arthritis. Neuroscience 37(1):143–153
Mapp PI, Walsh DA (2012) Mechanisms and targets of angiogenesis and nerve growth in osteoarthritis. Nat Rev Rheumatol 8(7):390–8. doi:10.1038/nrrheum.2012.80. Review.
Matthews PJ, Aziz Q, Facer P, Davis JB, Thompson DG, Anand P (2004) Increased capsaicin receptor TRPV1 nerve fibres in the inflamed human oesophagus. Eur J Gastroenterol Hepatol 16(9):897–902
Mauro LJ, Wenzel SJ, Sindberg GM (2010) Regulation of chick bone growth by leptin and catecholamines. Poult Sci 89(4):697–708
Mediero A, Cronstein BN (2013) Adenosine and bone metabolism. Trends Endocrinol Metab 24(6):290–300. doi:10.1016/j.tem.2013.02.001
Meyer LH, Franssen L, Pap T (2006) The role of mesenchymal cells in the pathophysiology of inflammatory arthritis. Best Pract Res Clin Rheumatol 20(5):969–981. doi:10.1016/j.berh.2006.06.005
Mignini F, Streccioni V, Amenta F (2003) Autonomic innervation of immune organs and neuroimmune modulation. Auton Autacoid Pharmacol 23(1):1–25
Miller LE, Justen HP, Scholmerich J, Straub RH (2000) The loss of sympathetic nerve fibers in the synovial tissue of patients with rheumatoid arthritis is accompanied by increased norepinephrine release from synovial macrophages. FASEB J 14(13):2097–2107
Miller LE, Grifka J, Scholmerich J, Straub RH (2002) Norepinephrine from synovial tyrosine hydroxylase positive cells is a strong indicator of synovial inflammation in rheumatoid arthritis. J Rheumatol 29(3):427–435
Miller LE, Weidler C, Falk W, Angele P, Schaumburger J, Scholmerich J, Straub RH (2004) Increased prevalence of semaphorin 3C, a repellent of sympathetic nerve fibers, in the synovial

tissue of patients with rheumatoid arthritis. Arthritis Rheum 50(4):1156–1163. doi:10.1002/art.20110

Millward-Sadler SJ, Wright MO, Davies LW, Nuki G, Salter DM (2000) Mechanotransduction via integrins and interleukin-4 results in altered aggrecan and matrix metalloproteinase 3 gene expression in normal, but not osteoarthritic, human articular chondrocytes. Arthritis Rheum 43(9):2091–2099. doi:10.1002/1529-0131(200009)43:9<2091::AID-ANR21>3.0.CO;2-C

Millward-Sadler SJ, Mackenzie A, Wright MO, Lee HS, Elliot K, Gerrard L, Fiskerstrand CE, Salter DM, Quinn JP (2003) Tachykinin expression in cartilage and function in human articular chondrocyte mechanotransduction. Arthritis Rheum 48(1):146–156

Mistry D, Chambers MG, Mason RM (2006) The role of adenosine in chondrocyte death in murine osteoarthritis and in a murine chondrocyte cell line. Osteoarthr Cartil OARS Osteoarthr Res Soc 14(5):486–495. doi:10.1016/j.joca.2005.11.015

Mitchell NS, Shepard N (1978) Changes in proteoglycan and collagen in cartilage in rheumatoid arthritis. J Bone Joint Surg Am 60(3):342–348

Mitchell J, Lai LP, Peralta F, Xu Y, Sugamori K (2011) Beta2-adrenergic receptors inhibit the expression of collagen type II in growth plate chondrocytes by stimulating the AP-1 factor Jun-B. Am J Physiol Endocrinol Metab 300(4):E633–E639

Molinoff PB (1984) Alpha- and beta-adrenergic receptor subtypes properties, distribution and regulation. Drugs 28(Suppl 2):1–15

Murakami K, Nakagawa H, Nishimura K, Matsuo S (2015) Changes in peptidergic fiber density in the synovium of mice with collagenase-induced acute arthritis. Can J Physiol Pharmacol 93(6):435–441. doi:10.1139/cjpp-2014-0446

Muschter D, Gottl C, Vogel M, Grifka J, Straub RH, Grassel S (2015a) Reactivity of rat bone marrow-derived macrophages to neurotransmitter stimulation in the context of collagen II-induced arthritis. Arthritis Res Ther 17(1):169. doi:10.1186/s13075-015-0684-4

Muschter D, Schafer N, Stangl H, Straub RH, Grassel S (2015b) Sympathetic neurotransmitters modulate osteoclastogenesis and osteoclast activity in the context of collagen-induced arthritis. PLoS ONE 10(10):e0139726. doi:10.1371/journal.pone.0139726

Nakasa T, Ishikawa M, Takada T, Miyaki S, Ochi M (2015) Attenuation of cartilage degeneration by calcitonin gene-related paptide receptor antagonist via inhibition of subchondral bone sclerosis in osteoarthritis mice. J Orthop Res Off Publ Orthop Res Soc. doi:10.1002/jor.23132

Napolitano A, Cesura AM, Da Prada M (1995) The role of monoamine oxidase and catechol O-methyltransferase in dopaminergic neurotransmission. J Neural Transm Suppl 45:35–45

Naukkarinen A, Nickoloff BJ, Farber EM (1989) Quantification of cutaneous sensory nerves and their substance P content in psoriasis. J Invest Dermatol 92(1):126–129

Niissalo S, Hukkanen M, Imai S, Tornwall J, Konttinen YT (2002) Neuropeptides in experimental and degenerative arthritis. Ann N Y Acad Sci 966:384–399

O'Sullivan J, D'Arcy S, Barry FP, Murphy JM, Coleman CM (2011) Mesenchymal chondroprogenitor cell origin and therapeutic potential. Stem Cell Res Ther 2(1):8. doi:10.1186/scrt49

Ohtori S, Inoue G, Koshi T, Ito T, Doya H, Moriya H, Takahashi K (2006) Substance P-saporin down-regulates substance P receptor immunoreactive sensory dorsal root ganglion neurons innervating the lumbar intervertebral discs in rats. Spine (Phila Pa 1976) 31(26):2987–2991. doi:10.1097/01.brs.0000250306.12996.fa

Okubo M, Kimura T, Fujita Y, Mochizuki S, Niki Y, Enomoto H, Suda Y, Toyama Y, Okada Y (2011) Semaphorin 3A is expressed in human osteoarthritic cartilage and antagonizes vascular endothelial growth factor 165-promoted chondrocyte migration: an implication for chondrocyte cloning. Arthritis Rheum 63(10):3000–3009. doi:10.1002/art.30482

Oliva F, Tarantino U, Maffulli N (2005) Immunohistochemical localization of calcitonin gene-related peptide and substance P in the rat knee cartilage at birth. Physiol Res 54(5):549–556

Opolka A, Straub RH, Pasoldt A, Grifka J, Grassel S (2012) Substance P and norepinephrine modulate murine chondrocyte proliferation and apoptosis. Arthritis Rheum 64(3):729–739

Otero M, Goldring MB (2007) Cells of the synovium in rheumatoid arthritis. Chondrocytes. Arthritis Res Ther 9(5):220. doi:10.1186/ar2292

Otero M, Lago R, Gomez R, Dieguez C, Lago F, Gomez-Reino J, Gualillo O (2006) Towards a pro-inflammatory and immunomodulatory emerging role of leptin. Rheumatology (Oxford) 45(8):944–950
Pongratz G, Straub RH (2013) Role of peripheral nerve fibres in acute and chronic inflammation in arthritis. Nat Rev Rheumatol 9(2):117–126. doi:10.1038/nrrheum.2012.181
Pongratz G, Straub RH (2014) The sympathetic nervous response in inflammation. Arthritis Res Ther 16(6):504
Popa C, Netea MG, Radstake TR, van Riel PL, Barrera P, van der Meer JW (2005) Markers of inflammation are negatively correlated with serum leptin in rheumatoid arthritis. Ann Rheum Dis 64(8):1195–1198
Poyner DR, Sexton PM, Marshall I, Smith DM, Quirion R, Born W, Muff R, Fischer JA, Foord SM (2002) International Union of Pharmacology. XXXII. The mammalian calcitonin gene-related peptides, adrenomedullin, amylin, and calcitonin receptors. Pharmacol Rev 54(2):233–246
Pretzel D, Linss S, Rochler S, Endres M, Kaps C, Alsalameh S, Kinne RW (2011) Relative percentage and zonal distribution of mesenchymal progenitor cells in human osteoarthritic and normal cartilage. Arthritis Res Ther 13(2):R64. doi:10.1186/ar3320
Purmessur D, Freemont AJ, Hoyland JA (2008) Expression and regulation of neurotrophins in the nondegenerate and degenerate human intervertebral disc. Arthritis Res Ther 10(4):R99. doi:10.1186/ar2487
Raj PP (2008) Intervertebral disc: anatomy-physiology-pathophysiology-treatment. Pain Pract 8(1):18–44. doi:10.1111/j.1533-2500.2007.00171.x
Rao AJ, Johnston TR, Harris AH, Smith RL, Costouros JG (2014) Inhibition of chondrocyte and synovial cell death after exposure to commonly used anesthetics: chondrocyte apoptosis after anesthetics. Am J Sports Med 42(1):50–58. doi:10.1177/0363546513507426
Ribatti D, Conconi MT, Nussdorfer GG (2007) Nonclassic endogenous novel [corrected] regulators of angiogenesis. Pharmacol Rev 59(2):185–205. doi:10.1124/pr.59.2.3
Richardson SM, Doyle P, Minogue BM, Gnanalingham K, Hoyland JA (2009) Increased expression of matrix metalloproteinase-10, nerve growth factor and substance P in the painful degenerate intervertebral disc. Arthritis Res Ther 11(4):R126. doi:10.1186/ar2793
Roemer FW, Guermazi A, Felson DT, Niu J, Nevitt MC, Crema MD, Lynch JA, Lewis CE, Torner J, Zhang Y (2011) Presence of MRI-detected joint effusion and synovitis increases the risk of cartilage loss in knees without osteoarthritis at 30-month follow-up: the MOST study. Ann Rheum Dis 70(10):1804–1809. doi:10.1136/ard.2011.150243
Romio M, Reinbeck B, Bongardt S, Huls S, Burghoff S, Schrader J (2011) Extracellular purine metabolism and signaling of CD73-derived adenosine in murine treg and teff cells. Am J Physiol Cell Physiol 301(2):C530–C539. doi:10.1152/ajpcell.00385.2010
Rosenfeld MG, Mermod JJ, Amara SG, Swanson LW, Sawchenko PE, Rivier J, Vale WW, Evans RM (1983) Production of a novel neuropeptide encoded by the calcitonin gene via tissue-specific RNA processing. Nature 304(5922):129–135
Ruff MR, Wahl SM, Pert CB (1985) Substance P receptor-mediated chemotaxis of human monocytes. Peptides 6(Suppl 2):107–111
Saba F, Soleimani M, Atashi A, Mortaz E, Shahjahani M, Roshandel E, Jaseb K, Saki N (2013) The role of the nervous system in hematopoietic stem cell mobilization. Lab Hematol 19(3):8–16. doi:10.1532/LH96.12013
Saito T (1992) Immunohistochemical study on the effect of high tibial osteotomy on the distribution pattern of neuropeptides in the synovium of the osteoarthritic knee. Nihon Seikeigeka Gakkai Zasshi 66(9):884–897
Saito T, Koshino T (2000) Distribution of neuropeptides in synovium of the knee with osteoarthritis. Clin Orthop Relat Res 376:172–182
Salo PT, Seeratten RA, Erwin WM, Bray RC (2002) Evidence for a neuropathic contribution to the development of spontaneous knee osteoarthrosis in a mouse model. Acta Orthop Scand 73(1):77–84

Sanders VM, Straub RH (2002) Norepinephrine, the beta-adrenergic receptor, and immunity. Brain Behav Immun 16(4):290–332

Sariola H (2001) The neurotrophic factors in non-neuronal tissues. Cell Mol Life Sci CMLS 58(8):1061–1066

Saxler G, Loer F, Skumavc M, Pfortner J, Hanesch U (2007) Localization of SP- and CGRP-immunopositive nerve fibers in the hip joint of patients with painful osteoarthritis and of patients with painless failed total hip arthroplasties. Eur J Pain 11(1):67–74. doi:10.1016/j.ejpain.2005.12.011

Schaible HG (2012) Mechanisms of chronic pain in osteoarthritis. Curr Rheumatol Rep 14(6):549–556. doi:10.1007/s11926-012-0279-x

Schaible HG, Straub RH (2014) Function of the sympathetic supply in acute and chronic experimental joint inflammation. Auton Neurosci 182:55–64. doi:10.1016/j.autneu.2013.12.004

Schinke T, Liese S, Priemel M, Haberland M, Schilling AF, Catala-Lehnen P, Blicharski D, Rueger JM, Gagel RF, Emeson RB, Amling M (2004) Decreased bone formation and osteopenia in mice lacking alpha-calcitonin gene-related peptide. J Bone Miner Res 19(12):2049–2056

Schubert T, Denk A, Magdefrau U, Kaufmann S, Bastone P, Lowin T, Schedel J, Bosserhoff AK (2009a) Role of the netrin system of repellent factors on synovial fibroblasts in rheumatoid arthritis and osteoarthritis. Int J Immunopathol Pharmacol 22(3):715–722

Schubert T, Kaufmann S, Wenke AK, Grassel S, Bosserhoff AK (2009b) Role of deleted in colon carcinoma in osteoarthritis and in chondrocyte migration. Rheumatology 48(11):1435–1441. doi:10.1093/rheumatology/kep245

Schwab W, Funk RH (1998) Innervation pattern of different cartilaginous tissues in the rat. Acta Anat (Basel) 163(4):184–190

Serafini T, Kennedy TE, Galko MJ, Mirzayan C, Jessell TM, Tessier-Lavigne M (1994) The netrins define a family of axon outgrowth-promoting proteins homologous to C. elegans UNC-6. Cell 78(3):409–424

Silva WA Jr, Covas DT, Panepucci RA, Proto-Siqueira R, Siufi JL, Zanette DL, Santos AR, Zago MA (2003) The profile of gene expression of human marrow mesenchymal stem cells. Stem Cells 21(6):661–669. doi:10.1634/stemcells.21-6-661

Simao AP, Mendonca VA, de Oliveira Almeida TM, Santos SA, Gomes WF, Coimbra CC, Lacerda AC (2014) Involvement of BDNF in knee osteoarthritis: the relationship with inflammation and clinical parameters. Rheumatol Int 34(8):1153–1157. doi:10.1007/s00296-013-2943-5

Sipos G, Sipos P, Altdorfer K, Pongor E, Feher E (2008) Correlation and immunolocalization of substance P nerve fibers and activated immune cells in human chronic gastritis. Anat Rec (Hoboken) 291(9):1140–1148. doi:10.1002/ar.20737

Sitkovsky MV (2003) Use of the A(2A) adenosine receptor as a physiological immunosuppressor and to engineer inflammation in vivo. Biochem Pharmacol 65(4):493–501

Smith CH, Barker JN, Morris RW, MacDonald DM, Lee TH (1993) Neuropeptides induce rapid expression of endothelial cell adhesion molecules and elicit granulocytic infiltration in human skin. J Immunol 151(6):3274–3282

Stangenberg L, Burzyn D, Binstadt BA, Weissleder R, Mahmood U, Benoist C, Mathis D (2014) Denervation protects limbs from inflammatory arthritis via an impact on the microvasculature. Proc Natl Acad Sci U S A 111(31):11419–11424. doi:10.1073/pnas.1410854111

Stanojevic S, Dimitrijevic M, Kustrimovic N, Mitic K, Vujic V, Leposavic G (2013) Adrenal hormone deprivation affects macrophage catecholamine metabolism and beta2-adrenoceptor density, but not propranolol stimulation of tumour necrosis factor-alpha production. Exp Physiol 98(3):665–678. doi:10.1113/expphysiol.2012.070524

Steele PA, Gibbins IL, Morris JL (1996) Projections of intrinsic cardiac neurons to different targets in the guinea-pig heart. J Auton Nerv Syst 56(3):191–200

Stolk L, van Meurs JB, Jhamai M, Arp PP, van Leeuwen JP, Hofman A, de Jong FH, Pols HA, Uitterlinden AG (2007) The catechol-O-methyltransferase Met158 low-activity allele and association with nonvertebral fracture risk in elderly men. J Clin Endocrinol Metab 92(8):3206–3212. doi:10.1210/jc.2006-2136

Strange-Vognsen HH, Laursen H (1997) Nerves in human epiphyseal uncalcified cartilage. J Pediatr Orthop B 6(1):56–58

Straub RH (2015) The origin of chronic inflammatory systemic diseases and their sequelae. San Diego Elsevier-Academic Press

Straub RH, Mayer M, Kreutz M, Leeb S, Scholmerich J, Falk W (2000) Neurotransmitters of the sympathetic nerve terminal are powerful chemoattractants for monocytes. J Leukoc Biol 67(4):553–558

Straub RH, Grum F, Strauch U, Capellino S, Bataille F, Bleich A, Falk W, Scholmerich J, Obermeier F (2008) Anti-inflammatory role of sympathetic nerves in chronic intestinal inflammation. Gut 57(7):911–921. doi:10.1136/gut.2007.125401

Straub RH, Bijlsma JW, Masi A, Cutolo M (2013) Role of neuroendocrine and neuroimmune mechanisms in chronic inflammatory rheumatic diseases – the 10-year update. Semin Arthritis Rheum 43(3):392–404. doi:10.1016/j.semarthrit.2013.04.008

Stucky CL, Lewin GR (1999) Isolectin B(4)-positive and -negative nociceptors are functionally distinct. J Neurosci Off J Soc Neurosci 19(15):6497–6505

Sun Y, Huang J, Xiang Y, Bastepe M, Juppner H, Kobilka BK, Zhang JJ, Huang XY (2007) Dosage-dependent switch from G protein-coupled to G protein-independent signaling by a GPCR. EMBO J 26(1):53–64. doi:10.1038/sj.emboj.7601502

Suri S, Gill SE, Massena DC, Wilson D, McWilliams DF, Walsh DA (2007) Neurovascular invasion at the osteochondral junction and in osteophytes in osteoarthritis. Ann Rheum Dis 66(11):1423–1428

Takahashi Y, Ohtori S, Takahashi K (2009) Peripheral nerve pathways of afferent fibers innervating the lumbar spine in rats. J Pain 10(4):416–425. doi:10.1016/j.jpain.2008.10.012

Takahata Y, Takarada T, Iemata M, Yamamoto T, Nakamura Y, Kodama A, Yoneda Y (2009) Functional expression of beta2 adrenergic receptors responsible for protection against oxidative stress through promotion of glutathione synthesis after Nrf2 upregulation in undifferentiated mesenchymal C3H10T1/2 stem cells. J Cell Physiol 218(2):268–275

Takao T, Iwaki T, Kondo J, Hiraki Y (2000) Immunohistochemistry of chondromodulin-I in the human intervertebral discs with special reference to the degenerative changes. Histochem J 32(9):545–550

Takarada T, Hojo H, Iemata M, Sahara K, Kodama A, Nakamura N, Hinoi E, Yoneda Y (2009) Interference by adrenaline with chondrogenic differentiation through suppression of gene transactivation mediated by Sox9 family members. Bone 45(3):568–578

Takeda S, Elefteriou F, Levasseur R, Liu X, Zhao L, Parker KL, Armstrong D, Ducy P, Karsenty G (2002) Leptin regulates bone formation via the sympathetic nervous system. Cell 111(3):305–17.

Takebayashi T, Cavanaugh JM, Kallakuri S, Chen C, Yamashita T (2006) Sympathetic afferent units from lumbar intervertebral discs. J Bone Joint Surg (Br) 88(4):554–557. doi:10.1302/0301-620X.88B4.17194

Tanabe T, Otani H, Mishima K, Ogawa R, Inagaki C (1996) Mechanisms of oxyradical production in substance P stimulated rheumatoid synovial cells. Rheumatol Int 16(4):159–167

Tessarollo L (1998) Pleiotropic functions of neurotrophins in development. Cytokine Growth Factor Rev 9(2):125–137

Tolofari SK, Richardson SM, Freemont AJ, Hoyland JA (2010) Expression of semaphorin 3A and its receptors in the human intervertebral disc: potential role in regulating neural ingrowth in the degenerate intervertebral disc. Arthritis Res Ther 12(1):R1. doi:10.1186/ar2898

Trzaska KA, King CC, Li KY, Kuzhikandathil EV, Nowycky MC, Ye JH, Rameshwar P (2009) Brain-derived neurotrophic factor facilitates maturation of mesenchymal stem cell-derived dopamine progenitors to functional neurons. J Neurochem 110(3):1058–1069. doi:10.1111/j.1471-4159.2009.06201.x

Van Linden A, Eltzschig HK (2007) Role of pulmonary adenosine during hypoxia: extracellular generation, signaling and metabolism by surface adenosine deaminase/CD26. Expert Opin Biol Ther 7(9):1437–1447. doi:10.1517/14712598.7.9.1437

Vega JA, Garcia-Suarez O, Hannestad J, Perez-Perez M, Germana A (2003) Neurotrophins and the immune system. J Anat 203(1):1–19

Vergroesen PP, Kingma I, Emanuel KS, Hoogendoorn RJ, Welting TJ, van Royen BJ, van Dieen JH, Smit TH (2015) Mechanics and biology in intervertebral disc degeneration: a vicious circle. Osteoarthr Cartil OARS Osteoarthr Res Soc 23(7):1057–1070. doi:10.1016/j.joca.2015.03.028

Vignon E, Broquet P, Mathieu P, Louisot P, Richard M (1990) Histaminergic H1, serotoninergic, beta adrenergic and dopaminergic receptors in human osteoarthritic cartilage. Biochem Int. 1990;20(2):251–5.

Wada N, Tanaka H, Ide H, Nohno T (2003) Ephrin-A2 regulates position-specific cell affinity and is involved in cartilage morphogenesis in the chick limb bud. Dev Biol 264(2):550–563

Walsh DA, McWilliams DF, Turley MJ, Dixon MR, Franses RE, Mapp PI, Wilson D (2010) Angiogenesis and nerve growth factor at the osteochondral junction in rheumatoid arthritis and osteoarthritis. Rheumatology 49(10):1852–1861. doi:10.1093/rheumatology/keq188

Wang L, Zhang L, Pan H, Peng S, Lv M, Lu WW (2014) Levels of neuropeptide Y in synovial fluid relate to pain in patients with knee osteoarthritis. BMC Musculoskelet Disord 15:319. doi:10.1186/1471-2474-15-319

Wang H, Zhang X, He JY, Zheng XF, Li D, Li Z, Zhu JF, Shen C, Cai GQ, Chen XD (2015) Increasing expression of substance P and calcitonin gene-related peptide in synovial tissue and fluid contribute to the progress of arthritis in developmental dysplasia of the hip. Arthritis Res Ther 17:4. doi:10.1186/s13075-014-0513-1

Weidler C, Holzer C, Harbuz M, Hofbauer R, Angele P, Scholmerich J, Straub RH (2005) Low density of sympathetic nerve fibres and increased density of brain derived neurotrophic factor positive cells in RA synovium. Ann Rheum Dis 64(1):13–20

Wheway J, Herzog H, Mackay F (2007) NPY and receptors in immune and inflammatory diseases. Curr Top Med Chem 7(17):1743–1752

Williams R, Khan IM, Richardson K, Nelson L, McCarthy HE, Analbelsi T, Singhrao SK, Dowthwaite GP, Jones RE, Baird DM, Lewis H, Roberts S, Shaw HM, Dudhia J, Fairclough J, Briggs T, Archer CW (2010) Identification and clonal characterisation of a progenitor cell sub-population in normal human articular cartilage. PLoS ONE 5(10):e13246. doi:10.1371/journal.pone.0013246

Wojtys EM, Beaman DN, Glover RA, Janda D (1990) Innervation of the human knee joint by substance-P fibers. Arthroscopy 6(4):254–263

Zhang Y, Lu L, Furlonger C, Wu GE, Paige CJ (2000) Hemokinin is a hematopoietic-specific tachykinin that regulates B lymphopoiesis. Nat Immunol 1(5):392–397. doi:10.1038/80826

Intervertebral Disc Degeneration

10

Akansha M. Shah, Sarah Yoon Ji Kwon, Wilson C.W. Chan, and Danny Chan

Abstract

The human spine consists of 23 intervertebral discs adjoining the vertebral bodies. These discs provide mechanical support and spinal motion and function to distribute loads from daily activities. Tissues of intervertebral discs share similarities to those of diarthrodial joints, such as a thin layer of cartilage that lines the interface between the joint and the bony elements and a central space rich in extracellular matrix molecules that promotes lubrication and maintains osmotic pressure. Like the pathophysiology of other cartilaginous joints, intervertebral discs undergo biomechanical and structural changes as a result of aging and mechanical insults. Due to higher mechanical loading, lumbar discs are more susceptible to degeneration, which can lead to symptomatic outcomes such as low back pain, sciatica, and other physical disabilities. These affect quality of life as we age and present a significant burden to the healthcare system globally. The etiology of intervertebral disc degeneration is not fully understood, but is a consequence of the changing structure and environment of its three interconnecting components, the nucleus pulposus, the annulus fibrosus, and the cartilaginous endplate, which function cooperatively to transmit load and regulate the cellular, biochemical, and nutritional properties of the disc. Abnormal changes in one or more of these disc compartments will compromise disc integrity and culminate in a degenerated disc state. Mechanistic insights into disc pathology can be gained through an understanding of the development and homeostasis of the intervertebral disc tissues. This chapter will provide an overview of the intervertebral disc in health and disease.

A.M. Shah, PhD • S.Y.J. Kwon, PhD • W.C.W. Chan, PhD • D. Chan, PhD (✉)
School of Biomedical Sciences, The University of Hong Kong, Pokfulam, Hong Kong
e-mail: aks.shah93@gmail.com; isyoonjis@gmail.com; cwilson@hku.hk; chand@hku.hk

S. Grässel, A. Aszódi (eds.), *Cartilage*, DOI 10.1007/978-3-319-45803-8_10

10.1 Evolution of the Vertebral Column and the Intervertebral Disc

Formation of the vertebral column, a metameric structure of vertebral bones and intervening intervertebral discs, marks a critical event in evolution. This was preceded by the presence of a rodlike midline structure (the notochord) in animals that lack a backbone. The lifelong notochord alone offers rigidity to the organism, protects the overlaying spinal cord, and enables the formation of an extended body. In higher organisms, the notochord exists only transiently in development as a primitive axial skeleton until the formation of the spinal elements (Stemple 2005). In the process, it functions as a signaling center, providing molecular signals that position and specify fates of surrounding tissues (Pourquie et al. 1993). This progression in evolution provides biological advantages that include the vertebral bones as attachment sites to other axial and appendicular bones and muscles and the intervertebral discs that allow flexibility and dissipate hydrodynamic loads along the spine.

10.2 Making of the Intervertebral Disc Tissues

While the nucleus pulposus is derived from the notochord, the annulus fibrosus and cartilaginous vertebrae arise from the sclerotome (Christ et al. 2000). Genetic studies of murine development and advances in time-lapse live imaging demonstrate that the three regions of the notochord along the anterior-posterior axis develop from the organizer tissue differently. In advancing notochord development, the anterior head process forms via a node-independent mechanism that involves the condensation of dispersed notochord progenitor cells located anterior to the forming node, the trunk notochord originates from the node and elongates by convergent extension, and the tail notochord arises from node-derived progenitors that migrate toward the posterior end (Yamanaka et al. 2007). These different cellular origins along the longitudinal body axis can introduce variations in the maintenance and function of cells in the future nucleus pulposus along different regions of the spine.

In early vertebral development, signals from the notochord pattern the surrounding paraxial mesoderm to form an unsegmented, presomitic mesoderm (PSM). The initial high levels of fibroblast growth factor (FGF) signaling maintain PSM cells in their unsegmented state (Dubrulle et al. 2001), but its reduction leads to segmentation of the PSM in the anterior-posterior direction sequentially, giving rise to bilateral pairs of somites on either side of the neural tube (Roughley 2004) (Fig. 10.1a). Somites are transient structures that consist of outer epithelial cells surrounding an inner mesenchymal core (the somitocoele) (Ferrer-Vaquer et al. 2010). Somitogenesis and establishment of boundaries between developing somites are tightly regulated by the periodic expression of cycling genes (segmentation clock) of the Notch and Wnt signaling pathways (Kageyama et al. 2007; Maroto and Pourquie 2001; Pourquie 2003; Saga and Takeda 2001). These two pathways cycle synchronously with somite formation but slightly out of phase with each other, with cyclic Wnt signaling acting upstream of cyclic Notch signaling. Genes that regulate Notch

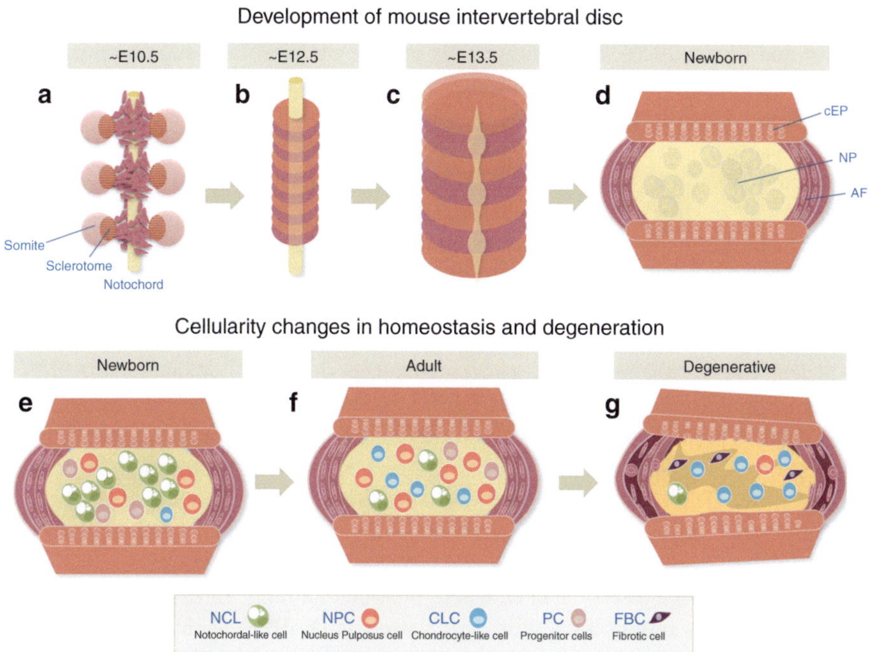

Fig. 10.1 Development and cellular changes in intervertebral discs. A diagrammatic representation of the embryonic development of intervertebral discs in mice (**a**–**d**), illustrating the role of the notochord in directing tissue formation and organization into the vertebral bodies and the intervertebral discs, and the formation of the nucleus pulposus from the notochord (**e**–**g**). The cellular changes in the nucleus pulposus with age and degeneration: notochordal-like cells are replaced by or progressively differentiate into nucleus pulposus and chondrocyte like cells, and fibrotic cells are present at later stages of degeneration. The presence of potential progenitor cells is also depicted. *Abbreviations*: *cEP* cartilaginous endplate, *NP* nucleus pulposus, *AF* annulus fibrosus

signaling (Mesp2, delta1, Hes1, Hes7, and lunatic fringe) turn on and off in a wave-like pattern in PSM cells, and each wave leads to the formation of a new somite (Bessho et al. 2003; Chen et al. 2005; Jiang et al. 2000; Morimoto et al. 2005; Pourquie 2011).

Once formed, cells in the somites differentiate with respect to their position starting with the anterior-most somite. The dorsal epithelium of the somite forms the dermomyotome, establishing the skeletal muscle and the dermis. Under the influence of bone morphogenetic protein (BMP) and Wnt signaling from the ectoderm and sonic hedgehog (Shh) signaling from the notochord and floor plate, cells in the somitocoele and ventral epithelium of the somite undergo epithelial to mesenchymal transition (EMT) to form mesenchymal sclerotomal cells that surround the notochord (Fig. 10.1a) (Borycki et al. 1998; Fan and Tessier-Lavigne 1994; Monsoro-Burq 2005). Subsequently, the sclerotome derived from one somite segregates into rostral and caudal halves, and re-segmentation occurs whereby the rostral half of one segment joins with the caudal half of a neighboring segment (Bagnall et al. 1988) (Fig. 10.1b). The sclerotome consists of multiple progenitor cells that differentiate into all connective tissues cell types (Brand-Saberi et al. 1996; Brand-Saberi and Christ 2000). Sclerotome cells specify into cartilage and annulus fibrosus lineages depending on the location of cells within the sclerotome and a complex network of growth factor signaling.

The rostral and caudal halves of adjacent sclerotome give rise to the cartilaginous vertebrae (Christ and Wilting 1992). Bone morphogenetic protein (BMP) activity concentrated in these regions act alongside Shh and Pax1/9-induced *Nkx3.2* transcriptional repressors to activate Sox9 expression and subsequent chondrocyte differentiation (Rodrigo et al. 2003; Tribioli and Lufkin 1999; Zeng et al. 2002). The annulus fibrosus, conversely, is derived from cells adjacent to the von Ebner's fissure, the border between the two sclerotome halves of a single somite. However, it is unclear whether the annulus fibrosus arises from the rostral or caudal side of this border (Bagnall and Sanders 1989; Goldstein and Kalcheim 1992; Huang et al. 1994), with a most recent dye-labeling study in avian discs (Bruggeman et al. 2012) suggesting that the annulus fibrosus originates from the rostral sclerotome. Nevertheless, TGF-β signaling plays an important role in annulus fibrosus fate determination. Targeted deletion of the TGF-β type II receptor (*Tgfbr2*) gene in *Col2a1*-expressing sclerotome cells (*Col2a1*-cre) in mice results in a reduced or absent intervertebral disc by E17.5, where the inner annulus fibrosus does not form and more closely resembles a wild-type vertebrae (Baffi et al. 2004, 2006). This suggests that TGF-β controls embryonic disc development by either precluding differentiation of vertebral cartilage in the disc space or by directly stimulating annulus fibrosus differentiation (Sohn et al. 2010).

At E13.5 in mouse embryonic development, the notochord also segments as it contracts in the periphery of the emerging vertebral bodies and expands in the developing intervertebral disc anlagen, giving rise to the nucleus pulposus (Fig. 10.1c) (Aszodi et al. 1998). As no cell death in regression regions or cell proliferation in expansion regions is observed during notochord segmentation, a possible mechanism of the pinching off process is that the developing vertebrae apply a biomechanical force on the notochord, squeezing the notochordal cells toward the less-compressed disc tissues while displacing them away from the vertebrae (Grotmol et al. 2003).

Evidence that the nucleus pulposus is derived from the notochord is provided through studies that utilize notochord-specific Cre mice to trace the fate of notochordal cells. Noteworthy are studies that used Shh-Cre/Shh-Cre^{ERT2} (Choi et al. 2008), a molecule highly expressed in the notochord and later in the nucleus pulposus, and Noto-Cre (McCann et al. 2012), a homeobox gene that is restricted to the node and posterior notochord during gastrulation (E7.5–E12.5) and is not detected subsequently. These studies demonstrate that many cells in the adult nucleus pulposus can be traced back to their developmental origin, up to a period of 9 months for the Noto-Cre (McCann et al. 2012) and 19 months for the Shh-Cre^{ERT2} (Choi et al. 2008). Whether other sources of cells contribute to the postnatal nucleus pulposus is not clear.

While the intervertebral disc structure is established embryonically, in mice, significant changes in cell and tissue morphology, such as the establishment of distinct annulus fibrosus lamellae, demarcation of the annulus-endplate boundary, endplate chondrocyte hypertrophy, and matrix gene expression, occur at early postnatal time points.

10.3 The Normal Disc: Working Together as a Functional Unit

Intervertebral discs are complex structures that connect adjacent vertebral segments along the spine. Each disc consists of a central, gelatinous nucleus pulpous, constrained by a circumferential ring of fibrocartilage, the annulus fibrosus, and is situated between two cartilaginous endplates that are contiguous with adjacent vertebral bodies (Fig. 10.1d). These components of the intervertebral disc are distinct in composition and structure, conferring mechanical properties that permit the intervertebral disc to distribute load while maintaining flexibility of the vertebral column. Thus, different structural compartments of the disc act synergistically to produce a functional unit.

10.3.1 Nucleus Pulposus

The nucleus pulposus is at the center of the intervertebral disc. Cells of the nucleus pulposus produce the necessary proteoglycans, mainly aggrecan, to maintain a high osmotic pressure in the disc environment (Roughley et al. 2006a, Roughley and Mort 2014). The abundant water content gives the nucleus pulposus its gelatinous appearance and enables distribution of compressive loads and maintenance of a functional disc height (Boxberger et al. 2008). In early stages of life, the nucleus pulposus is highly cellular, but with age, cell density decreases (Urban et al. 2000) and changes in cell morphology are observed (Cappello et al. 2006) (Fig. 10.1e–g).

Although cells in the nucleus pulposus are heterogeneous, early in life, it majorly consists of notochordal-like cells (Fig. 10.1e). The notochordal-like (NCL) cells appear in clusters, they exhibit a well-defined Golgi and extensive endoplasmic reticulum, have very few mitochondria, and possess actin filament-bounded vacuoles that contain an osmotically active material (possibly unsecreted glycosaminoglycans) (Adams et al. 1990; Smith et al. 2011; Trout et al. 1982a, b). Given that cells in the nucleus pulposus originate from the developing notochord, they express BrachyuryT, a known marker of the notochord and notochord tumors, and sonic hedgehog (Shh). In addition, co-expression of cytokeratin 8 and vimentin are observed in bovine notochordal-like cells (Gilson et al. 2010).

However, with growth and aging, the notochordal-like cells are replaced by or progressively differentiate into mature nucleus pulposus cells (NPCs) that produce the bulk of the extracellular matrix for the functionality of the nucleus pulposus (Fig. 10.1f–g). This coincides with the appearance of chondrocyte-like cells (CLCs) that exhibit morphology similar to chondrocytes, with a thin cortex actin surrounding the nucleus pulposus and a small cytoplasmic volume (Fig. 10.1b). Markers common to chondrocyte-like cells include CD24, cytokeratin 18, cytokeratin 19, and cadherin 2, and they can be distinguished from articular chondrocyte using markers such as PAX1, FOXF1, CA12, and OVOS2 (Choi et al. 2015; Minogue et al. 2010; Power et al. 2011). Since chondrocyte-like cells are reportedly less metabolically active and more resistant to nutritional stress, a transition to chondrocyte-like cells with age may be induced by factors such as reduced nutrient supply to the avascular nucleus pulposus upon calcification of the cartilaginous endplates (Guehring et al. 2009). Finally, at a later stage in adult life, fibrotic cells populate the disc nucleus pulposus (Arkesteijn et al. 2015; Ishii et al. 1991).

With different species and aging, the relative proportions of cells in the nucleus pulposus vary. In humans and chondrodystrophic dogs, the notochordal cells decrease rapidly with establishment of the mature spinal structure and aging (Cappello et al. 2006; Hunter et al. 2003, 2004; Smolders et al. 2013). On the other hand, in rodents and rabbits, pools of notochordal cells are maintained through adulthood (Miyazaki et al. 2009; Poiraudeau et al. 1999). A disc that retains greater number of notochordal-like cells through adulthood (i.e., non-chondrodystrophic dogs) appears to sustain a healthy nucleus pulposus with a well-hydrated structure (Hunter et al. 2004; Smolders et al. 2013). Conversely, disappearance of notochordal-like cells causes a more fibrous nucleus pulposus and early degenerative changes (Melrose et al. 1996).

A recent study of the human and mouse nucleus pulposus identified tyrosine-protein kinase receptor (Tie2) and disialoganglioside 2 (GD2) as potential cell surface markers for cells in the nucleus pulposus (Sakai et al. 2012). A hierarchal expression pattern of progenitor cell markers was proposed, where $Tie2^+/GD2^-/CD24^-$ cells behave as dormant stem cells. Upon activation they become $Tie2^+/GD2^+/CD24^-$ with self-renewal potential and are potential progenitor cells. Upon differentiation, Tie2 expression is lost, and the $Tie2^-/GD2^+/CD24^-$ cells are characterized as cells of the early nucleus pulposus. Subsequently, cells lose expression of both Tie2 and GD2 but maintain CD24 expression ($Tie2^-/GD2^-/CD24^+$) that defines a mature nucleus pulposus cell phenotype of chondrocyte-like cells in the disc nucleus pulposus. This study suggests a chronological differentiation of cells in the nucleus pulposus and supports the presence of progenitors that are able to commit to the nucleus pulposus cell lineage. However, a decline in progenitor potential (reduction of $Tie2^+$ cells) is observed with aging, correlating with the onset of disc degeneration in humans (Sakai et al. 2012).

10.3.2 Annulus Fibrosus

At E13.5–E14.5, the lamellar structure of the annulus fibrosus is not well defined in mouse development, but resident cells in the region are oriented concentrically around the newly formed nucleus pulposus (Peacock 1951). Organization of this template sheet structure during development may be a result of pressures in the nucleus pulposus that trigger the formation of actin stress fibers in annular fibroblasts, connecting the cadherin junctions of adjacent cells (Hayes et al. 1999). This initial cellular organization may be necessary for the formation of tough circumferential angle-ply layers and highly concerted matrix deposition that are progressively established during postnatal growth. However, the molecular signals that direct annulus fibrosus cells to assemble into a required lamellar orientation remain unclear.

This microstructural composition and arrangement enable the annulus fibrosus to contain the radial bulge of the nucleus pulposus and uniformly distribute torsional and compressive loadings in the disc space. Furthermore, it facilitates joint flexibility by distention and rotation. In humans, the number of fiber bundles in the annulus fibrosus can range from 20 to 62 (Marchand and Ahmed 1990), with the lamellar thickness increasing with age, both circumferentially and radially. The inner and outer layers of a fully developed annulus fibrosus differ substantially in their biochemical and cellular composition and thus have potentially distinct biomechanical roles. Cells in the outer annulus fibrosus are elongated, fibroblast-like, and extend in the long axis of the fibrils, whereas those in the inner annulus fibrosus have a spherical morphology similar to chondrocytes (Bruehlmann et al. 2002). Being a tension-bearing tissue, the annulus fibrosus has a high collagen content, with collagen fibers in adjacent lamellae oriented in opposite directions. The progression from a predominantly collagen I matrix in the outer annulus fibrosus to a collagen II matrix in the inner annulus reflects changes in the loading environment from more tension in the outer regions to greater compression in the inner compartment.

The cellular origin of the annulus fibrosus is less well defined. Lineage tracing in mice with Tbx18-Cre established that a pool of labeled cells in the anterior part of the annulus fibrosus are derived from the sclerotome (Bruggeman et al. 2012). Furthermore, using fluorescent lipophilic dyes, DiI and DiA, to differentially label the rostral (upper) and caudal (lower) regions of the sclerotome in chick development, it was shown that cells from the rostral half of the sclerotome contribute to caudal half of vertebral body as well as the intervertebral space, while the caudal half of sclerotome contributes to rostral half of adjacent vertebrae (Bruggeman et al. 2012). However, these studies cannot account for all the cells in the annulus fibrosus, and alternative sources of progenitor cells may contribute to the formation, maintenance, and renewal of cells in the annulus fibrosus in postnatal life.

10.3.3 Cartilaginous Endplate

The cartilaginous endplate is the least studied intervertebral disc compartment, despite its important role in disc nutrition and mechanics. The endplate represents a transition structure of hyaline cartilage that physically and biomechanically separates the disc region from the vertebrae, barring the motion segment from applying pressure onto the vertebral bone. In humans, its thickness varies from about 0.2 mm in the center to 0.9 mm in the periphery and is thickest at birth and thins with age (Moon et al. 2013). The endplate exhibits structural similarity with the articular cartilage of synovial joints; thus, it consists of subpopulations of chondrocytes embedded in an extracellular matrix of aggrecan and collagen II fibrils that are oriented parallel to the longitudinal axis (Moore 2006; Sahlman et al. 2001). Collagen

X is expressed in the pericellular matrix of hypertrophic chondrocytes (Aigner et al. 1998). However, the origin of the collagen X-producing hypertrophic chondrocytes remains unclear.

Calcification of the endplate initiates at around 6 human years of age and ossifies by about 13 human years of age (forming a secondary ossification center) (Bick and Copel 1951). In humans, the secondary ossification center adjacent to the endplate fuses with the primary ossification center of the vertebral bone, terminating skeletal growth and producing a "closed" vertebral growth plate at puberty. In goats and rabbits, the primary and secondary ossification centers remain distinct, and the osseous layer of the endplate is termed bony endplate (Zhang et al. 2014). Although the vertebral growth plate remains "open" through adult life in these animals, skeletal growth ceases. Unlike rabbits and goats that contain only one to three layers of chondrocytes in their cartilaginous endplate, mice contain three to seven layers of cells (Zhang et al. 2014). The cartilaginous endplate of rodents manifests as a continuation of the growth plate with pockets of secondary ossification centers in spines of elder mice (Zhang et al. 2014). Nevertheless, the extent of calcification and growth plate – cartilaginous endplate continuous or discontinuous morphology – varies with age of the animal as well as position of the disc along the spine.

Like tendon inserts into bone to form a contiguous tissue, the cartilaginous endplate and annulus fibrosus are tightly linked. The less organized inner annulus fibrosus fiber bundles insert obliquely into the cartilaginous endplate, divide into sub-bundles, and merge with the hyaline cartilage matrix. The junction at which the two tissues intersect is defined as the inner annulus-endplate interface, which contains a mixed cell population comprised of chondrocytes and cells with morphology intermediate of elongated annulus fibrosus cells and round hypertrophic chondrocytes. This histological feature suggests that cells in the two structures may be closely related. Likewise, organized lamellar fibers of the outer annulus fibrosus penetrate directly into the vertebrae with "Sharpey's fibers" bridging the two tissues (Hashizume 1980; Inoue 1981) or curve laterally at a 90° angle and merge with the periosteum along the vertebrae (Nosikova et al. 2012). These interfaces, where soft tissue integrates into hard tissue, develop to withstand the wide range of mechanical forces (shear, compression, tensile, and torsion) experienced by the intervertebral disc.

10.4 Clinical Features of Intervertebral Disc Degeneration

Low back pain is a debilitating condition that is estimated to affect 84 % of the population at some point in their life, with 10 % suffering from chronic disability (Walker 2000). While there may be many causes of back pain, intervertebral disc degeneration is a major contributing factor in 40 % of the cases (Cheung et al. 2009). Degenerated lumbar discs have been observed in patients as early as 11–16 years of age (Boos et al. 2002). In fact, 20 % of individuals in teen years

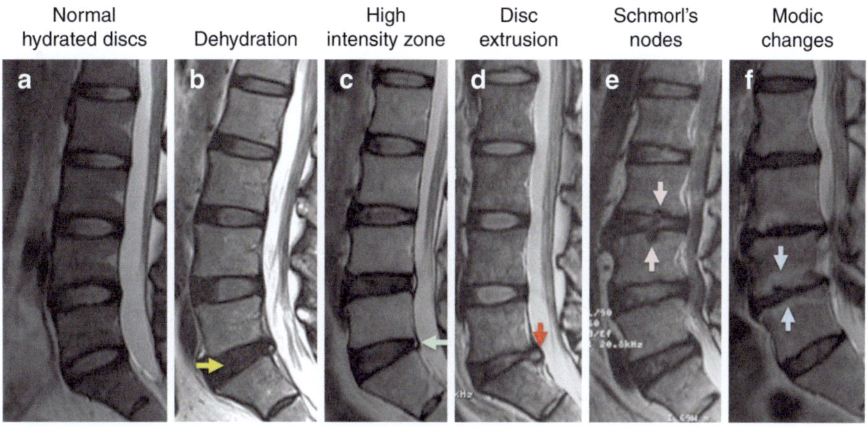

Fig. 10.2 Common clinical feature of intervertebral changes by MRI. Selected human MRI of the lumbar region showing common clinical features observed in the intervertebral disc with age and degeneration. The *colored arrows* highlight the respective changes in each category. (**a**) An individual with normal hydrated intervertebral discs. (**b**) An individual with a dehydrated nucleus pulposus (*yellow arrow*). (**c**) An image depicting tears within the annulus fibrosus shown as a brighter region (*light green arrow*). (**d**) Deformation of the annulus fibrosus and extrusion of the nucleus pulposus impinging on the sciatic nerve (*red arrow*). (**e**) Abnormal structural changes the cartilaginous endplates (*light pink arrows*). (**f**) Signal intensity changes in the bone marrow regions of the vertebral bodies adjacent to a degenerative disc (*light blue arrows*)

procure mild signs of degeneration, which increases sharply with age such that 10 % of 50-year-old discs and 60 % of 70-year-old discs are severely degenerate (Miller et al. 1988). Despite its wide prevalence, a unified definition of disc degeneration is lacking. In comprehending the scope of the degenerative condition, a thorough evaluation of the structural and biological changes during degeneration is essential.

Magnetic resonance imaging (MRI) is a useful technique to assess the quality of the intervertebral disc by means of a T2-weighted sagittal scan (Fig. 10.2). The Schneiderman (Schneiderman et al. 1987) or Pfirrmann (Pfirrmann et al. 2001) classifications are the commonly used grading protocols for assessing degenerative changes using MRI. An MRI scan of a healthy human lumbar disc is presented in Fig. 10.2a. The bright nucleus pulposus represents a fully hydrated disc of normal height. The most marked change in disc degeneration is a decrease in signal intensity (a dark disc), indicating a loss of water content (Fig. 10.2b). Other features of degeneration include tears in the annulus fibrosus shown as "high intensity zones" (Fig. 10.2c), extrusion of the nucleus pulposus from the annulus fibrosus and its subsequent compression on the sciatic nerve (sciatica) (Fig. 10.2d), and herniation of the nucleus pulposus to the cartilaginous endplates which form structures called Schmorl's nodes (Fig. 10.2e). In some cases, Modic changes can be detected as signal variations in the marrow regions adjacent to the cartilaginous endplates (Fig. 10.2f) (Hu et al. 2009; Yu et al. 2012). The relevance of Modic changes is not clear but many studies have shown a high

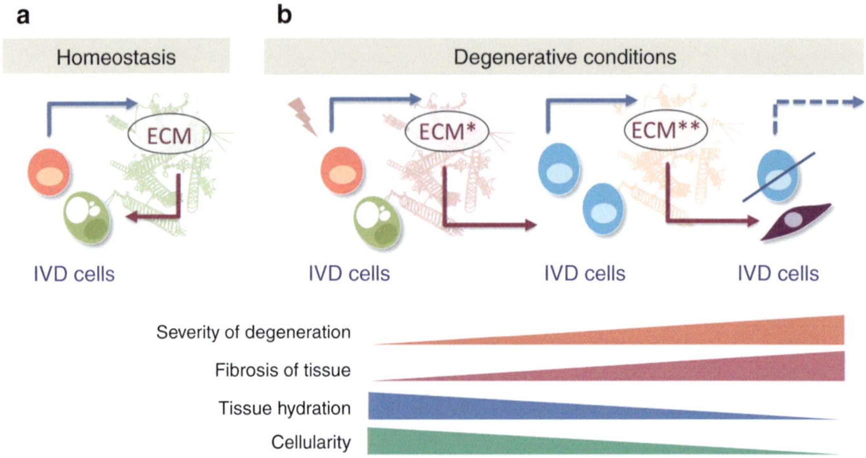

Fig. 10.3 Extracellular matrix (*ECM*) changes influencing progression of intervertebral disc (*IVD*) degeneration. Diagrammatic representation depicting the relationship between the extracellular matrix and cells in a healthy disc (**a**) where a homeostatic environment can be maintained. (**b**) Insults to cells or matrix initiate a cascade of events that lead to progressive changes in matrix production or degradation and cellular differentiation to that of a degenerating intervertebral disc

association with symptomatic outcomes such as back pain (Albert et al. 2008). The features described above can be presented in singleton or in combination.

10.5 The Etiology of Intervertebral Disc Degeneration

Cell tracing and fate mapping in mice suggest that cells in the nucleus pulposus are long lived, but undergo biochemical and phenotypic changes upon natural aging and degeneration. Degeneration has been described as an abnormal cell-mediated response that leads to a gradual failure in tissue architecture (Fig. 10.3). Hallmarks of disc degeneration include an overall breakdown of extracellular matrix, abnormal matrix synthesis, and changes in cell number, cell morphology, and cell behavior. These features alter the nutritional, biomechanical, and structural properties of the nucleus pulposus, annulus fibrosus, and cartilaginous endplate, causing the disc to function less efficiently in a vicious cycle that leads to disc degeneration (Fig. 10.3). Degeneration is also marked by a reduced capacity for tissue maintenance and repair, which possibly involves the mobilization of progenitor cell pools found within the intervertebral disc tissues. The clinical features presented in MRIs suggest that degeneration of the disc is complex, and it is likely that the anomalies described previously are interconnected. However, a direct relationship between these changes and their potential sequence of manifestation is not well understood. Furthermore, a condition brought on by development, intervertebral "dysgeneration," has been proposed to describe cases where subtle differences exist in the disc from early on, causing an atypical cascade of events with age (Luk and Samartzis 2015).

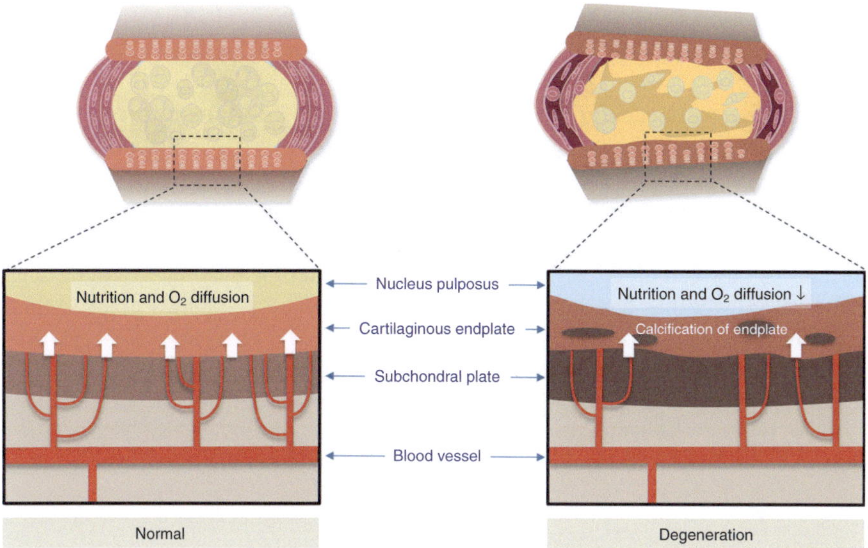

Fig. 10.4 Nutrient and oxygen supply in the intervertebral disc. Diagrammatic representation of the blood supply to the intervertebral disc with blood vessels ending at the junction of the cartilaginous endplate in normal and degenerate stages. The intervertebral disc proper is normally avascular and the cartilaginous endplate presents as selective barrier and the major route of nutrient and oxygen supply. Conditions with impaired blood vessel formation and calcification of the cartilaginous endplate will have an impact on nutrient supply and movement of solutes through the cartilaginous endplate

10.6 Living in the Harsh Environment of an Intervertebral Disc

Disc cells require a sufficient supply of nutrients and adequate removal of metabolic waste for normal function and survival. An imbalance between nutrient availability and consumption may lead to aberrant cellular behaviors that contribute to disc degeneration (Grunhagen et al. 2011; Urban et al. 2004). This is a challenge for cells in the intervertebral disc as the tissue is avascular, and cells at the center of the adult human disc are approximately 8 mm away from the nearest blood supply that terminates at the cartilaginous endplates (Fig. 10.4) (Urban et al. 2004). Nutrients such as glucose and oxygen and substrates for matrix production such as amino acids enter via bloody supply in the cartilaginous endplates and the margins of the annulus fibrosus (Naresh-Babu et al. 2016). Cells in the outer annulus fibrosus engage in metabolite exchange via blood vessels within the soft tissues at the disc periphery, as well as small capillaries that penetrate 1–2 mm into the outermost lamella in adult humans (Hassler 1969; Rudert and Tillmann 1993). The remaining majority of disc cells rely on nutrient supply via capillaries that penetrate channels in the subchondral plate and terminate at the border of the cartilaginous endplate (Holm et al. 1981). Small solutes are transported from capillaries to cells across the disc matrix by diffusion (Holm et al. 1981; Naresh-Babu et al. 2016), and cellular wastes such as lactic acid are removed in the opposite direction (Magnier et al. 2009).

The dense extracellular matrix of collagens and polyanionic proteoglycans restricts the size and charge of solutes transported to the disc (Magnier et al. 2009; Roberts et al. 1996). The negative charge of aggrecan enhances the penetration of small cations (sodium), whereas anions (sulfate and chloride) are partially excluded (Urban and Maroudas 1979). Moreover, a high aggrecan concentration minimizes the spaces between glycosaminoglycan chains and sterically excludes large solute molecules (cytokines and growth factors), which make use of load-induced fluid movement for their transport (Chandran and Horkay 2012; Ferguson et al. 2004).

The rate of solute transport depends on many parameters, from the blood supply to compartmental permeability and cellular metabolic demand. A fine balance between the rate nutrient supply and consumption is a key for tissue homeostasis (Urban et al. 2004). A sustained imbalance will lead to cellular and biochemical changes.

10.6.1 The Role of Cartilaginous Endplate in Nutritional Supply

The cartilaginous endplate acts as a selective barrier, and this property influences the movement of solutes into and out of the intervertebral disc (Magnier et al. 2009; Roberts et al. 1996). A major risk factor for intervertebral disc degeneration is calcification of the endplate (Roberts et al. 1996) or occlusion of endplate channels (Benneker et al. 2005), which have been demonstrated to disturb transport into the disc region (Rajasekaran et al. 2004; Roberts et al. 1996) that could lead to cell death (Galbusera et al. 2013). MRI studies of human intervertebral discs using contrast medium showed that diffusion into discs via the cartilaginous endplate was lower in early degenerative discs (Nguyen-minh et al. 1998; Rajasekaran et al. 2004), suggesting impaired solute transport upon degeneration. Furthermore, a gradual change of the endplate to fibrocartilage and focal necrosis upon aging may also impede diffusion of solutes (Boos et al. 2002), although this has not been experimentally demonstrated.

10.6.2 Impact of the Vasculature on Nutrient Supply

The impact of systemic blood circulation on disc cell health is evident in the correlation between disc degeneration and disorders that obstruct blood supply to the vertebral column or the capillary beds of the endplate. These include atherosclerotic calcification of the abdominal aorta (Kurunlahti et al. 1999), Gaucher's disease and Caisson's disease (Jones 1997), sickle cell anemia (Babhulkar 1997; Jones and Engleman 1966), lipidemia, and hyperuricemia (Nerlich and Boos 2016). Furthermore, post-contrast MRI studies of human intervertebral discs found that signal intensity increased inside the disc after administration of vasoactive drugs, suggesting that adjustment in the rate of blood flow via muscarinic receptors of vascular channels in the subchondral plate affects the delivery of nutrients to the disc (Rajasekaran et al. 2008). Reduction of vascularity in the endplate also occurs with aging (Boos et al. 2002; Edelson and Nathan 1988) resulting in general reduction of blood supply to the disc, which may explain a higher predisposition of aged discs to degeneration. Obliteration of blood vessels in the endplate is paralleled by structural changes such as cartilage disorganization and an increase in cartilage cracks and microfractures (Boos et al. 2002).

10.6.3 Factors Influencing Nutrient Demand

Factors such as endplate calcification or changes in blood flow patterns can reduce nutrient supply leading to disc degeneration. However, increasing the rate of nutrient utilization per cell (the cellular demand) can also add to the risk. Cellular demand can be increased by growth factors or cytokines that enhance lactic acid production and glucose utilization; these factors are shown to be elevated in herniated human lumbar discs (Andrade et al. 2013; Gronblad et al. 1994; Weiler et al. 2005). Moderate levels of growth factors are essential for maintaining a healthy disc by stimulating matrix production (Li et al. 2004). However, upon loss of proteoglycans and an increased permeability of the matrix in degeneration increased levels of growth factors (and cytokines) infiltrate, affecting cellular metabolism and the balance between nutrient supply and demand (Grunhagen et al. 2011).

Addition of exogenous growth factors (GDF-5, TGF-β, FGF, or EGF) to bovine, canine, or human cell cultures in vitro (Chujo et al. 2006; Kim et al. 2003; Gruber et al. 1997; Thompson et al. 1991) or injection into mouse tail discs in vivo (Walsh et al. 2004) can stimulate cell proliferation and cell activity. Increased cell activity and cell numbers can boost glucose consumption rates, which may deplete nutrients from the intervertebral disc niche (Mokhbi Soukane et al. 2009). Furthermore, upregulated proinflammatory cytokines in the disc, such as tumor necrosis factor alpha (TNFα) and interleukin-1 (IL-1) (Johnson et al. 2015), propagate local inflammatory signals via prostaglandin E2 (PGE_2) (Bachmeier et al. 2007; Rannou et al. 2000; Shinmei et al. 1988; Takahashi et al. 1996), stimulate angiogenesis (Doita et al. 1996), and the production of matrix degrading enzymes in disc cells (Le Maitre et al. 2005). These changes will result in enhanced metabolic activity, although this has not been addressed directly.

Mechanical stresses can also increase nutritional requirements of cells beyond diffusional capacity of the disc. For instance, as the disc deforms under mechanical loads, the hydrostatic pressure inside the disc increases and fluid moves out of the disc. This change in fluid dynamics increases the interstitial fixed charge density, attracts more ions into the tissue, and elevates osmotic pressure in the disc (Boos et al. 1993; Chan et al. 2011; Ishihara et al. 1997; Urban et al. 1994). The hyperosmotic extracellular environment leads to a decrease in cell volume (Pritchard et al. 2002; Verkman et al. 1996) and triggers recovery mechanisms such as activation of ion channels to restore intracellular water and chemical composition (Erickson et al.

2001). Since a vital sodium gradient is maintained across plasma membranes by the Na^+/K^+ pump, which utilizes 50–70 % of the cellular adenosine triphosphate (ATP), changes in ionic distribution can increase Na^+/K^+-ATPase activity in a cell, thereby increasing cellular demand of energy (Horner et al. 2001; Mobasheri et al. 1997). Thus, alterations in the physical environment surrounding cells due to loading can cause changes in cellular metabolism, resulting in a different biosynthetic activity of the cell.

10.6.4 Balancing Nutrient Supply and Demand for Tissue Homeostasis

The concentration of nutrient available to disc cells (the steepness of its concentration gradient) is determined by the balance between rate of influx (e.g., rate of blood flow) and the rate of consumption by cells (Urban et al. 2004). Thus, a high cell density and rate of cell metabolism will increase the net nutrient consumption, thereby the rate of nutrient diffusion (Magnier et al. 2009). However, cell number and metabolism are also regulated in a negative feedback loop, where a decreased nutrient supply will restrict cell mass and metabolic rates, thus maintaining homeostasis among different regions of the disc. In the intervertebral disc, concentrations of nutrients are highest at the disc periphery and lowest at the center (Magnier et al. 2009), and cell density in the disc is also highest at the outer edge of the annulus fibrosus, closest to the nutrient supply, and falls rapidly toward the inner annulus fibrosus and the nucleus pulposus (Maroudas et al. 1975; Urban et al. 2004). This correlation between nutrient levels and cell density across the disc suggests that cell numbers may be regulated by nutritional constraints.

Cells in the nucleus pulposus and inner annulus fibrosus are often exposed to suboptimal microenvironments with low oxygen tension, limited glucose availability, relatively low pH (due to lactic acid buildup), and high metabolic waste accumulation (Magnier et al. 2009). In this avascular environment, disc cells predominantly rely on glycolysis for energy production, which requires glucose and produces lactic acid (Holm et al. 1981). Therefore, disc cells are particularly sensitive to changes in glucose and lactic acid concentrations. A reduction in glucose concentration to below 0.5 mmol/L or a decrease in pH to below 6.4 comprises cell viability of bovine cells in vitro (Bibby et al. 2005; Horner and Urban 2001). Moreover in a low pH environment, bovine disc cells continue to produce active matrix metalloproteinase (MMPs) that break down matrix macromolecules, while matrix production is inhibited (Razaq et al. 2003).

On the other hand, the effect of oxygen concentration and hypoxia on disc cell function is currently under debate. Studies in canine and bovine discs have demonstrated sulfate incorporation for matrix production is highest in the nucleus pulposus at 5 % O_2 and at pH 7.0. In vitro studies suggest that disc cells have been adapted to hypoxic environments (Agrawal et al. 2007; Feng et al. 2013; Mwale et al. 2011; Risbud et al. 2005) in part by activating hypoxia-inducible factors (HIF), particularly HIF-1. HIF-1 enhances the expression of matrix genes such as aggrecan in nucleus pulposus cells in vitro (Agrawal et al. 2007) and also promotes cell survival by modulating apoptosis and regulating glucose metabolism (Risbud et al. 2005) However, an oxygen concentration below 5 % causes cellular inactivity (Horner and Urban 2001) and reduced matrix synthesis (Ishihara and Urban 1999). Thus, the effects of depleted nutrients and increased metabolites have a cumulative detrimental effect on metabolism and survival of cells in the intervertebral disc niche, as well as on the integrity of the disc matrix.

10.7 The Extracellular Matrix Environment of an Intervertebral Disc

The extracellular matrix is a dynamic network of insoluble fibers (collagens and elastin), soluble polymers (glycosaminoglycans and proteoglycans), and adhesive glycoproteins (e.g., fibronectin, COMP, matrillin, laminin) (Sivan et al. 2014). Each disc compartment consists of a different combination of extracellular matrix proteins for proper function, and cells maintain a balance between anabolic and catabolic processes for tissue homeostasis. Cells regulate their behavior by sensing their changing microenvironment via cell surface receptors such as integrins that physically link the matrix proteins with intracellular cytoskeletal elements (Gilchrist et al. 2007).

10.7.1 Extracellular Matrix Organization in Tissues of the Intervertebral Disc

The nucleus pulposus is composed of large amounts of proteoglycans, predominantly aggrecan (30–50 % of dry weight) (Adams and Muir 1976; Eyre 1979), which generates a swelling pressure inside the disc to resist compressive loads. Aggrecan is retained in the nucleus pulposus by interacting with hyaluronic acid to form supramolecular aggregates (Morgelin et al. 1988), and the sulfate groups on its glycosaminoglycan chains (chondroitin and keratin sulfate) attract water molecules (70–85 % of total weight) by osmosis (Watanabe et al. 1998). Human chondroitin sulfate may also serve to restrict neural ingrowth, according to recent in vitro studies (Johnson et al. 2002; Purmessur et al. 2015). In humans, aggrecan content increases in juvenile years, peaks at late adolescence, but subsequently declines in adult life so that the nucleus pulposus becomes less hydrated with aging (Roughley et al. 2006a). In the nucleus pulposus, proteoglycans are entrapped in an unorganized network of fibril-forming collagens (collagen II and small amounts of collagens XI) that are decorated with fibril-associated collagen with interruptions in the triple helix (FACITs) (collagen IX) (Wu et al. 1992; van der Rest and Mayne 1988). These collagen copolymers (20 % of dry weight) form a structural framework that contains the swelling pressure of the nucleus pulposus. Collagens IX and XI also exist to possibly link collagen molecules to other matrix components.

In a healthy human annulus fibrosus, collagens and proteoglycans make up 70 % and 10 % of the dry weight, respectively, and the water content is 50 % (Eyre 1979; Eyre and Muir 1976). The collagen II fibers in the inner annulus fibrosus have wider interfibrillar spaces that retain a greater volume of water than collagen I in the outer annulus fibrosus (Han et al. 2013; Schollmeier et al. 2000). This confers a superior ability of the inner annulus fibrosus region to undergo deformation and endure high mechanical pressures. A co-localized network of elastin fibers concentrated between lamellae may also contribute to restore collagen and lamellar organization upon distortion (Yu et al. 2005, 2007a). Collagen I forms heterotypic fibrils with collagens III and V, and the translamellar bridging network of the outer annulus fibrosus connects fibers of adjacent lamellae to provide a structural alignment that limits interlamellar movement (Melrose et al. 2008; Smith and Elliott 2011).

Like the nucleus pulposus, the cartilaginous endplate is also composed of a network of collagens II, IX, and XI and aggrecan in the pericellular microenvironment of chondrocytes, although their relative quantities differ. Total collagens make up 60–80 % of the dry weight, whereas glycosaminoglycans constitute 17 % of the dry weight (Setton et al. 1993). Alternative splicing of the *COL2A1* gene produces procollagen mRNA that include (IIA) or exclude (IIB) an exon encoding a cysteine-rich N-terminal domain. In mice pro-IIA mRNA exist in non-chondrogenic and pre-chondrogenic cells, and expression switches to collagen IIB upon chondrocyte differentiation (Sandell et al. 1991). Moreover, collagen X, encoded by the *COL10A1* gene, is mainly expressed in the central region of the endplate in hypertrophic chondrocytes and regulates mineralization along with collagen II, by entrapping matrix vesicles and facilitating calcium influx (Lammi et al. 1998; Shen 2005). The pericellular matrix is believed to influence physiochemical signals sensed by chondrocytes and augment their protective response to mechanical compression.

10.7.2 The Role of Proteinases in Matrix Remodeling

The matrix is a dynamic structure that is routinely broken down by proteinases such as matrix metalloproteinases (MMPs) and a disintegrin and metalloproteinase with thrombospondin motifs (ADAMTS) produced by disc cells. MMPs are capable of cleaving most matrix constituents, with MMPs 1, 8, and 13 degrading intact collagens I and II and MMPs 2 and 9 (gelatinases) cleaving partially degraded collagen molecules (Nagase and Woessner 1999). The ability of MMPs to degrade aggrecan, however, is significantly lower than that of the ADAMTS aggrecanases (ADAMTSs 1, 4, 5, 9, and 15) (Durigova et al. 2011; Nagase and Kashiwagi 2003; Pockert et al. 2009). These molecules along with other proteolytic enzymes are implicated in the homeostatic turnover of disc extracellular matrices. Equilibrium between synthesis, breakdown, and accumulation of matrix macromolecules determines matrix integrity and influences the mechanical behavior of the intervertebral disc as well as the maintenance of an avascular/aneural microenvironment of a healthy disc.

10.7.3 The Changing Matrix Environment in Disc Degeneration

Intracellular events affecting glycosaminoglycan synthesis or the extracellular degradation of aggrecan can cause disc dehydration, a common feature of degeneration. In response to excessive mechanical loading (Neidlinger-Wilke et al. 2006) or infiltration of inflammatory cytokines such as TNFα and IL-1 (Hoyland et al. 2008; Le Maitre et al. 2005, 2007a), disc cells produce MMPs and ADAMTS proteinases (Adams and Roughley 2006; Roberts et al. 2000). Their proteolytic cleavage of aggregated and non-aggregated aggrecan causes an extensive decrease in fragment size and untimely loss from the disc tissue. Changes in disc hydration may also result from shortening of chondroitin sulfate chains. The increasing avascularity of degenerating discs can further reduce chondroitin sulfate synthesis, which requires the oxidation of hexose to uronic acid. These changes, in addition to the steady decline of aggrecan with aging, cause dehydration of the nucleus pulposus and an increasing inability to recover from compressive deformation (Lyons et al. 1981; Urban and McMullin 1988). Moreover, as the hydrostatic pressure of the disc falls, mechanical loading may apply inappropriate stress concentrations along the endplate or annulus fibrosus forming fissures and tears between collagen fibrils, disrupting the lamellar structure (Adams et al. 1996). Prolonged stresses on the disc can cause the nucleus pulposus to protrude through the weakened annulus fibrosus (herniation) or though the endplate (Schmorl's nodes) (Adams and Dolan 2012; Urban and Roberts 2003).

High concentration and charge of aggrecan in the normal disc also prevents movement of large uncharged molecules (growth factors and cytokines) into the disc microenvironment (Maroudas et al. 1975). Upon loss of aggrecan, infiltration of these molecules may advance degeneration by affecting cellular behavior and matrix catabolism. For instance, IL-1 induces fibrotic changes that include an upregulation of MMP3 and 13 and ADAMTS4 in the nucleus pulposus, a shift in expression from collagen II to collagen I in nucleus pulposus and annulus fibrosus cells (Le Maitre et al. 2005, 2007b), and further degradation of aggrecan. It is also considered to induce apoptosis and senescence of disc cells. Breakdown of normal disc extracellular matrix and formation of annular clefts and tears in turn stimulate neovascularization and accelerate the degenerative processes (Carreon et al. 1997; Karamouzian et al. 2010; Vernon-Roberts et al. 2007).

Finally, ectopic expression of collagen X has been identified in the nucleus pulposus and annulus fibrosus at late stages of degeneration in humans (Boos et al. 1997). Induction of collagen X signifies cell hypertrophy and can be accompanied by increased expression of alkaline phosphatase (ALP) and Runt-related transcription factor 2 (Runx2), causing abnormal calcification (Boos et al. 1997; Rutges et al 2010).

Although the exact sequence of changes in degenerative cascade has not been clearly defined, changes in matrix composition in the disc, particularly the loss of aggrecan, impedes its ability to maintain osmotic pressure and remain intact under mechanical loading. Changes in the extracellular matrix also hinder cellular maintenance and activity, resulting in a downward spiral of destructive events.

10.8 Mechanical Loading and Intervertebral Disc Degenerative Events

Association between the mechanical environment and mechanical properties of disc cells with degeneration has been demonstrated through in vivo studies that use external loading devices, in vitro organ culture studies, and application of hydrostatic pressure in in vitro cell cultures. A disc cell's response to stresses depends on the type, magnitude, frequency, and duration of the loading. While physiological loading conditions (<1 MPa magnitude, <3 Hz frequency, and <24-h duration) are anabolic, loads outside this range are catabolic. Outside the "safe window" of tissue mechanics, initially overload and eventually immobilization affect tissue homeostasis and cell mechanobiological response (Stokes and Iatridis 2004).

The "wear-and-tear" or overload hypothesis represents a single, excessive loading cycle or a frequent number of moderate loading cycles that accumulate microscopic damage. The damage heals slowly due slow tissue turnover rates and progressively weakens the ability of the disc tissues to withstand mechanical pressure, increasing their susceptibility to fatigue failure. For instance, the laminated structure of the annulus fibrosus requires damage propagation through multiple micro-failure (matrix cracking, delaminations, and fiber failure). Damage is propagated mainly by delamination in presence of high interlaminar shear stresses after initial tears in the annulus, while fiber failure occurs under severe loading conditions. Structural changes such as the decrease in the number of layers and an expanded layer from increased spacing of collagen fibers may also alter disc cell mechano-responsiveness and advance the degenerative process.

Various mechanical interventions on cadaver discs and animal motion segments have been demonstrated to cause structural and compositional changes in the discs. Administration of high compressive loading with bending on normal cadaver discs (aged 40–50) produced disc prolapse or herniation (Adams and Hutton 1982; McNally et al. 1993), and application of cyclic compressive loading caused endplate fractures, Schmorl's nodes, and a decreased disc height (Brinckmann et al. 1988). Endplate fractures and Schmorl's nodes reduce pressure in the nucleus pulposus and generate compressive stress peaks in the annulus fibrosus (Adams et al. 2000). Repetitive loading on endplate-damaged discs further intensifies stresses on the annulus fibrosus, causing inversion of the inner annulus, extreme outward bulging of the outer annulus, radial fissures, and disc prolapse.

The zonal variations in mechanical environment, cell population, and cell mechanical properties between the annulus fibrosus and nucleus pulposus result in a distinct cascade of load-induced behaviors. An Ilizarov-type device used to apply dynamic loading on skeletally mature Wistar rat tail discs demonstrated that while changes in gene expression in the annulus fibrosus is dependent on magnitude of stress, the nucleus pulposus is more frequency dependent (Maclean et al. 2004). In the nucleus pulposus, low frequencies increased aggrecan and collagen mRNA, whereas higher frequencies stimulated aggrecanase and collagenase expression. In contrast, the annulus fibrosus exhibited increased aggrecanase and collagenase mRNA levels with higher magnitudes of compression regardless of the frequency of the applied compression.

While the consequences of cyclic loading are determined by both the magnitude and frequency of the applied load, mechanosensitivity of disc cells to static loading is largely magnitude dependent. An in vivo study analyzed mouse tail discs after 1 week of continuous loading at different stress levels (Lotz et al. 1998). Increasing loading induced disorganization of lamellar structures in the inner annulus fibrosus and replacement of fibroblast-like cells with chondrocyte-like cells, whereas the outer annulus fibrosus was able to maintain its organization. Collagen II expression was reduced in all areas of the disc regardless of magnitude of compression, whereas aggrecan mRNA decreased in areas of highest loading. In the nucleus pulposus, large cell clusters were dissociated into smaller islands of cells.

Static bending of mouse tail discs for 1 week also presented lamellar inversion at the concave (compressed) side (Court et al. 2001), and as in human scoliotic discs, the concave annulus fibrosus contained a greater number of dying cells. When the bending is released, chondrocyte-like cells are detected in the concave annulus, suggesting three possible differentiations of fibroblasts into chondrocytes, selective loss of fibroblast-like cells, or chondrocyte migration from the endplate. Comparable observations can be made using a compression bending mouse model induced by tail looping that can be used to study cellular response to stretch and compressive forces (Sakai et al. 2015).

When the structural interrelationship of disc compartments fails at late stages of degeneration, tissue stiffness and hypomobility result. In vivo animal models demonstrate that the immobilization condition shows reduced protein synthesis due to lack of mechanical stimulus or lack of nutritional supply. Yet further investigation on the effects of hypomobility on the progression of disc disease is necessary.

10.9 Genetic Risk Factors of Disc Degeneration

Intervertebral disc degeneration is considered to be a complex, multifactorial disease involving many genetic risk factors that interact with one another and the environment. This is supported by a study of an Arabic pedigree which established familial inheritance as a risk factor for intervertebral disc degeneration and suggested that lumbar disc degeneration is a polygenic disease with a complex mode of inheritance (Livshits et al. 2001). Indeed, lumbar disc degeneration (LDD) is estimated to have a heritability rate from 47 to 74 % in familial and twin studies (Battie et al. 2008; Sambrook et al. 1999). To understand the cause of intervertebral disc degeneration, genetic studies provide the most direct and informative route to elucidate the genes and cellular pathways involved, followed by functional testing using established model systems.

Much effort has been made to identify the genetic risk factors associated with intervertebral disc degeneration in the lumbar spine. The initial focus was based on the candidate-gene approach with genetic association studies assessing specific genes as potential risk factors. Many genes in the inflammatory pathways and components of the extracellular matrix were tested based on our initial understanding of the biology of disc degeneration. Positive associations were identified, such as collagen IX genes (*COL9A2* and *COL9A3*) (Annunen et al. 1999; Jim et al. 2005; Paassilta et al. 2001); aggrecan (*ACAN*) (Kawaguchi et al. 1999; Roughley et al. 2006b); MMPs-1, -2, -3, and -9 (Dong et al. 2007; Song et al. 2008a; Sun et al. 2009; Takahashi et al. 2001); and IL-1 (Solovieva et al. 2004a, b, 2006).

However, many of the studies have not been replicated and only showed low levels of association, likely due to small cohort size and variable phenotype definition. A recent review of 52 association analyses (Eskola et al. 2012) suggested moderate levels of evidence for six candidate genes, *ASPN* (asporin) (Song et al. 2008b), *COL11A1* (collagen XI α1) (Mio et al. 2007), *GDF5* (growth differentiation factor 5) (Williams et al. 2011), *SKT* (Sickle tail) (Karasugi et al. 2009), *THBS2* (thrombospondin 2) (Hirose et al. 2008), and *MMP9* (matrix metalloproteinase 9) (Hirose et al. 2008).

A popular alternative approach for the study of complex diseases is performing genome-wide association studies (GWAS) to identity candidate loci across the genome in an unbiased assessment (Hirschhorn and Daly 2005). However, this requires a very large cohort size to reach statistical significance, as thousands of single nucleotide polymorphisms (SNPs) must undergo multiple testing corrections. Due to the lack of large cohorts that provide GWAS data for intervertebral disc degeneration, an alternative is to perform a meta-analysis of smaller cohorts to increase the genetic power. In a meta-analysis study totaling 4,683 subjects, a strong association ($p = 2.8 \times 10^{-8}$) for rs926849 which is located in an intron of the Parkinson protein 2, E3 ubiquitin protein ligase (*PARK2*) gene on chromosome 6 was detected (Williams et al. 2013). The relevance of this variant in lumbar disc degeneration is not clear, and further confirmation and functional assessment are needed. More recently, an approach combining linkage analysis of families with early onset lumbar disc degeneration and GWAS studies totaling 32,642 subjects across three ethnic populations identified chondroitin 6-O-sulfotransferase (CHST3) as a risk factor (Song et al. 2013). The associated functional SNP is on a miRNA binding site and may possibly alter the level of CHST3 in affected individuals. CHST3 is enzyme required for proper chondroitin sulfate synthesis, which is essential for water retention and inhibition of neural ingrowth.

10.10 Concluding Remarks

In the last decade, we have gained significant insight into the pathobiology of intervertebral disc degeneration. From detailed studies on humans and animal models, we are beginning to decipher the nutritional, mechanical, cellular, and biochemical contributions to the degenerative process. While there are differences in the development, anatomy, cellular characteristics, and mechanical properties between animal species and humans, research is most effective when an optimal animal model is used to address well-defined and specific research questions, and subsequent validations are carried out in human cells or tissues. It is also essential to look beyond the dehydrating nucleus pulposus as the major cause of disc degeneration. Cellular and biochemical changes in the nucleus pulposus could be the consequence of earlier changes in the surrounding tissues and disc environment.

The topics discussed in this chapter highlight the complexity of disc degeneration, and to comprehensively study its causes and consequences, we must take advantage of the most recent advances in technology. For instance, single cell transcriptomics (Tang et al. 2009) has proved a powerful approach for dissecting cellular heterogeneity. Thus, it can be readily applied to recognize the identity of cells in the nucleus pulposus and surrounding disc tissues, to understand the cellular changes in a degenerating disc, and also to identify potential local progenitor cells that may be used to stimulate endogenous repair or exogenous cell therapy strategies. Furthermore, a comparative deep proteomic analysis of degenerate and non-degenerate human intervertebral discs can provide supplemental information to better understand how cells in different disc compartments may respond to a changing environment.

Findings from genetic studies are vital for understanding the cause of human disc degeneration. However, as of yet, very few genes have been identified and the effect size is small. Since variations in phenotypic parameters that define the degenerative status of the disc confounds genetic studies, an accurate assessment of the phenotype and development of a common classification system for disc degeneration is necessary. This improvement, alongside the collection of larger cohorts and the advent of next-generation sequencing, will enable meaningful meta-analysis studies to increase the statistical power of detection and identify rare genes with bigger effect size.

Combining human genetics with in vitro or in vivo model systems for functional assessment studies provides an ideal scenario. Upon identifying individuals that harbor specific risk factors or even protective factors of disc degeneration, induced pluripotent stem (iPS) cells (Yu et al. 2007a, b) can be generated from blood or urine samples (Loh et al. 2009; Zhou et al. 2012) and then differentiated into the desired cell type(s) for further analysis of disease outcome. These are all feasible technologies and approaches to consider for future studies. Understanding the etiology of intervertebral disc degeneration could offer potential therapeutic strategies that not only relieve procuring symptoms but also hope to restore the form and function of intervertebral discs.

References

Adams MA, Dolan P (2012) Intervertebral disc degeneration: evidence for two distinct phenotypes. J Anat 221(6):497–506. doi:10.1111/j.1469-7580.2012.01551.x

Adams MA, Hutton WC (1982) Prolapsed intervertebral disc. A hyperflexion injury. 1981 Volvo Award in Basic Science. Spine 7(3):184–191

Adams P, Muir H (1976) Qualitative changes with age of proteoglycans of human lumbar discs. Ann Rheum Dis 35(4):289–296

Adams MA, Roughley PJ (2006) What is intervertebral disc degeneration, and what causes it? Spine (Phila Pa 1976) 31(18):2151–2161. doi:10.1097/01.brs.0000231761.73859.2c

Adams DS, Keller R, Koehl MA (1990) The mechanics of notochord elongation, straightening and stiffening in the embryo of Xenopus laevis. Development 110(1):115–130

Adams MA, McNally DS, Dolan P (1996) 'Stress' distributions inside intervertebral discs. The effects of age and degeneration. J Bone Joint Surg Br 78(6):965–972

Adams MA, Freeman BJ, Morrison HP, Nelson IW, Dolan P (2000) Mechanical initiation of intervertebral disc degeneration. Spine 25(13):1625–1636

Agrawal A, Guttapalli A, Narayan S, Albert TJ, Shapiro IM, Risbud MV (2007) Normoxic stabilization of HIF-1alpha drives glycolytic metabolism and regulates aggrecan gene expression in nucleus pulposus cells of the rat intervertebral disk. Am J Physiol Cell Physiol 293(2):C621–C631. doi:10.1152/ajpcell.00538.2006

Aigner T, Gresk-otter KR, Fairbank JC, von der Mark K, Urban JP (1998) Variation with age in the pattern of type X collagen expression in normal and scoliotic human intervertebral discs. Calcif Tissue Int 63(3):263–268

Albert HB, Kjaer P, Jensen TS, Sorensen JS, Bendix T, Manniche C (2008) Modic changes, possible causes and relation to low back pain. Med Hypotheses 70(2):361–368. doi:10.1016/j.mehy.2007.05.014

Andrade P, Hoogland G, Garcia MA, Steinbusch HW, Daemen MA, Visser-Vandewalle V (2013) Elevated IL-1beta and IL-6 levels in lumbar herniated discs in patients with sciatic pain. Eur Spine J 22(4):714–720. doi:10.1007/s00586-012-2502-x

Annunen S, Paassilta P, Lohiniva J, Perala M, Pihlajamaa T, Karppinen J, Tervonen O, Kroger H, Lahde S, Vanharanta H, Ryhanen L, Goring HH, Ott J, Prockop DJ, Ala-Kokko L (1999) An allele of COL9A2 associated with intervertebral disc disease. Science 285(5426):409–412

Arkesteijn IT, Smolders LA, Spillekom S, Riemers FM, Potier E, Meij BP, Ito K, Tryfonidou MA (2015) Effect of coculturing canine notochordal, nucleus pulposus and mesenchymal stromal cells for intervertebral disc regeneration. Arthritis Res Ther 17:60. doi:10.1186/s13075-015-0569-6

Aszodi A, Chan D, Hunziker E, Bateman JF, Fassler R (1998) Collagen II is essential for the removal of the notochord and the formation of intervertebral discs. J Cell Biol 143(5):1399–1412

Babhulkar S (1997) Osteonecrosis in sickle cell hemoglobinopathy. In: Urbaniak JR, Jones JPJ (eds) Osteonecrosis: etiology, diagnosis, and treatment. American Orthopaedic Association, Rosemont, pp 131–133

Bachmeier BE, Nerlich AG, Weiler C, Paesold G, Jochum M, Boos N (2007) Analysis of tissue distribution of TNF-alpha, TNF-alpha-receptors, and the activating TNF-alpha-converting enzyme suggests activation of the TNF-alpha system in the aging intervertebral disc. Ann N Y Acad Sci 1096:44–54. doi:10.1196/annals.1397.069

Baffi MO, Slattery E, Sohn P, Moses HL, Chytil A, Serra R (2004) Conditional deletion of the TGF-beta type II receptor in Col2a expressing cells results in defects in the axial skeleton without alterations in chondrocyte differentiation or embryonic development of long bones. Dev Biol 276(1):124–142. doi:10.1016/j.ydbio.2004.08.027

Baffi MO, Moran MA, Serra R (2006) Tgfbr2 regulates the maintenance of boundaries in the axial skeleton. Dev Biol 296(2):363–374. doi:10.1016/j.ydbio.2006.06.002

Bagnall KM, Sanders EJ (1989) The binding pattern of peanut lectin associated with sclerotome migration and the formation of the vertebral axis in the chick embryo. Anat Embryol (Berl) 180(5):505–513

Bagnall KM, Higgins SJ, Sanders EJ (1988) The contribution made by a single somite to the vertebral column: experimental evidence in support of resegmentation using the chick-quail chimaera model. Development 103(1):69–85

Battie MC, Levalahti E, Videman T, Burton K, Kaprio J (2008) Heritability of lumbar flexibility and the role of disc degeneration and body weight. J Appl Physiol (1985) 104(2):379–385. doi:10.1152/japplphysiol.01009.2007

Benneker LM, Heini PF, Alini M, Anderson SE, Ito K (2005) 2004 Young Investigator Award Winner: vertebral endplate marrow contact channel occlusions and intervertebral disc degeneration. Spine 30(2):167–173
Bessho Y, Hirata H, Masamizu Y, Kageyama R (2003) Periodic repression by the bHLH factor Hes7 is an essential mechanism for the somite segmentation clock. Genes Dev 17(12):1451–1456. doi:10.1101/gad.1092303
Bibby SR, Jones DA, Ripley RM, Urban JP (2005) Metabolism of the intervertebral disc: effects of low levels of oxygen, glucose, and pH on rates of energy metabolism of bovine nucleus pulposus cells. Spine 30(5):487–496
Bick EM, Copel JW (1951) The ring apophysis of the human vertebra; contribution to human osteogeny. II. J Bone Joint Surg Am 33-A(3):783–787
Boos N, Wallin A, Gbedegbegnon T, Aebi M, Boesch C (1993) Quantitative MR imaging of lumbar intervertebral disks and vertebral bodies: influence of diurnal water content variations. Radiology 188(2):351–354. doi:10.1148/radiology.188.2.8327677
Boos N, Nerlich AG, Wiest I, von der Mark K, Aebi M (1997) Immunolocalization of type X collagen in human lumbar intervertebral discs during ageing and degeneration. Histochem Cell Biol 108(6):471–480
Boos N, Weissbach S, Rohrbach H, Weiler C, Spratt KF, Nerlich AG (2002) Classification of age-related changes in lumbar intervertebral discs: 2002 Volvo Award in basic science. Spine 27(23):2631–2644. doi:10.1097/01.BRS.0000035304.27153.5B
Borycki AG, Mendham L, Emerson CP Jr (1998) Control of somite patterning by Sonic hedgehog and its downstream signal response genes. Development 125(4):777–790
Boxberger JI, Auerbach JD, Sen S, Elliott DM (2008) An in vivo model of reduced nucleus pulposus glycosaminoglycan content in the rat lumbar intervertebral disc. Spine 33(2):146–154. doi:10.1097/BRS.0b013e31816054f8
Brand-Saberi B, Christ B (2000) Evolution and development of distinct cell lineages derived from somites. Curr Top Dev Biol 48:1–42
Brand-Saberi B, Muller TS, Wilting J, Christ B, Birchmeier C (1996) Scatter factor/hepatocyte growth factor (SF/HGF) induces emigration of myogenic cells at interlimb level in vivo. Dev Biol 179(1):303–308. doi:10.1006/dbio.1996.0260
Brinckmann P, Biggemann M, Hilweg D (1988) Fatigue fracture of human lumbar vertebrae. Clin Biomech 3(Suppl 1):i-S23. doi:10.1016/S0268-0033(88)80001-9
Bruehlmann SB, Rattner JB, Matyas JR, Duncan NA (2002) Regional variations in the cellular matrix of the annulus fibrosus of the intervertebral disc. J Anat 201(2):159–171
Bruggeman BJ, Maier JA, Mohiuddin YS, Powers R, Lo Y, Guimaraes-Camboa N, Evans SM, Harfe BD (2012) Avian intervertebral disc arises from rostral sclerotome and lacks a nucleus pulposus: implications for evolution of the vertebrate disc. Dev Dyn 241(4):675–683. doi:10.1002/dvdy.23750
Cappello R, Bird JL, Pfeiffer D, Bayliss MT, Dudhia J (2006) Notochordal cell produce and assemble extracellular matrix in a distinct manner, which may be responsible for the maintenance of healthy nucleus pulposus. Spine 31(8):873–882. doi:10.1097/01.brs.0000209302.00820.fd; discussion 883
Carreon LY, Ito T, Yamada M, Uchiyama S, Takahashi HE (1997) Neovascularization induced by anulus and its inhibition by cartilage endplate. Its role in disc absorption. Spine (Phila Pa 1976) 22(13):1429–1434; discussion 1446–1427
Chan SC, Ferguson SJ, Gantenbein-Ritter B (2011) The effects of dynamic loading on the intervertebral disc. Eur Spine J 20(11):1796–1812. doi:10.1007/s00586-011-1827-1
Chandran PL, Horkay F (2012) Aggrecan, an unusual polyelectrolyte: review of solution behavior and physiological implications. Acta Biomater 8(1):3–12. doi:10.1016/j.actbio.2011.08.011
Chen E, Stringer SE, Rusch MA, Selleck SB, Ekker SC (2005) A unique role for 6-O sulfation modification in zebrafish vascular development. Dev Biol 284(2):364–376. doi:10.1016/j.ydbio.2005.05.032

Cheung KM, Karppinen J, Chan D, Ho DW, Song YQ, Sham P, Cheah KS, Leong JC, Luk KD (2009) Prevalence and pattern of lumbar magnetic resonance imaging changes in a population study of one thousand forty-three individuals. Spine 34(9):934–940. doi:10.1097/BRS.0b013e3181a01b3f

Choi KS, Cohn MJ, Harfe BD (2008) Identification of nucleus pulposus precursor cells and notochordal remnants in the mouse: implications for disk degeneration and chordoma formation. Dev Dyn 237(12):3953–3958. doi:10.1002/dvdy.21805

Choi H, Johnson ZI, Risbud MV (2015) Understanding nucleus pulposus cell phenotype: a prerequisite for stem cell based therapies to treat intervertebral disc degeneration. Curr Stem Cell Res Ther 10(4):307–316

Christ B, Wilting J (1992) From somites to vertebral column. Ann Anat 174(1):23–32

Christ B, Huang R, Wilting J (2000) The development of the avian vertebral column. Anat Embryol (Berl) 202(3):179–194

Chujo T, An HS, Akeda K, Miyamoto K, Muehleman C, Attawia M, Andersson G, Masuda K (2006) Effects of growth differentiation factor-5 on the intervertebral disc – in vitro bovine study and in vivo rabbit disc degeneration model study. Spine (Phila Pa 1976) 31(25):2909–2917. doi:10.1097/01.brs.0000248428.22823.86

Court C, Colliou OK, Chin JR, Liebenberg E, Bradford DS, Lotz JC (2001) The effect of static in vivo bending on the murine intervertebral disc. Spine J 1(4):239–245

Doita M, Kanatani T, Harada T, Mizuno K (1996) Immunohistologic study of the ruptured intervertebral disc of the lumbar spine. Spine (Phila Pa 1976) 21(2):235–241

Dong DM, Yao M, Liu B, Sun CY, Jiang YQ, Wang YS (2007) Association between the -1306C/T polymorphism of matrix metalloproteinase-2 gene and lumbar disc disease in Chinese young adults. Eur Spine J 16(11):1958–1961. doi:10.1007/s00586-007-0454-3

Dubrulle J, McGrew MJ, Pourquie O (2001) FGF signaling controls somite boundary position and regulates segmentation clock control of spatiotemporal Hox gene activation. Cell 106(2):219–232

Durigova M, Nagase H, Mort JS, Roughley PJ (2011) MMPs are less efficient than ADAMTS5 in cleaving aggrecan core protein. Matrix Biol 30(2):145–153. doi:10.1016/j.matbio.2010.10.007

Edelson JG, Nathan H (1988) Stages in the natural history of the vertebral end-plates. Spine 13(1):21–26

Erickson GR, Alexopoulos LG, Guilak F (2001) Hyper-osmotic stress induces volume change and calcium transients in chondrocytes by transmembrane, phospholipid, and G-protein pathways. J Biomech 34(12):1527–1535

Eskola PJ, Lemmela S, Kjaer P, Solovieva S, Mannikko M, Tommerup N, Lind-Thomsen A, Husgafvel-Pursiainen K, Cheung KM, Chan D, Samartzis D, Karppinen J (2012) Genetic association studies in lumbar disc degeneration: a systematic review. PLoS One 7(11):e49995. doi:10.1371/journal.pone.0049995

Eyre DR (1979) Biochemistry of the intervertebral disc. Int Rev Connect Tissue Res 8:227–291

Eyre DR, Muir H (1976) Types I and II collagens in intervertebral disc. Interchanging radial distributions in annulus fibrosus. Biochem J 157(1):267–270

Fan CM, Tessier-Lavigne M (1994) Patterning of mammalian somites by surface ectoderm and notochord: evidence for sclerotome induction by a hedgehog omolog. Cell 79(7):1175–1186

Feng G, Li L, Liu H, Song Y, Huang F, Tu C, Shen B, Gong Q, Li T, Liu L, Zeng J, Kong Q, Yi M, Gupte M, Ma PX, Pei F (2013) Hypoxia differentially regulates human nucleus pulposus and annulus fibrosus cell extracellular matrix production in 3D scaffolds. Osteoarthritis Cartilage 21(4):582–588. doi:10.1016/j.joca.2013.01.001

Ferguson SJ, Ito K, Nolte LP (2004) Fluid flow and convective transport of solutes within the intervertebral disc. J Biomech 37(2):213–221

Ferrer-Vaquer A, Viotti M, Hadjantonakis AK (2010) Transitions between epithelial and mesenchymal states and the morphogenesis of the early mouse embryo. Cell Adh Migr 4(3):447–457

Galbusera F, Mietsch A, Schmidt H, Wilke HJ, Neidlinger-Wilke C (2013) Effect of intervertebral disc degeneration on disc cell viability: a numerical investigation. Comput Methods Biomech Biomed Engin 16(3):328–337. doi:10.1080/10255842.2011.619184

Gilchrist CL, Witvoet-Braam SW, Guilak F, Setton LA (2007) Measurement of intracellular strain on deformable substrates with texture correlation. J Biomech 40(4):786–794. doi:10.1016/j.jbiomech.2006.03.013

Gilson A, Dreger M, Urban JP (2010) Differential expression level of cytokeratin 8 in cells of the bovine nucleus pulposus complicates the search for specific intervertebral disc cell markers. Arthritis Res Ther 12(1):R24. doi:10.1186/ar2931

Goldstein RS, Kalcheim C (1992) Determination of epithelial half-somites in skeletal morphogenesis. Development 116(2):441–445

Gronblad M, Virri J, Tolonen J, Seitsalo S, Kaapa E, Kankare J, Myllynen P, Karaharju EO (1994) A controlled immunohistochemical study of inflammatory cells in disc herniation tissue. Spine (Phila Pa 1976) 19(24):2744–2751

Grotmol S, Kryvi H, Nordvik K, Totland GK (2003) Notochord segmentation may lay down the pathway for the development of the vertebral bodies in the Atlantic salmon. Anat Embryol (Berl) 207(4–5):263–272. doi:10.1007/s00429-003-0349-y

Gruber HE, Fisher EC Jr, Desai B, Stasky AA, Hoelscher G, Hanley EN Jr (1997) Human intervertebral disc cells from the annulus: three-dimensional culture in agarose or alginate and responsiveness to TGF-beta1. Exp Cell Res 235(1):13–21

Grunhagen T, Shirazi-Adl A, Fairbank JC, Urban JP (2011) Intervertebral disk nutrition: a review of factors influencing concentrations of nutrients and metabolites. Orthop Clin North Am 42(4):465–477. doi:10.1016/j.ocl.2011.07.010, vii

Guehring T, Wilde G, Sumner M, Grunhagen T, Karney GB, Tirlapur UK, Urban JP (2009) Notochordal intervertebral disc cells: sensitivity to nutrient deprivation. Arthritis Rheum 60(4):1026–1034. doi:10.1002/art.24407

Han WM, Heo SJ, Driscoll TP, Smith LJ, Mauck RL, Elliott DM (2013) Macro- to microscale strain transfer in fibrous tissues is heterogeneous and tissue-specific. Biophys J 105(3):807–817. doi:10.1016/j.bpj.2013.06.023

Hashizume Y (1980) Pathological studies on the ossification of the posterior longitudinal ligament (opll). Acta Pathol Jpn 30(2):255–273

Hassler O (1969) The human intervertebral disc. A micro-angiographical study on its vascular supply at various ages. Acta Orthop Scand 40(6):765–772

Hayes AJ, Benjamin M, Ralphs JR (1999) Role of actin stress fibres in the development of the intervertebral disc: cytoskeletal control of extracellular matrix assembly. Dev Dyn 215(3):179–189. doi:10.1002/(SICI)1097-0177(199907)215:3<179::AID-AJA1>3.0.CO;2-Q

Hirose Y, Chiba K, Karasugi T, Nakajima M, Kawaguchi Y, Mikami Y, Furuichi T, Mio F, Miyake A, Miyamoto T, Ozaki K, Takahashi A, Mizuta H, Kubo T, Kimura T, Tanaka T, Toyama Y, Ikegawa S (2008) A functional polymorphism in THBS2 that affects alternative splicing and MMP binding is associated with lumbar-disc herniation. Am J Hum Genet 82(5):1122–1129. doi:10.1016/j.ajhg.2008.03.013

Hirschhorn JN, Daly MJ (2005) Genome-wide association studies for common diseases and complex traits. Nat Rev Genet 6(2):95–108. doi:10.1038/nrg1521

Holm S, Maroudas A, Urban JP, Selstam G, Nachemson A (1981) Nutrition of the intervertebral disc: solute transport and metabolism. Connect Tissue Res 8(2):101–119

Horner HA, Urban JP (2001) 2001 Volvo Award Winner in Basic Science Studies: effect of nutrient supply on the viability of cells from the nucleus pulposus of the intervertebral disc. Spine 26(23):2543–2549

Hoyland JA, Le Maitre C, Freemont AJ (2008) Investigation of the role of IL-1 and TNF in matrix degradation in the intervertebral disc. Rheumatology (Oxford) 47(6):809–814. doi:10.1093/rheumatology/ken056

Hu ZJ, Zhao FD, Fang XQ, Fan SW (2009) Modic changes, possible causes and promotion to lumbar intervertebral disc degeneration. Med Hypotheses 73(6):930–932. doi:10.1016/j.mehy.2009.06.038

Huang R, Zhi Q, Wilting J, Christ B (1994) The fate of somitocoele cells in avian embryos. Anat Embryol (Berl) 190(3):243–250

Hunter CJ, Matyas JR, Duncan NA (2003) The three-dimensional architecture of the notochordal nucleus pulposus: novel observations on cell structures in the canine intervertebral disc. J Anat 202(Pt 3):279–291

Hunter CJ, Matyas JR, Duncan NA (2004) The functional significance of cell clusters in the notochordal nucleus pulposus: survival and signaling in the canine intervertebral disc. Spine 29(10):1099–1104

Inoue H (1981) Three-dimensional architecture of lumbar intervertebral discs. Spine (Phila Pa 1976) 6(2):139–146

Ishihara H, Urban JP (1999) Effects of low oxygen concentrations and metabolic inhibitors on proteoglycan and protein synthesis rates in the intervertebral disc. J Orthop Res 17(6):829–835. doi:10.1002/jor.1100170607

Ishihara H, Warensjo K, Roberts S, Urban JP (1997) Proteoglycan synthesis in the intervertebral disk nucleus: the role of extracellular osmolality. Am J Physiol 272(5 Pt 1):C1499–C1506

Ishii T, Tsuji H, Sano A, Katoh Y, Matsui H, Terahata N (1991) Histochemical and ultrastructural observations on brown degeneration of human intervertebral disc. J Orthop Res 9(1):78–90. doi:10.1002/jor.1100090111

Jiang YJ, Aerne BL, Smithers L, Haddon C, Ish-Horowicz D, Lewis J (2000) Notch signalling and the synchronization of the somite segmentation clock. Nature 408(6811):475–479. doi:10.1038/35044091

Jim JJ, Noponen-Hietala N, Cheung KM, Ott J, Karppinen J, Sahraravand A, Luk KD, Yip SP, Sham PC, Song YQ, Leong JC, Cheah KS, Ala-Kokko L, Chan D (2005) The TRP2 allele of COL9A2 is an age-dependent risk factor for the development and severity of intervertebral disc degeneration. Spine 30(24):2735–2742

Johnson WE, Caterson B, Eisenstein SM, Hynds DL, Snow DM, Roberts S (2002) Human intervertebral disc aggrecan inhibits nerve growth in vitro. Arthritis Rheum 46(10):2658–2664. doi:10.1002/art.10585

Johnson ZI, Schoepflin ZR, Choi H, Shapiro IM, Risbud MV (2015) Disc in flames: roles of TNF-alpha and IL-1beta in intervertebral disc degeneration. Eur Cell Mater 30:104–116; discussion 116–107

Jones JPJ (1997) Subchondral osteonecrosis can conceivably cause disk degeneration and primary osteoarthritis. In: Urbaniak JR, Jones JPJ (eds) Osteonecrosis: etiology, diagnosis, and treatment. American Orthopaedic Association, Rosemont, pp 135–142

Jones JP Jr, Engleman EP (1966) Osseous avascular necrosis associated with systemic abnormalities. Arthritis Rheum 9(5):728–736

Kageyama R, Masamizu Y, Niwa Y (2007) Oscillator mechanism of Notch pathway in the segmentation clock. Dev Dyn 236(6):1403–1409. doi:10.1002/dvdy.21114

Karamouzian S, Eskandary H, Faramarzee M, Saba M, Safizade H, Ghadipasha M, Malekpoor AR, Ohadi A (2010) Frequency of lumbar intervertebral disc calcification and angiogenesis, and their correlation with clinical, surgical, and magnetic resonance imaging findings. Spine (Phila Pa 1976) 35(8):881–886. doi:10.1097/BRS.0b013e3181b9c986

Karasugi T, Semba K, Hirose Y, Kelempisioti A, Nakajima M, Miyake A, Furuichi T, Kawaguchi Y, Mikami Y, Chiba K, Kamata M, Ozaki K, Takahashi A, Makela P, Karppinen J, Kimura T, Kubo T, Toyama Y, Yamamura K, Mannikko M, Mizuta H, Ikegawa S (2009) Association of the tag SNPs in the human SKT gene (KIAA1217) with lumbar disc herniation. J Bone Miner Res 24(9):1537–1543. doi:10.1359/jbmr.090314

Kawaguchi Y, Osada R, Kanamori M, Ishihara H, Ohmori K, Matsui H, Kimura T (1999) Association between an aggrecan gene polymorphism and lumbar disc degeneration. Spine 24(23):2456–2460

Kim DJ, Moon SH, Kim H, Kwon UH, Park MS, Han KJ, Hahn SB, Lee HM (2003) Bone morphogenetic protein-2 facilitates expression of chondrogenic, not osteogenic, phenotype of

human intervertebral disc cells. Spine (Phila Pa 1976) 28(24):2679–2684. doi:10.1097/01.BRS.0000101445.46487.16
Kurunlahti M, Tervonen O, Vanharanta H, Ilkko E, Suramo I (1999) Association of atherosclerosis with low back pain and the degree of disc degeneration. Spine 24(20):2080–2084
Lammi P, Inkinen RI, von der Mark K, Puustjarvi K, Arokoski J, Hyttinen MM, Lammi MJ (1998) Localization of type X collagen in the intervertebral disc of mature beagle dogs. Matrix Biol 17(6):449–453
Le Maitre CL, Freemont AJ, Hoyland JA (2005) The role of interleukin-1 in the pathogenesis of human intervertebral disc degeneration. Arthritis Res Ther 7(4):R732–R745. doi:10.1186/ar1732
Le Maitre CL, Hoyland JA, Freemont AJ (2007a) Catabolic cytokine expression in degenerate and herniated human intervertebral discs: IL-1beta and TNFalpha expression profile. Arthritis Res Ther 9(4):R77. doi:10.1186/ar2275
Le Maitre CL, Pockert A, Buttle DJ, Freemont AJ, Hoyland JA (2007b) Matrix synthesis and degradation in human intervertebral disc degeneration. Biochem Soc Trans 35(Pt 4):652–655. doi:10.1042/BST0350652
Li X, Leo BM, Beck G, Balian G, Anderson GD (2004) Collagen and proteoglycan abnormalities in the GDF-5-deficient mice and molecular changes when treating disk cells with recombinant growth factor. Spine (Phila Pa 1976) 29(20):2229–2234
Livshits G, Cohen Z, Higla O, Yakovenko K (2001) Familial history, age and smoking are important risk factors for disc degeneration disease in Arabic pedigrees. Eur J Epidemiol 17(7):643–651
Loh YH, Agarwal S, Park IH, Urbach A, Huo H, Heffner GC, Kim K, Miller JD, Ng K, Daley GQ (2009) Generation of induced pluripotent stem cells from human blood. Blood 113(22):5476–5479. doi:10.1182/blood-2009-02-204800
Lotz J, Colliou OK, Chin JR, Duncan NA, Liebenberg E (1998) 1998 Volvo Award Winner in Biomechanical Studies: compression-induced degeneration of the intervertebral disc: an in vivo mouse model and finite-element study. Spine 23(23):2493–2506
Luk KD, Samartzis D (2015) Intervertebral disc "dysgeneration". Spine J 15(9):1915–1918. doi:10.1016/j.spinee.2014.07.020
Lyons G, Eisenstein SM, Sweet MB (1981) Biochemical changes in intervertebral disc degeneration. Biochim Biophys Acta 673(4):443–453
Maclean JJ, Lee CR, Alini M, Iatridis JC (2004) Anabolic and catabolic mRNA levels of the intervertebral disc vary with the magnitude and frequency of in vivo dynamic compression. J Orthop Res 22(6):1193–1200. doi:10.1016/j.orthres.2004.04.004
Magnier C, Boiron O, Wendling-Mansuy S, Chabrand P, Deplano V (2009) Nutrient distribution and metabolism in the intervertebral disc in the unloaded state: a parametric study. J Biomech 42(2):100–108. doi:10.1016/j.jbiomech.2008.10.034
Marchand F, Ahmed AM (1990) Investigation of the laminate structure of lumbar disc anulus fibrosus. Spine 15(5):402–410
Maroto M, Pourquie O (2001) A molecular clock involved in somite segmentation. Curr Top Dev Biol 51:221–248
Maroudas A, Stockwell RA, Nachemson A, Urban J (1975) Factors involved in the nutrition of the human lumbar intervertebral disc: cellularity and diffusion of glucose in vitro. J Anat 120(Pt 1):113–130
McCann MR, Tamplin OJ, Rossant J, Seguin CA (2012) Tracing notochord-derived cells using a Noto-cre mouse: implications for intervertebral disc development. Dis Model Mech 5(1):73–82. doi:10.1242/dmm.008128
McNally DS, Adams MA, Goodship AE (1993) Can intervertebral disc prolapse be predicted by disc mechanics? Spine 18(11):1525–1530
Melrose J, Taylor TK, Ghosh P (1996) Variation in intervertebral disc serine proteinase inhibitory proteins with ageing in a chondrodystrophoid (beagle) and a non-chondrodystrophoid (greyhound) canine breed. Gerontology 42(6):322–329

Melrose J, Smith SM, Appleyard RC, Little CB (2008) Aggrecan, versican and type VI collagen are components of annular translamellar crossbridges in the intervertebral disc. Eur Spine J 17(2):314–324. doi:10.1007/s00586-007-0538-0

Miller JA, Schmatz C, Schultz AB (1988) Lumbar disc degeneration: correlation with age, sex, and spine level in 600 autopsy specimens. Spine 13(2):173–178

Minogue BM, Richardson SM, Zeef LA, Freemont AJ, Hoyland JA (2010) Characterization of the human nucleus pulposus cell phenotype and evaluation of novel marker gene expression to define adult stem cell differentiation. Arthritis Rheum 62(12):3695–3705. doi:10.1002/art.27710

Mio F, Chiba K, Hirose Y, Kawaguchi Y, Mikami Y, Oya T, Mori M, Kamata M, Matsumoto M, Ozaki K, Tanaka T, Takahashi A, Kubo T, Kimura T, Toyama Y, Ikegawa S (2007) A functional polymorphism in COL11A1, which encodes the alpha 1 chain of type XI collagen, is associated with susceptibility to lumbar disc herniation. Am J Hum Genet 81(6):1271–1277. doi:10.1086/522377

Miyazaki T, Kobayashi S, Takeno K, Meir A, Urban J, Baba H (2009) A phenotypic comparison of proteoglycan production of intervertebral disc cells isolated from rats, rabbits, and bovine tails; which animal model is most suitable to study tissue engineering and biological repair of human disc disorders? Tissue Eng Part A 15(12):3835–3846. doi:10.1089/ten.tea.2009.0250

Mobasheri A, Hall AC, Urban JP, France SJ, Smith AL (1997) Immunologic andautoradiographic localisation of the Naþ, K(þ)-ATPase in articular cartilage: upregulationin response to changes in extracellular Naþ concentration. Int J Biochem Cell Biol 29:649–657

Mokhbi Soukane D, Shirazi-Adl A, Urban JP (2009) Investigation of solute concentrations in a 3D model of intervertebral disc. Eur Spine J 18(2):254–262. doi:10.1007/s00586-008-0822-7

Monsoro-Burq AH (2005) Sclerotome development and morphogenesis: when experimental embryology meets genetics. Int J Dev Biol 49(2–3):301–308. doi:10.1387/ijdb.041953am

Moon SM, Yoder JH, Wright AC, Smith LJ, Vresilovic EJ, Elliott DM (2013) Evaluation of intervertebral disc cartilaginous endplate structure using magnetic resonance imaging. Eur Spine J 22(8):1820–1828. doi:10.1007/s00586-013-2798-1

Moore RJ (2006) The vertebral endplate: disc degeneration, disc regeneration. Eur Spine J 15(Suppl 3):S333–S337. doi:10.1007/s00586-006-0170-4

Morgelin M, Paulsson M, Hardingham TE, Heinegard D, Engel J (1988) Cartilage proteoglycans. Assembly with hyaluronate and link protein as studied by electron microscopy. Biochem J 253(1):175–185

Morimoto M, Takahashi Y, Endo M, Saga Y (2005) The Mesp2 transcription factor establishes segmental borders by suppressing Notch activity. Nature 435(7040):354–359. doi:10.1038/nature03591

Mwale F, Ciobanu I, Giannitsios D, Roughley P, Steffen T, Antoniou J (2011) Effect of oxygen levels on proteoglycan synthesis by intervertebral disc cells. Spine 36(2):E131–E138. doi:10.1097/BRS.0b013e3181d52b9e

Nagase H, Kashiwagi M (2003) Aggrecanases and cartilage matrix degradation. Arthritis Res Ther 5(2):94–103

Nagase H, Woessner JF Jr (1999) Matrix metalloproteinases. J Biol Chem 274(31):21491–21494

Naresh-Babu J, Neelima G, Reshma Begum S, Siva-Leela V (2016) Diffusion characteristics of human annulus fibrosus. – A study documenting the dependence of annulus fibrosus on endplate for diffusion. Spine J. doi:10.1016/j.spinee.2016.03.046

Neidlinger-Wilke C, Wurtz K, Urban JP, Borm W, Arand M, Ignatius A, Wilke HJ, Claes LE (2006) Regulation of gene expression in intervertebral disc cells by low and high hydrostatic pressure. Eur Spine J 15(Suppl 3):S372–S378. doi:10.1007/s00586-006-0112-1

Nerlich AG, Boos N (2016) Advances in lumbar degenerative disk disease pathophysiology comprehension. In: João Luiz Pinheiro-Franco, Alexander R. Vaccaro, Edward C. Benzel and H. Michael Mayer (eds) Advanced concepts in lumbar degenerative disk disease. Springer, Berlin, pp 41–60. doi:10.1007/978-3-662-47756-4_4

Nguyen-minh C, Haughton VM, Papke RA, An H, Censky SC (1998) Measuring diffusion of solutes into intervertebral disks with MR imaging and paramagnetic contrast medium. AJNR Am J Neuroradiol 19(9):1781–1784

Nosikova YS, Santerre JP, Grynpas M, Gibson G, Kandel RA (2012) Characterization of the annulus fibrosus-vertebral body interface: identification of new structural features. J Anat 221(6):577–589. doi:10.1111/j.1469-7580.2012.01537.x

Paassilta P, Lohiniva J, Goring HH, Perala M, Raina SS, Karppinen J, Hakala M, Palm T, Kroger H, Kaitila I, Vanharanta H, Ott J, Ala-Kokko L (2001) Identification of a novel common genetic risk factor for lumbar disk disease. JAMA 285(14):1843–1849

Peacock A (1951) Observations on the prenatal development of the intervertebral disc in man. J Anat 85(3):260–274

Pfirrmann CW, Metzdorf A, Zanetti M, Hodler J, Boos N (2001) Magnetic resonance classification of lumbar intervertebral disc degeneration. Spine 26(17):1873–1878

Pockert AJ, Richardson SM, Le Maitre CL, Lyon M, Deakin JA, Buttle DJ, Freemont AJ, Hoyland JA (2009) Modified expression of the ADAMTS enzymes and tissue inhibitor of metalloproteinases 3 during human intervertebral disc degeneration. Arthritis Rheum 60(2):482–491. doi:10.1002/art.24291

Poiraudeau S, Monteiro I, Anract P, Blanchard O, Revel M, Corvol MT (1999) Phenotypic characteristics of rabbit intervertebral disc cells. Comparison with cartilage cells from the same animals. Spine 24(9):837–844

Pourquie O (2003) The segmentation clock: converting embryonic time into spatial pattern. Science 301(5631):328–330. doi:10.1126/science.1085887

Pourquie O (2011) Vertebrate segmentation: from cyclic gene networks to scoliosis. Cell 145(5):650–663. doi:10.1016/j.cell.2011.05.011

Pourquie O, Coltey M, Teillet MA, Ordahl C, Le Douarin NM (1993) Control of dorsoventral patterning of somitic derivatives by notochord and floor plate. Proc Natl Acad Sci U S A 90(11):5242–5246

Power KA, Grad S, Rutges JP, Creemers LB, van Rijen MH, O'Gaora P, Wall JG, Alini M, Pandit A, Gallagher WM (2011) Identification of cell surface-specific markers to target human nucleus pulposus cells: expression of carbonic anhydrase XII varies with age and degeneration. Arthritis Rheum 63(12):3876–3886. doi:10.1002/art.30607

Pritchard S, Erickson GR, Guilak F (2002) Hyperosmotically induced volume change and calcium signaling in intervertebral disk cells: the role of the actin cytoskeleton. Biophys J 83(5):2502–2510. doi:10.1016/S0006-3495(02)75261-2

Purmessur D, Cornejo MC, Cho SK, Roughley PJ, Linhardt RJ, Hecht AC, Iatridis JC (2015) Intact glycosaminoglycans from intervertebral disc-derived notochordal cell-conditioned media inhibit neurite growth while maintaining neuronal cell viability. Spine J 15(5):1060–1069. doi:10.1016/j.spinee.2015.02.003

Rajasekaran S, Babu JN, Arun R, Armstrong BR, Shetty AP, Murugan S (2004) ISSLS prize winner: a study of diffusion in human lumbar discs: a serial magnetic resonance imaging study documenting the influence of the endplate on diffusion in normal and degenerate discs. Spine 29(23):2654–2667

Rajasekaran S, Venkatadass K, Naresh Babu J, Ganesh K, Shetty AP (2008) Pharmacological enhancement of disc diffusion and differentiation of healthy, ageing and degenerated discs: results from in-vivo serial post-contrast MRI studies in 365 human lumbar discs. Eur Spine J 17(5):626–643. doi:10.1007/s00586-008-0645-6

Rannou F, Corvol MT, Hudry C, Anract P, Dumontier MF, Tsagris L, Revel M, Poiraudeau S (2000) Sensitivity of anulus fibrosus cells to interleukin 1 beta. Comparison with articular chondrocytes. Spine (Phila Pa 1976) 25(1):17–23

Razaq S, Wilkins RJ, Urban JP (2003) The effect of extracellular pH on matrix turnover by cells of the bovine nucleus pulposus. Eur Spine J 12(4):341–349. doi:10.1007/s00586-003-0582-3

Risbud MV, Fertala J, Vresilovic EJ, Albert TJ, Shapiro IM (2005) Nucleus pulposus cells upregulate PI3K/Akt and MEK/ERK signaling pathways under hypoxic conditions and resist apoptosis induced by serum withdrawal. Spine 30(8):882–889

Roberts S, Urban JP, Evans H, Eisenstein SM (1996) Transport properties of the human cartilage endplate in relation to its composition and calcification. Spine 21(4):415–420

Roberts S, Caterson B, Menage J, Evans EH, Jaffray DC, Eisenstein SM (2000) Matrix metalloproteinases and aggrecanase: their role in disorders of the human intervertebral disc. Spine (Phila Pa 1976) 25(23):3005–3013

Rodrigo I, Hill RE, Balling R, Munsterberg A, Imai K (2003) Pax1 and Pax9 activate Bapx1 to induce chondrogenic differentiation in the sclerotome. Development 130(3):473–482

Roughley PJ (2004) Biology of intervertebral disc aging and degeneration: involvement of the extracellular matrix. Spine 29(23):2691–2699

Roughley PJ, Mort JS (2014) The role of aggrecan in normal and osteoarthritic cartilage. J Exp Orthop 1(1):8. doi:10.1186/s40634-014-0008-7

Roughley PJ, Melching LI, Heathfield TF, Pearce RH, Mort JS (2006a) The structure and degradation of aggrecan in human intervertebral disc. Eur Spine J 15(Suppl 3):S326–S332. doi:10.1007/s00586-006-0127-7

Roughley PJ, Martens D, Rantakokko J, Alini M, Mwale F, Antoniou J (2006b) The involvement of aggrecan polymorphism in degeneration of human intervertebral disc and articular cartilage. Eur Cell Mater 11:1–7; discussion 7

Rudert M, Tillmann B (1993) Detection of lymph and blood vessels in the human intervertebral disc by histochemical and immunohistochemical methods. Ann Anat 175(3):237–242

Rutges JP, Duit RA, Kummer JA, Oner FC, van Rijen MH, Verbout AJ, Castelein RM, Dhert WJ, Creemers LB (2010) Hypertrophic differentiation and calcification during intervertebral disc degeneration. Osteoarthritis Cartilage 18(11):1487–1495. doi:10.1016/j.joca.2010.08.006

Saga Y, Takeda H (2001) The making of the somite: molecular events in vertebrate segmentation. Nat Rev Genet 2(11):835–845. doi:10.1038/35098552

Sahlman J, Inkinen R, Hirvonen T, Lammi MJ, Lammi PE, Nieminen J, Lapvetelainen T, Prockop DJ, Arita M, Li SW, Hyttinen MM, Helminen HJ, Puustjarvi K (2001) Premature vertebral endplate ossification and mild disc degeneration in mice after inactivation of one allele belonging to the Col2a1 gene for Type II collagen. Spine (Phila Pa 1976) 26(23):2558–2565

Sakai D, Nakamura Y, Nakai T, Mishima T, Kato S, Grad S, Alini M, Risbud MV, Chan D, Cheah KS, Yamamura K, Masuda K, Okano H, Ando K, Mochida J (2012) Exhaustion of nucleus pulposus progenitor cells with ageing and degeneration of the intervertebral disc. Nat Commun 3:1264. doi:10.1038/ncomms2226

Sakai D, Nishimura K, Tanaka M, Nakajima D, Grad S, Alini M, Kawada H, Ando K, Mochida J (2015) Migration of bone marrow-derived cells for endogenous repair in a new tail-looping disc degeneration model in the mouse: a pilot study. Spine J 15(6):1356–1365. doi:10.1016/j.spinee.2013.07.491

Sambrook PN, MacGregor AJ, Spector TD (1999) Genetic influences on cervical and lumbar disc degeneration: a magnetic resonance imaging study in twins. Arthritis Rheum 42(2):366–372. doi:10.1002/1529-0131(199902)42:2<366::AID-ANR20>3.0.CO;2-6

Sandell LJ, Morris N, Robbins JR, Goldring MB (1991) Alternatively spliced type II procollagen mRNAs define distinct populations of cells during vertebral development: differential expression of the amino-propeptide. J Cell Biol 114(6):1307–1319

Schneiderman G, Flannigan B, Kingston S, Thomas J, Dillin WH, Watkins RG (1987) Magnetic resonance imaging in the diagnosis of disc degeneration: correlation with discography. Spine 12(3):276–281

Schollmeier G, Lahr-Eigen R, Lewandrowski KU (2000) Observations on fiber-forming collagens in the anulus fibrosus. Spine (Phila Pa 1976) 25(21):2736–2741

Setton LA, Zhu W, Weidenbaum M, Ratcliffe A, Mow VC (1993) Compressive properties of the cartilaginous end-plate of the baboon lumbar spine. J Orthop Res 11(2):228–239. doi:10.1002/jor.1100110210

Shen G (2005) The role of type X collagen in facilitating and regulating endochondral ossification of articular cartilage. Orthod Craniofac Res 8(1):11–17. doi:10.1111/j.1601-6343.2004.00308.x
Shinmei M, Kikuchi T, Yamagishi M, Shimomura Y (1988) The role of interleukin-1 on proteoglycan metabolism of rabbit annulus fibrosus cells cultured in vitro. Spine (Phila Pa 1976) 13(11):1284–1290
Sivan SS, Hayes AJ, Wachtel E, Caterson B, Merkher Y, Maroudas A, Brown S, Roberts S (2014) Biochemical composition and turnover of the extracellular matrix of the normal and degenerate intervertebral disc. Eur Spine J 23(Suppl 3):S344–S353. doi:10.1007/s00586-013-2767-8
Smith LJ, Elliott DM (2011) Formation of lamellar cross bridges in the annulus fibrosus of the intervertebral disc is a consequence of vascular regression. Matrix Biol 30(4):267–274. doi:10.1016/j.matbio.2011.03.009
Smith LJ, Chiaro JA, Nerurkar NL, Cortes DH, Horava SD, Hebela NM, Mauck RL, Dodge GR, Elliott DM (2011) Nucleus pulposus cells synthesize a functional extracellular matrix and respond to inflammatory cytokine challenge following long-term agarose culture. Eur Cell Mater 22:291–301
Smolders LA, Bergknut N, Grinwis GC, Hagman R, Lagerstedt AS, Hazewinkel HA, Tryfonidou MA, Meij BP (2013) Intervertebral disc degeneration in the dog. Part 2: chondrodystrophic and non-chondrodystrophic breeds. Vet J 195(3):292–299. doi:10.1016/j.tvjl.2012.10.011
Sohn P, Cox M, Chen D, Serra R (2010) Molecular profiling of the developing mouse axial skeleton: a role for Tgfbr2 in the development of the intervertebral disc. BMC Dev Biol 10:29. doi:10.1186/1471-213X-10-29
Solovieva S, Kouhia S, Leino-Arjas P, Ala-Kokko L, Luoma K, Raininko R, Saarela J, Riihimaki H (2004a) Interleukin 1 polymorphisms and intervertebral disc degeneration. Epidemiology 15(5):626–633
Solovieva S, Leino-Arjas P, Saarela J, Luoma K, Raininko R, Riihimaki H (2004b) Possible association of interleukin 1 gene locus polymorphisms with low back pain. Pain 109(1–2):8–19. doi:10.1016/j.pain.2003.10.020
Solovieva S, Lohiniva J, Leino-Arjas P, Raininko R, Luoma K, Ala-Kokko L, Riihimaki H (2006) Intervertebral disc degeneration in relation to the COL9A3 and the IL-1ss gene polymorphisms. Eur Spine J 15(5):613–619. doi:10.1007/s00586-005-0988-1
Song YQ, Ho DW, Karppinen J, Kao PY, Fan BJ, Luk KD, Yip SP, Leong JC, Cheah KS, Sham P, Chan D, Cheung KM (2008a) Association between promoter -1607 polymorphism of MMP1 and lumbar disc disease in Southern Chinese. BMC Med Genet 9:38. doi:10.1186/1471-2350-9-38
Song YQ, Cheung KM, Ho DW, Poon SC, Chiba K, Kawaguchi Y, Hirose Y, Alini M, Grad S, Yee AF, Leong JC, Luk KD, Yip SP, Karppinen J, Cheah KS, Sham P, Ikegawa S, Chan D (2008b) Association of the asporin D14 allele with lumbar-disc degeneration in Asians. Am J Hum Genet 82(3):744–747. doi:10.1016/j.ajhg.2007.12.017
Song YQ, Karasugi T, Cheung KM, Chiba K, Ho DW, Miyake A, Kao PY, Sze KL, Yee A, Takahashi A, Kawaguchi Y, Mikami Y, Matsumoto M, Togawa D, Kanayama M, Shi D, Dai J, Jiang Q, Wu C, Tian W, Wang N, Leong JC, Luk KD, Yip SP, Cherny SS, Wang J, Mundlos S, Kelempisioti A, Eskola PJ, Mannikko M, Makela P, Karppinen J, Jarvelin MR, O'Reilly PF, Kubo M, Kimura T, Kubo T, Toyama Y, Mizuta H, Cheah KS, Tsunoda T, Sham PC, Ikegawa S, Chan D (2013) Lumbar disc degeneration is linked to a carbohydrate sulfotransferase 3 variant. J Clin Invest 123(11):4909–4917. doi:10.1172/JCI69277
Stemple DL (2005) Structure and function of the notochord: an essential organ for chordate development. Development 132(11):2503–2512. doi:10.1242/dev.01812
Stokes IA, Iatridis JC (2004) Mechanical conditions that accelerate intervertebral disc degeneration: overload versus immobilization. Spine 29(23):2724–2732
Sun ZM, Miao L, Zhang YG, Ming L (2009) Association between the -1562 C/T polymorphism of matrix metalloproteinase-9 gene and lumbar disc disease in the young adult population in North China. Connect Tissue Res 50(3):181–185. doi:10.1080/03008200802585630

Takahashi H, Suguro T, Okazima Y, Motegi M, Okada Y, Kakiuchi T (1996) Inflammatory cytokines in the herniated disc of the lumbar spine. Spine (Phila Pa 1976) 21(2):218–224
Takahashi M, Haro H, Wakabayashi Y, Kawa-uchi T, Komori H, Shinomiya K (2001) The association of degeneration of the intervertebral disc with 5a/6a polymorphism in the promoter of the human matrix metalloproteinase-3 gene. J Bone Joint Surg Br 83(4):491–495
Tang F, Barbacioru C, Wang Y, Nordman E, Lee C, Xu N, Wang X, Bodeau J, Tuch BB, Siddiqui A, Lao K, Surani MA (2009) mRNA-Seq whole-transcriptome analysis of a single cell. Nat Methods 6(5):377–382. doi:10.1038/nmeth.1315
Thompson JP, Oegema TR Jr, Bradford DS (1991) Stimulation of mature canine intervertebral disc by growth factors. Spine (Phila Pa 1976) 16(3):253–260
Tribioli C, Lufkin T (1999) The murine Bapx1 homeobox gene plays a critical role in embryonic development of the axial skeleton and spleen. Development 126(24):5699–5711
Trout JJ, Buckwalter JA, Moore KC, Landas SK (1982a) Ultrastructure of the human intervertebral disc. I. Changes in notochordal cells with age. Tissue Cell 14(2):359–369
Trout JJ, Buckwalter JA, Moore KC (1982b) Ultrastructure of the human intervertebral disc: II. Cells of the nucleus pulposus. Anat Rec 204(4):307–314. doi:10.1002/ar.1092040403
Urban JP, Maroudas A (1979) The measurement of fixed charged density in the intervertebral disc. Biochim Biophys Acta (BBA) – Gen Subj 586(1):166–178. doi:10.1016/0304-4165(79)90415-X
Urban JP, McMullin JF (1988) Swelling pressure of the lumbar intervertebral discs: influence of age, spinal level, composition, and degeneration. Spine (Phila Pa 1976) 13(2):179–187
Urban JP, Roberts S (2003) Degeneration of the intervertebral disc. Arthritis Res Ther 5(3):120–130
Urban JP, Ishihara H, Mouat MJ (1994) Differential metabolic responses of the nucleus and outer annulus to changes in extracellular physical factors. Trans Orthop Res Soc 19:134
Urban JP, Roberts S, Ralphs JR (2000) The nucleus of the intervertebral disc from development to degeneration. Am Zool 40(1):53–61. doi:10.1093/icb/40.1.53
Urban JP, Smith S, Fairbank JC (2004) Nutrition of the intervertebral disc. Spine 29(23):2700–2709
van der Rest M, Mayne R (1988) Type IX collagen proteoglycan from cartilage is covalently cross-linked to type II collagen. J Biol Chem 263(4):1615–1618
Verkman AS, van Hoek AN, Ma T, Frigeri A, Skach WR, Mitra A, Tamarappoo BK, Farinas J (1996) Water transport across mammalian cell membranes. Am J Physiol 270(1 Pt 1):C12–C30
Vernon-Roberts B, Moore RJ, Fraser RD (2007) The natural history of age-related disc degeneration: the pathology and sequelae of tears. Spine (Phila Pa 1976) 32(25):2797–2804. doi:10.1097/BRS.0b013e31815b64d2
Walker BF (2000) The prevalence of low back pain: a systematic review of the literature from 1966 to 1998. J Spinal Disord 13(3):205–217
Walsh AJ, Bradford DS, Lotz JC (2004) In vivo growth factor treatment of degenerated intervertebral discs. Spine (Phila Pa 1976) 29(2):156–163. doi:10.1097/01.BRS.0000107231.67854.9F
Watanabe H, Yamada Y, Kimata K (1998) Roles of aggrecan, a large chondroitin sulfate proteoglycan, in cartilage structure and function. J Biochem 124(4):687–693
Weiler C, Nerlich AG, Bachmeier BE, Boos N (2005) Expression and distribution of tumor necrosis factor alpha in human lumbar intervertebral discs: a study in surgical specimen and autopsy controls. Spine (Phila Pa 1976) 30(1):44–53; discussion 54
Williams FM, Popham M, Hart DJ, de Schepper E, Bierma-Zeinstra S, Hofman A, Uitterlinden AG, Arden NK, Cooper C, Spector TD, Valdes AM, van Meurs J (2011) GDF5 single-nucleotide polymorphism rs143383 is associated with lumbar disc degeneration in Northern European women. Arthritis Rheum 63(3):708–712. doi:10.1002/art.30169
Williams FM, Bansal AT, van Meurs JB, Bell JT, Meulenbelt I, Suri P, Rivadeneira F, Sambrook PN, Hofman A, Bierma-Zeinstra S, Menni C, Kloppenburg M, Slagboom PE, Hunter DJ, MacGregor AJ, Uitterlinden AG, Spector TD (2013) Novel genetic variants associated with

lumbar disc degeneration in northern Europeans: a meta-analysis of 4600 subjects. Ann Rheum Dis 72(7):1141–1148. doi:10.1136/annrheumdis-2012-201551

Wu JJ, Woods PE, Eyre DR (1992) Identification of cross-linking sites in bovine cartilage type IX collagen reveals an antiparallel type II-type IX molecular relationship and type IX to type IX bonding. J Biol Chem 267(32):23007–23014

Yamanaka Y, Tamplin OJ, Beckers A, Gossler A, Rossant J (2007) Live imaging and genetic analysis of mouse notochord formation reveals regional morphogenetic mechanisms. Dev Cell 13(6):884–896. doi:10.1016/j.devcel.2007.10.016

Yu J, Fairbank JC, Roberts S, Urban JP (2005) The elastic fiber network of the anulus fibrosus of the normal and scoliotic human intervertebral disc. Spine (Phila Pa 1976) 30(16):1815–1820

Yu J, Tirlapur U, Fairbank J, Handford P, Roberts S, Winlove CP, Cui Z, Urban J (2007a) Microfibrils, elastin fibres and collagen fibres in the human intervertebral disc and bovine tail disc. J Anat 210(4):460–471. doi:10.1111/j.1469-7580.2007.00707.x

Yu J, Vodyanik MA, Smuga-Otto K, Antosiewicz-Bourget J, Frane JL, Tian S, Nie J, Jonsdottir GA, Ruotti V, Stewart R, Slukvin II, Thomson JA (2007b) Induced pluripotent stem cell lines derived from human somatic cells. Science 318(5858):1917–1920. doi:10.1126/science.1151526

Yu SW, Yen CY, Wu CH, Kao FC, Kao YH, Tu YK (2012) Radiographic and clinical results of posterior dynamic stabilization for the treatment of multisegment degenerative disc disease with a minimum follow-up of 3 years. Arch Orthop Trauma Surg 132(5):583–589. doi:10.1007/s00402-012-1460-4

Zeng L, Kempf H, Murtaugh LC, Sato ME, Lassar AB (2002) Shh establishes an Nkx3.2/Sox9 autoregulatory loop that is maintained by BMP signals to induce somitic chondrogenesis. Genes Dev 16(15):1990–2005. doi:10.1101/gad.1008002

Zhang Y, Lenart BA, Lee JK, Chen D, Shi P, Ren J, Muehleman C, Chen D, An HS (2014) Histological features of endplates of the mammalian spine: from mice to men. Spine 39(5):E312–E317. doi:10.1097/BRS.0000000000000174

Zhou T, Benda C, Dunzinger S, Huang Y, Ho JC, Yang J, Wang Y, Zhang Y, Zhuang Q, Li Y, Bao X, Tse HF, Grillari J, Grillari-Voglauer R, Pei D, Esteban MA (2012) Generation of human induced pluripotent stem cells from urine samples. Nat Protoc 7(12):2080–2089. doi:10.1038/nprot.2012.115

MIX
Papier aus verantwortungsvollen Quellen
Paper from responsible sources
FSC® C105338

If you have any concerns about our products,
you can contact us on
ProductSafety@springernature.com

In case Publisher is established outside the EU,
the EU authorized representative is:
Springer Nature Customer Service Center GmbH
Europaplatz 3, 69115 Heidelberg, Germany

Printed by Libri Plureos GmbH
in Hamburg, Germany